Vibrations and Acoustic Radiation of Thin Structures

Vibrations and Acoustic Radiation of Thin Structures

Physical Basis, Theoretical Analysis and Numerical Methods

Paul J.T. Filippi

Series Editor
Société Française d'Acoustique

ISTE

WILEY

First published in Great Britain and the United States in 2008 by ISTE Ltd and John Wiley & Sons, Inc.

ISTE Ltd
27-37 St George's Road
London SW19 4EU
UK

www.iste.co.uk

John Wiley & Sons, Inc.
111 River Street
Hoboken, NJ 07030
USA

www.wiley.com

Library of Congress Cataloging-in-Publication Data

Filippi, Paul J.T.
Vibrations and acoustic radiation of thin structures : physical basis, theoretical analysis and numerical methods / Paul J.T. Filippi.
p. cm.
Includes bibliographical references and index.
ISBN 978-1-84821-056-1
1. Sound--Transmission. 2. Sound-waves. 3. Thin-walled structures--Vibration. 4. Radiation sources. I. Title.
QC243.F56 2008
620.2--dc22

2008019708

British Library Cataloguing-in-Publication Data
A CIP record for this book is available from the British Library
ISBN: 978-1-84821-056-1

Printed and bound in Great Britain by CPI Antony Rowe, Chippenham, Wiltshire.

FSC
Mixed Sources
Product group from well-managed forests and other controlled sources
Cert no. SGS-COC-2953
www.fsc.org
© 1996 Forest Stewardship Council

To my wife Dominique,
who inspires in all ways

Contents

Preface

This monograph is the result of lectures given by the author at the *Université d'Aix-Marseille (France)* to students of the *Diplôme d'Etudes Approfondies de Mécanique* (which is now the *Master de Mécanique*). It is aimed mainly at postgraduate students, PhD students and practicing acoustical scientists and engineers.

Among the most important sources of noise pollution are transport means, that is, cars, trucks, trains, planes, boats, etc. All these vehicles are essentially composed of thin vibrating structures. This is the reason why the present book is devoted to the vibrations and vibro-acoustics of thin structures only.

The simplest thin structure is the thin plate, then comes the circular cylindrical thin shell and the spherical thin shell. These basic structures provide a set of examples which make it possible to understand the basis of the physical phenomena of vibrations and sound radiation. Of course, most of the practical situations involve more complex structures, but their vibratory and acoustic behaviors are very similar to those of the simple structures described here, and the mathematical and numerical tools necessary to predict their response are much the same as those used for the simple examples.

Another aim of this monograph is to propose a homogenous theoretical approach to plates and shells.

Chapter 1 is devoted to equations which describe a good approximation of the vibrations of thin solids, and more precisely: plates, circular cylindrical shells and spherical shells. Analytical or numerical solutions of the mechanics equations are always based on a variational principle, which is, of course, the mathematical transcription of the conservation of energy principle which governs any phenomenon in physics. Thus, to establish the approximate equations governing the vibrations of thin structures we start from the expressions of the potential and kinetic energies of three-dimensional elastic solids, written in a convenient coordinate system: a Cartesian system for the plate, a cylindrical or spherical system for the shells. The hypothesis "thin structure" makes it possible to expand the three components of the displacement and the six independent components of the stresses as Taylor-like series of the transverse variable, leading to an approximate system of equations. We adopt the simplest approximations which are quite sufficient for a good understanding of the physical phenomena. Nevertheless, the method which is used can easily provide more accurate equations as they are proposed in the basic treatises cited in the bibliography.

Chapter 2 deals with the vibrations of *in vacuo* thin structures. The most important part concerns beams and plates. The classical method, based on the separation of variables, used to solve the vibration equation of simple plates of constant thickness (circular and rectangular) is developed in detail. Then, similar methods are applied to plates with a non-constant thickness. Finally, the Boundary Element Method (BEM) is described in some detail and illustrated by a comparison between numerical predictions and experimental results.

The chapter then continues with the problem of shell vibrations. For circular cylindrical shells, some of the existing analytical methods are proposed which enable us to give the expression of the resonance modes and of the response to a harmonic excitation. The Boundary Element Method is also described. For spherical shells, it seems that no analytical method exists. The main reason is that the equations are not separable. Thus, the presentation is limited to the variational equations which govern the resonance modes and the forced harmonic regime and to a general method for solving them is briefly outlined.

The third and last chapter deals with the important problems of acoustical engineering of sound generation by vibrating structures and sound transmission through elastic structures. It starts with a very simple academic one-dimensional example: the transmission of acoustic energy through a spring supported piston in a wave guide and the radiation of sound by such a system. Although this system is not realistic – we do not see how an experiment could be conducted – its simplicity makes it possible to develop an exhaustive study: the equations which describe the system can be solved analytically, both in the frequency and time domains. The interest of such an example is that it points out clearly the main aspects of the phenomena involved in sound transmission and sound radiation by vibrating structures.

After a short section, in which the basic concepts and equations of acoustics are recalled, several vibro-acoustics problems are examined in some detail. These concern plates and circular cylindrical shells. The important notion of "fluid-loaded resonance modes" is introduced: these modes are characteristics of the structure-fluid system and can be used to predict the response to any excitation (harmonic, transient, random). Numerical methods for computing either the resonance modes of a fluid-loaded structure or its response to an external excitation are described. Numerical results are given and, as far as possible, compared with experiments which have been selected from recent PhD theses.

At the end of the three chapters, a few exercises are proposed as complements of the text itself. At the end of this monograph, the bibliography proposes two kinds of references: basic textbooks in which the reader can find much more detail on the different aspects which are developed; specialized papers on the topics, and particularly those from which numerical and experimental results have been used to illustrate the theoretical developments.

The aim of this monograph is to present the basic concepts and methods necessary for the study of vibro-acoustics phenomena. As such, only classical analytical and numerical methods are described: separation of variables, series expansions in terms of special functions, matched asymptotic expansions, Boundary Element Methods (BEM). Nowadays, much more powerful numerical methods have been developed, for example, Statistical Energy Analysis (SEA), Finite Element Methods (FEM) and mixed methods such as various BEM–FEM methods, medium and high frequency approximations, numerical techniques for improving the performances of BEM and FEM computer programs (in particular the Fast Multi-pole Method), etc. Several specialized books have already been published on these topics. Several pieces of software for acoustics and vibro-acoustics engineering are now available.

Chapter 1

Equations Governing the Vibrations of Thin Structures

1.1. Introduction

1.1.1. *General Considerations on Thin Structures*

Thin structures are commonly called thin plates, thin shells, beams or rings (the term *thin* is generally omitted when there is no ambiguity). A plate or a shell is a solid in which one of its dimensions – called its *thickness* – is small compared to the other two; beams and rings are solids which have two dimensions which are small compared to the third one. The term *small* means that some approximations of the general equations of elasticity are sufficient to describe stresses and strains accurately enough: thus, depending on the accuracy which is required to predict the physical phenomenon, different equations are used. The present chapter deals with the simplest approximation. More elaborated approximations can be found in several textbooks, such as those mentioned in the bibliography [FLU 90, LAN 67, LEI 69, LEI 73, LOV 44].

The geometry of a shell (or a plate) is described by three variables: two of them – say ξ_1 and ξ_2 – are the parametric coordinates of a surface Σ (a plane in the case of a plate); the third one, ξ_3, in the direction normal to Σ and sometimes called the *transverse variable*, is a function of the first two which varies within two bounds, $-h^-(\xi_1, \xi_2)$ and $+h^+(\xi_1, \xi_2)$, which remain small compared to any characteristic dimension of Σ. Beams and rings are described in a similar way.

Approximate equations governing the vibrations of thin structures are based on several hypotheses. The main one is that the expansions of every mechanical quantity (displacement, forces, momentums, etc.) into a Taylor series of the transverse variable can be truncated at a low order. The second hypothesis is that the two boundaries $\xi_3 = -h^-(\xi_1, \xi_2)$ and $\xi_3 = +h^+(\xi_1, \xi_2)$ can be considered as free of any constraint: this means that the constraints exerted on these surfaces are small compared to the volume constraints. When vibrations are concerned, it is necessary to assume that wavelengths involved are large compared to the maximum thickness of the structure.

There are essentially two methods to establish approximate equations for the vibrations of thin structures. The most ancient one consists of approximating the forces and momentums exerted of an elementary volume of solid. This leads immediately to a system of partial differential equations and, then, the energy equations can be deduced.

The second method – which we can call the *energy method* – starts with the energy equation of the three-dimensional solid and approximations are made: this leads to the energy equation of the thin structure from which the partial differential equations are deduced. This second approach is adopted here. The main reason is that it leads to a variational form of the problem which is perfectly suitable for numerical computation (expansion of the solution in terms of a set of basis functions or finite element methods).

1.1.2. *Overview of the Energy Method*

Let Σ, with boundary $\partial\Sigma$, be a surface which can be parametrized by a coordinate system (ξ_1, ξ_2). It is assumed that a unit normal vector $\vec{\xi}_3$ exists everywhere on this surface. A point in the neighborhood of Σ can be defined by local coordinates (ξ_1, ξ_2, ξ_3), where ξ_3 is counted along the normal vector $\vec{\xi}_3$. For simplicity, it is assumed that this coordinate system is an orthogonal system.

Let us define two regular functions $h^-(\xi_1, \xi_2) < 0$ and $h^+(\xi_1, \xi_2) > 0$ with $|h^-|$ and h^+ small compared to the domains of variations of ξ_1 and ξ_2. Space domain Ω defined by $\{(\xi_1, \xi_2) \in \Sigma, h^-(\xi_1, \xi_2) \leq \xi_3 \leq h^+(\xi_1, \xi_2)\}$ is occupied by an elastic (or visco-elastic) solid. It is assumed that boundaries $\xi_3 = h^-(\xi_1, \xi_2)$ and $\xi_3 = h^+(\xi_1, \xi_2)$ are free (or submitted to loads which, in a first approximation, are negligible).

Let $\mathcal{D}_{ij}$ be the strain tensor and $\mathcal{S}_{ij}$ the stress tensor, where the subscripts i and j take the values 1, 2 and 3. The potential energy of the solid is given by the integral:

$$\mathcal{E}_p = \frac{1}{2} \int_{\Sigma} \int_{h^-}^{h^+} \mathcal{S}_{ij}\mathcal{D}_{ij} \, \mathrm{d}\xi_3 \, \mathrm{d}\Sigma$$

In this equation, as well as throughout this chapter, the convention of *summation over repeated subscripts* is adopted, that is:

$$\mathcal{S}_{ij}\mathcal{D}_{ij} = \sum_{i=1}^{3}\sum_{j=1}^{3} \mathcal{S}_{ij}\mathcal{D}_{ij}$$

Because of the hypothesis that the wavelengths of the vibratory waves are large compared with the thickness of domain Ω, the strain and stress tensors are expected to vary slowly with respect to variable ξ_3. Thus, it is reasonable to expand each of them into a Taylor series of this variable:

$$\mathcal{D}_{ij} = \mathcal{D}^0_{ij} + \xi_3\mathcal{D}^1_{ij} + \cdots \qquad\qquad \mathcal{S}_{ij} = \mathcal{S}^0_{ij} + \xi_3\mathcal{S}^1_{ij} + \cdots$$

The hypothesis of free boundaries for $\xi_3 = h^-$ and $\xi_3 = h^+$ is written as:

$$\mathcal{S}^0_{ij} + h^-\mathcal{S}^1_{ij} + \cdots = 0\,, \qquad \mathcal{S}^0_{ij} + h^+\mathcal{S}^1_{ij} + \cdots = 0 \qquad \forall(\xi_1, \xi_2)$$

This provides relationships between the terms of the stress tensor expansion, in particular, we obtain:

$$\mathcal{S}^0_{ij} = 0 \qquad \mathcal{S}^1_{ij} = 0$$

The stress–strain relationship (here, Hooke's law) is then applied and relationships between the $\mathcal{D}^k_{ij}$ are obtained. All these results are introduced into the expression of the potential energy. The quantity to be integrated is thus a Taylor series with respect to transverse variable ξ_3 and, as a consequence, the integral over this variable can be performed analytically. Finally, the potential energy is expressed by a two-dimensional integral over the mean surface Σ. The same approximation is made to express the kinetic energy.

To obtain the variational form of the approximated equation governing the vibrations of the thin body, the *virtual works theorem* is applied. As is usually done, an integration by parts leads to the corresponding partial differential equations and provides boundary conditions along $\partial\Sigma$.

1.2. Thin Plates

Let Σ be a domain of the plane (x_1, x_2), with boundary $\partial\Sigma$. It is assumed that there exists almost everywhere a unit vector $\vec{n}$ normal to $\partial\Sigma$ and pointing outward; there also exists a unit tangent vector $\vec{s}$ which makes an angle $\pi/2$ with $\vec{n}$. Let Ω be the cylindrical domain with basis Σ and extending from $x_3 = -h/2$ to $x_3 = h/2$, where h remains small compared to any characteristic dimension of Σ: this means that the thickness of Ω is a few percent of this characteristic length. In section 1.2.1, it is assumed that h is constant; the equations for a plate of variable thickness are given section 1.2.2.

A homogenous isotropic elastic solid occupies Ω: it has a density μ_s, a Young's modulus E and a Poisson's ratio ν. The boundaries $x_3 = -h/2$ and $x_3 = h/2$ are free (external forces applied to the plate are zero or negligible). It is assumed that there is no in-plane external force.

As is commonly done in mechanics, in the following, the derivation of a function f with respect to variable x_i is denoted by $f,_i$.

1.2.1. *Plate with Constant Thickness*

Let (U_1, U_2, U_3) be the components of the displacement of a point of the solid. The strain tensor $\mathcal{D}_{ij}$ is defined by:

$$\mathcal{D}_{ij} = \frac{1}{2}(U_i,_j + U_j,_i)$$

Let $\mathcal{S}_{ij}$ be the stress tensor. Assuming that Hooke's law is valid, the strain–stress relationship is expressed by:

$$\begin{aligned}
\mathcal{S}_{11} &= \frac{E}{(1+\nu)(1-2\nu)}\left[(1-\nu)\mathcal{D}_{11} + \nu(\mathcal{D}_{22} + \mathcal{D}_{33})\right] \\
\mathcal{S}_{22} &= \frac{E}{(1+\nu)(1-2\nu)}\left[(1-\nu)\mathcal{D}_{22} + \nu(\mathcal{D}_{33} + \mathcal{D}_{11})\right] \\
\mathcal{S}_{33} &= \frac{E}{(1+\nu)(1-2\nu)}\left[(1-\nu)\mathcal{D}_{33} + \nu(\mathcal{D}_{11} + \mathcal{D}_{22})\right] \\
\mathcal{S}_{12} &= \frac{E}{1+\nu}\mathcal{D}_{12} = \mathcal{S}_{21}\ ,\ \mathcal{S}_{13} = \frac{E}{1+\nu}\mathcal{D}_{13} = \mathcal{S}_{31} \\
\mathcal{S}_{23} &= \frac{E}{1+\nu}\mathcal{D}_{23} = \mathcal{S}_{32}
\end{aligned} \tag{1.1}$$

We look for approximations of the displacement and the stress tensor as truncated Taylor series in x_3, that is:

$$\begin{aligned}
U_i(x_1, x_2, x_3) &= U_i^0(x_1, x_2) + x_3 U_i^1(x_1, x_2) + \mathcal{O}(x_3{}^2) \\
\mathcal{S}_{ij}(x_1, x_2, x_3) &= \mathcal{S}_{ij}^0(x_1, x_2) + x_3 \mathcal{S}_{ij}^1(x_1, x_2) + \mathcal{O}(x_3{}^2)
\end{aligned}$$

The free boundary condition at $x_3 = \pm h/2$ is written as:

$$\mathcal{S}_{i3}(x_1, x_2, \pm h/2) = 0 \ \forall (x_1, x_2)$$

This implies the following equalities:

$$\mathcal{S}^0_{i3}(x_1, x_2) = \mathcal{S}^1_{i3}(x_1, x_2) = 0 \ \forall (x_1, x_2) \ \Rightarrow \ \mathcal{S}_{i3}(x_1, x_2, x_3) = \mathcal{O}({x_3}^2)$$

Introducing this result into Hooke's law, it appears that all the components of the displacement can be expressed in terms of component $w = U^0_3$ only; more precisely, we obtain:

$$\begin{gathered} U_1 \simeq -x_3 w_{,1} \quad , \quad U_2 \simeq -x_3 w_{,2} \\ \mathcal{D}_{11} \simeq -x_3 w_{,11} = d_{11} \quad , \mathcal{D}_{22} \simeq -x_3 w_{,22} = d_{22} \\ \mathcal{D}_{12} \simeq -x_3 w_{,12} = d_{12} \quad , \quad \mathcal{D}_{13} \simeq 0 = d_{13} \quad , \quad \mathcal{D}_{23} \simeq 0 = d_{23} \\ \mathcal{D}_{33} \simeq x_3 \frac{\nu}{1-\nu}\left(w_{,11} + w_{,22}\right) = d_{33} \end{gathered} \tag{1.2}$$

The potential energy of the solid is the integral over Ω of quantity $\mathcal{S}_{ij}\mathcal{D}_{ij}$; it is approximated by the following positive quantity:

$$\begin{aligned} \mathcal{E}_p &= \frac{E}{2(1-\nu^2)} \int_\Sigma \int_{-h/2}^{+h/2} {x_3}^2 \, \mathrm{d}x_3 \Big[\nu\left(w_{,11} + w_{,22}\right)^2 \\ &\qquad + (1-\nu)\left(w_{,12}^2 + w_{,21}^2 + w_{,11}^2 + w_{,22}^2\right)\Big] \mathrm{d}\Sigma \\ &= \frac{E}{2(1-\nu^2)} \frac{h^3}{12} \int_\Sigma \Big[\nu\left(w_{,11} + w_{,22}\right)^2 \\ &\qquad + (1-\nu)\left(w_{,12}^2 + w_{,21}^2 + w_{,11}^2 + w_{,22}^2\right)\Big] \mathrm{d}\Sigma \end{aligned} \tag{1.3}$$

The same approximations of the displacement leads to the following approximation for the kinetic energy:

$$\mathcal{E}_c = \frac{\mu_s h}{2} \int_\Sigma \dot{w}^2 \, \mathrm{d}\Sigma \tag{1.4}$$

where $\dot{w}$ is the time derivative of w.

Let us now assume that a force, normal to Σ, with density f is exerted on the plate. The *virtual works theorem* implies that the work of the external force corresponding to a virtual displacement δw obtained within a time interval δt

is equal to the variation of the total energy of the solid, that is:

$$\int_{\Sigma}\left\{\frac{Eh^3}{12(1-\nu^2)}\Big[\nu\big(w,_{11}+w,_{22}\big)\big(\delta w,_{11}+\delta w,_{22}\big) + (1-\nu)\big(w,_{11}\delta w,_{11}+w,_{22}\delta w,_{22}+w,_{12}\delta w,_{12}+w,_{21}\delta w,_{21}\big)\Big] + \mu_s h\,\ddot{w}\,\delta w\right\}\mathrm{d}\Sigma = \int_{\Sigma} f\,\delta w\;\mathrm{d}\Sigma \quad (1.5)$$

or equivalently:

$$\int_{\Sigma}\left\{\frac{Eh^3}{12(1-\nu^2)}\Big[\big(w,_{11}+w,_{22}\big)\big(\delta w,_{11}+\delta w,_{22}\big) + (1-\nu)\big(2w,_{12}\delta w,_{12}-w,_{11}\delta w,_{22}-w,_{22}\delta w,_{11}\big)\Big] + \mu_s h\,\ddot{w}\,\delta w\right\}\mathrm{d}\Sigma = \int_{\Sigma} f\,\delta w\;\mathrm{d}\Sigma \quad (1.5')$$

The variation of the kinetic energy is obtained using the following equality:

$$\delta\dot{w}^2 = 2\dot{w}\,\delta\dot{w} = 2\dot{w}\,\ddot{w}\,\delta t = 2\ddot{w}\,\delta w$$

Integrations by parts lead to:

$$\int_{\Sigma}\left\{\frac{Eh^3}{12(1-\nu^2)}\Delta^2 w + \mu_s h\,\ddot{w}\right\}\delta w\,\mathrm{d}\Sigma + \frac{Eh^3}{12(1-\nu^2)}\int_{\partial\Sigma}\big[\ell_1(w)\ \mathrm{Tr}\,\partial_n\delta w - \mathrm{Tr}\,\partial_n\Delta w\ \mathrm{Tr}\,\delta w + \ell_2(w)\ \mathrm{Tr}\,\partial_s\delta w\big]\,\mathrm{d}\bar{s} = \int_{\Sigma} f\,\delta w\;\mathrm{d}\Sigma \quad (1.6)$$

where $\bar{s}$ is the curvilinear abscissae along $\partial\Sigma$. The different operators in (1.6) are defined as follows:

$$\Delta^2 = \frac{\partial^4}{\partial x_1^4} + 2\frac{\partial^4}{\partial x_1^2\partial x_2^2} + \frac{\partial^4}{\partial x_2^4}$$

$$\mathrm{Tr}\,w(M) = \lim_{P\in\Sigma\to M\in\partial\Sigma} w(P)$$

$$\mathrm{Tr}\,\partial_n w = \lim_{P\in\Sigma\to M\in\partial\Sigma} \vec{n}(M)\cdot\vec{\nabla}_P\,w(P)\,,\quad \mathrm{Tr}\,\partial_s w = \lim_{P\in\Sigma\to M\in\partial\Sigma} \vec{s}(M)\cdot\vec{\nabla}_P\,w(P)$$

$$\mathrm{Tr}\,\partial_{s^2} w = \lim_{P\in\Sigma\to M\in\partial\Sigma} \vec{s}(M)\cdot\vec{\nabla}_P[\vec{s}(M)\cdot\vec{\nabla}_P\,w(P)]$$

$$\mathrm{Tr}\,\partial_n\partial_s w = \lim_{P\in\Sigma\to M\in\partial\Sigma} \vec{n}(M)\cdot\vec{\nabla}_P[\vec{s}(M)\cdot\vec{\nabla}_P\,w(P)]$$

$$\ell_1(w) = \mathrm{Tr}\,\Delta w - (1-\nu)\ \mathrm{Tr}\,\partial_{s^2} w \quad , \quad \ell_2(w) = (1-\nu)\ \mathrm{Tr}\,\partial_n\partial_s w$$

REMARK.– An elementary calculation shows that $\operatorname{Tr} \partial_s w = \partial_{\bar{s}} \operatorname{Tr} w$, where $\partial_{\bar{s}}$ is the derivation with respect to the curvilinear abscissae.

PROOF OF EQUATION (1.6).– Let θ be the angle between the axis x_1 and the normal $\vec{n}$ at a point P of $\partial\Sigma$. The differential operators with respect to (x_1, x_2) and to the directions (n, s) are related as follows (see Figure 1.1):

$$
\begin{aligned}
\partial_{x_1} &= \cos(\theta)\partial_n - \sin(\theta)\partial_s & \qquad \partial_n &= \cos(\theta)\partial_{x_1} + \sin(\theta)\partial_{x_2} \\
\partial_{x_2} &= \sin(\theta)\partial_n + \cos(\theta)\partial_s & \qquad \partial_s &= -\sin(\theta)\partial_{x_1} + \cos(\theta)\partial_{x_2}
\end{aligned}
$$

Let us consider the first integral in equation (1.5′)

$$
I_1 = \int_{\Sigma} (w_{,11} + w_{,22})(\delta w_{,11} + \delta w_{,22})\, \mathrm{d}\Sigma
$$

Its first component is integrated by parts with respect to x_1 and we obtain:

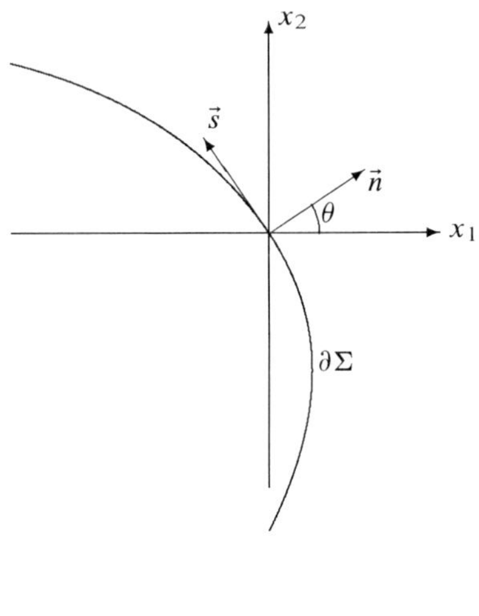

Figure 1.1. *Orientations of the normal and tangent unit vectors with respect to the coordinate axes*

$$I_{11} = \int_\Sigma (w_{,11} + w_{,22})\delta w_{,11}\,\mathrm{d}\Sigma = \int \mathrm{d}x_2 \int_{x_1^-(x_2)}^{x_1^+(x_2)} (w_{,11} + w_{,22})\delta w_{,11}\,\mathrm{d}x_1$$

$$= \int_\Sigma (w_{,1111} + w_{,2211})\delta w \mathrm{d}\Sigma$$

$$+ \int (w_{,11} + w_{,22})\delta w_{,1}\Big|_{x_1^-(x_2)}^{x_1^+(x_2)} \mathrm{d}x_2 - \int (w_{,11} + w_{,22})_{,1}\,\delta w\Big|_{x_1^-(x_2)}^{x_1^+(x_2)} \mathrm{d}x_2$$

$$= \int_\Sigma (w_{,1111} + w_{,2211})\delta w\,\mathrm{d}\Sigma$$

$$+ \int_{\partial\Sigma} \big[\operatorname{Tr} \Delta w \operatorname{Tr} \delta w_{,1} \cos(\theta) - \operatorname{Tr} \Delta w_{,1} \cos(\theta) \operatorname{Tr} \delta w\big] \mathrm{d}\bar{s}$$

In the same way, we obtain:

$$I_{12} = \int_\Sigma (w_{,11} + w_{,22})\delta w_{,22}\,\mathrm{d}\Sigma$$

$$= \int_\Sigma (w_{,1122} + w_{,2222})\delta w\,\mathrm{d}\Sigma + \int_{\partial\Sigma} \big[\operatorname{Tr} \Delta w \operatorname{Tr} \delta w_{,2} \sin(\theta)$$

$$- \operatorname{Tr} \Delta w_{,2} \sin(\theta) \operatorname{Tr} \delta w\big] \mathrm{d}\bar{s}$$

Gathering these results, we have that the first integral in (1.5′) becomes:

$$I_1 = \int_\Sigma \Delta^2 w\,\mathrm{d}\Sigma + \int_{\partial\Sigma} \big[\operatorname{Tr} \Delta w \operatorname{Tr} \partial_n \delta w - \operatorname{Tr} \partial_n \Delta w \operatorname{Tr} \delta w\big] \mathrm{d}\bar{s}$$

Let us now consider the terms with a factor $(1 - \nu)$. The calculation method being the same, we give the results only. In order to preserve the symmetric roles played by the variables x_1 and x_2, the first integral is split into two equal terms: the integrations by parts are performed with respect to x_1 and, then, to x_2 on the first term, and in the reverse order on the second term. The result is:

$$I_2 = 2\int_\Sigma w_{,12}\,\delta w_{,12}\,\mathrm{d}\Sigma = \int_\Sigma w_{,12}\,\delta w_{,12}\,\mathrm{d}\Sigma + \int_\Sigma w_{,21}\,\delta w_{,21}\,\mathrm{d}\Sigma$$

$$= 2\int_\Sigma w_{,1122}\,\delta w\,\mathrm{d}\Sigma + \int_{\partial\Sigma} \big[\operatorname{Tr} w_{,12} \operatorname{Tr} \delta w_{,1} \sin\theta - \operatorname{Tr} w_{,122} \operatorname{Tr} \delta w \cos\theta\big] \mathrm{d}\bar{s}$$

$$+ \int_{\partial\Sigma} \big[\operatorname{Tr} w_{,12} \operatorname{Tr} \delta w_{,2} \cos\theta - \operatorname{Tr} w_{,112} \operatorname{Tr} \delta w \sin\theta\big] \mathrm{d}\bar{s}$$

The third and fourth terms become:

$$\begin{aligned}
I_3 =& -\int_{\Sigma} w_{,11}\,\delta w_{,22}\,\mathrm{d}\Sigma \\
=& -\int_{\Sigma} w_{,1122}\,\delta w\,\mathrm{d}\Sigma - \int_{\partial\Sigma}\big[\operatorname{Tr} w_{,11}\ \operatorname{Tr}\delta w_{,2}\sin(\theta) - \operatorname{Tr} w_{,112}\sin(\theta)\operatorname{Tr}\delta w\big]\,\mathrm{d}\bar{s} \\
I_4 =& -\int_{\Sigma} w_{,22}\,\delta w_{,11}\,\mathrm{d}\Sigma \\
=& -\int_{\Sigma} w_{,1122}\,\delta w\,\mathrm{d}\Sigma - \int_{\partial\Sigma}\big[\operatorname{Tr} w_{,22}\ \operatorname{Tr}\delta w_{,1}\cos(\theta) - \operatorname{Tr} w_{,122}\cos(\theta)\operatorname{Tr}\delta w\big]\,\mathrm{d}\bar{s}
\end{aligned}$$

Summing up these results leads to:

$$\begin{aligned}
I_2 + I_3 + I_4 = \int_{\partial\Sigma} \Big[&\cos(\theta)\big(-\operatorname{Tr} w_{,22}\ \operatorname{Tr}\delta w_{,1} + \operatorname{Tr} w_{,12}\ \operatorname{Tr}\delta w_{,2}\big)+ \\
&\sin(\theta)\big(-\operatorname{Tr} w_{,11}\ \operatorname{Tr}\delta w_{,2} + \operatorname{Tr} w_{,12}\ \operatorname{Tr}\delta w_{,1}\big)\Big]\,\mathrm{d}\bar{s}
\end{aligned}$$

To end the proof, the derivation operators with respect to variables (x_1, x_2) are expressed in terms of the derivation operators with respect to (n, s). This is a simple, but nevertheless, tedious calculation which is left to the reader. □

The boundary integral in expression (1.6) represents the work of the different forces and moments that the plate exerts on its support. The physical meaning of the various terms is, thus, easy:

- ◇ $-Eh^3/12(1-\nu^2)\ \operatorname{Tr}\partial_n \Delta w$ is the factor of δw: it represents the density of shearing forces that the plate boundary exerts on its support.
- ◇ $-\ell_1(w) = -Eh^3/12(1-\nu^2)[\operatorname{Tr}\Delta w - (1-\nu)\ \operatorname{Tr}\partial_{s^2} w]$ is the factor of $\operatorname{Tr}\partial_n \delta w$: it represents the density of bending moments (rotation around the tangential direction).
- ◇ $-\ell_2(w) = -(1-\nu)Eh^3/12(1-\nu^2)\ \operatorname{Tr}\partial_n\partial_s w$ is the factor of $\operatorname{Tr}\partial_s\delta w$: it represents the density of twisting moments (rotation around the normal direction).

Finally, in the case of a regular boundary $\partial\Sigma$, that is, a boundary without angular points, the term $\ell_2(w)$ in equation (1.6) is continuous and an integration by parts of the term $\ell_2(w)\operatorname{Tr}\partial_s\delta w = \ell_2(w)\partial_{\bar{s}}\operatorname{Tr}\delta w$ can be performed without any caution. We obtain:

$$\begin{aligned}\int_{\Sigma}\left\{\frac{Eh^3}{12(1-\nu^2)}\Delta^2 w+\mu_s h\,\ddot{w}\right\}\delta w\,\mathrm{d}\Sigma \\ +\frac{Eh^3}{12(1-\nu^2)}\int_{\partial\Sigma}\Big\{\big[\operatorname{Tr}\Delta w-(1-\nu)\operatorname{Tr}\partial_{s^2}w\big]\operatorname{Tr}\partial_n\delta w \\ -\big[(1-\nu)\,\partial_{\bar{s}}\operatorname{Tr}\partial_n\partial_s w+\operatorname{Tr}\partial_n\Delta w\big]\operatorname{Tr}\delta w\Big\}\,\mathrm{d}\bar{s}=\int_{\Sigma}f\,\delta w\,\mathrm{d}\Sigma\end{aligned} \quad (1.7)$$

The term $-(1-\nu)Eh^3/12(1-\nu^2)\,\partial_s\mathrm{Tr}\,\partial_n\partial_s w$ is the tangential derivative of the twisting moment density. The coefficient of $\mathrm{Tr}\,\delta w$ is called the *Kelvin-Kirchhoff edge reaction*.

Integral relationship (1.7) must be satisfied for any virtual displacement δw; thus, the integrals over Σ and over $\partial\Sigma$ must cancel separately. The cancellation of the integral over Σ leads to the well-known *thin plate equation*:

$$\begin{aligned}&\left(D\Delta^2+\mu_s h\frac{\partial^2}{\partial t^2}\right)w=f\\ &\text{with :}\quad D=\frac{Eh^3}{12(1-\nu^2)}\end{aligned} \quad (1.8)$$

where D is called the *plate flexural rigidity*. If a harmonic time dependence of the form $\mathrm{e}^{-\imath\omega t}$ is assumed, this equation becomes:

$$\left(\Delta^2-\lambda^4\right)w=\frac{f}{D}\,,\quad \text{with } \lambda^4=\frac{\mu_s h\omega^2}{D} \quad (1.9)$$

(because no confusion can occur, we have used the same symbols f and w for the amplitudes of the harmonic excitation and the corresponding displacement: this avoids needless heavy notations).

The cancellation of the boundary integrals provides what is called *the natural boundary conditions* (which, of course, are a mathematical idealization of the physical conditions which can be imposed geometrically or mechanically); their expressions are the same for a transient or a harmonic excitation:

◇ Clamped boundary:

$$\mathrm{Tr}\,w=0\quad,\quad \mathrm{Tr}\,\partial_n w=0$$

◇ Free boundary:

$$\mathrm{Tr}\,\Delta w - (1-\nu)\,\mathrm{Tr}\,\partial_{s^2} w = 0$$
$$\mathrm{Tr}\,\partial_n \Delta w + (1-\nu)\partial_{\vec{s}}\mathrm{Tr}\,\partial_n \partial_s w = 0$$

The second of these two conditions is known as *Kirchhoff's condition* (Kirchhoff's contribution to the plates theory has been essential).

◇ Simply supported boundary:

$$\mathrm{Tr}\, w = 0 \quad , \quad \mathrm{Tr}\,\Delta w - (1-\nu)\ \mathrm{Tr}\,\partial_{s^2} w = 0$$

These boundary conditions imply that there is no energy loss across the plate boundaries. For that reason they are called *conservative boundary conditions.*

To conclude this section, let us mention that the plate equation obtained here is the simplest one. Many authors have developed more accurate equations which are valid for plates whose thickness is not very small; equally, equations for plates made of non-isotropic material and for sandwich plates can be found in the literature.

1.2.2. *Plate with Variable Thickness*

Accounting for a thickness variation does not present any extra difficulty. Following exactly the same steps as in the preceding section, we obtain:

$$\int_\Sigma \Big\{\Delta(D\Delta w) + (1-\nu)\big[2(Dw,_{12}),_{12}$$
$$- (Dw,_{11}),_{22} - (Dw,_{22}),_{11}\big] + \mu_s h\, \ddot{w}\Big\}\,\delta w\,\mathrm{d}\Sigma$$
$$+ \int_{\partial\Sigma} \big[\ell_1^v(w)\ \mathrm{Tr}\,\partial_n\delta w + \ell_2^v(w)\,\partial_{\vec{s}}\mathrm{Tr}\,\delta w - \ell_3^v(w)\ \mathrm{Tr}\,\delta w\big]\,\mathrm{d}s$$
$$= \int_\Sigma f\,\delta w\ \mathrm{d}\Sigma \qquad (1.10)$$

with:

$$D = \frac{Eh^3}{12(1-\nu^2)}$$
$$\ell_1^v(w) = \mathrm{Tr}\,(D\Delta w) - (1-\nu)\,\mathrm{Tr}\,(D\partial_{s^2} w)$$
$$\ell_2^v(w) = (1-\nu)\,\mathrm{Tr}\,(D\partial_n\partial_s w)$$
$$\ell_3^v(w) = \mathrm{Tr}\,\partial_n(D\Delta w) - (1-\nu)\big[\,\mathrm{Tr}\,\partial_s(D\partial_n\partial_s w) - \mathrm{Tr}\,\partial_n(D\partial_{s^2} w)\big]$$

These terms are related to the physical efforts exerted by the plate boundary on its support:

- $-\ell_3^v(w) =$ density of shearing forces;
- $-\ell_1^v(w) =$ density of bending moments;
- $-\ell_2^v(w) =$ density of twisting moments.

If $\partial\Sigma$ is a regular curve (no angular point), an integration by parts of the third term of the boundary integral can be performed, which leads to:

$$\int_\Sigma \Big\{ \Delta(D\Delta w) + (1-\nu)\big[2(Dw_{,12})_{,12} - (Dw_{,11})_{,22} - (Dw_{,22})_{,11}\big] + \mu_s h\,\ddot{w} \Big\}\,\delta w\,\mathrm{d}\Sigma + \int_{\partial\Sigma} \big\{\ell_1^v(w)\ \mathrm{Tr}\,\partial_n \delta w - \big[\partial_{\bar{s}}\ell_2^v(w) + \ell_3^v(w)\big]\,\mathrm{Tr}\,\delta w\big\}\,\mathrm{d}s = \int_\Sigma f\ \delta w\ \mathrm{d}\Sigma \quad (1.11)$$

This integral relationship must be satisfied for any virtual displacement, so the surface integral and the boundary integral must cancel separately. Thus, the plate displacement w satisfies the following partial-differential equation:

$$\Delta(D\Delta w) + (1-\nu)\big[2(Dw_{,12})_{,12} - (Dw_{,11})_{,22} - (Dw_{,22})_{,11}\big] + \mu_s h\,\ddot{w} = f \quad (1.12)$$

The cancellation of the boundary integral is obtained by the boundary conditions satisfied by w. As for the plate with constant thickness, there are three classical boundary conditions:

- Clamped boundary:

$$\mathrm{Tr}\,w = 0 \quad , \quad \mathrm{Tr}\,\partial_n w = 0$$

- Free boundary:

$$\mathrm{Tr}\,\big(D\Delta w\big) - (1-\nu)\ \mathrm{Tr}\,\big(D\partial_{s^2} w\big) = 0$$

$$\mathrm{Tr}\,\partial_n\big(D\Delta w\big) + (1-\nu)\big[\,\mathrm{Tr}\,\partial_n\big(D\partial_{s^2} w\big) - \mathrm{Tr}\,\partial_s\big(D\partial_n\partial_s w\big) + \partial_{\bar{s}}\mathrm{Tr}\,\big(D\partial_n\partial_s w\big)\big] = 0$$

◇ Simply supported boundary:

$$\operatorname{Tr} w = 0 \quad , \quad \operatorname{Tr}\big(D\Delta w\big) - (1-\nu)\ \operatorname{Tr}\big(D\partial_{s^2} w\big) = 0$$

The "clamped boundary" condition, which is purely geometrical, is identical to the result obtained for a plate of constant thickness. Conversely, the "free boundary" and the "simply supported boundary" conditions, which are essentially mechanical conditions, involve the variations in plate rigidity. These boundary conditions are often called *natural boundary conditions* because they appear naturally when the partial differential equation of the plate is established. However, in practice, the engineer is often faced with more complex boundary conditions which are more difficult to describe mathematically.

1.2.3. *Boundary with an Angular Point*

In the last two sections, equations (1.6) and (1.10) present no difficulty for performing an integration by parts because we are assured that the terms $\ell_2(w)$ and $\ell_2^v(w)$ are continuous. If the boundary has an angular point, these terms are not *a priori* continuous: indeed, they involve derivatives with respect to normal and tangent vectors. At an angular point, there are two normal vectors and two tangent vectors (see Figure 1.2). Assume that $\partial\Sigma$ has an angular point Q. Let $(\vec{n}_1, \vec{s}_1)$ and $(\vec{n}_2, \vec{s}_2)$ be the two sets of normal and tangent unit vectors. Let Q_1 – resp. Q_2 – be the limit of a point P belonging to the arc (1) – resp. (2) – tending to Q: of course Q_1 and Q_2 are geometrically the same point (they coincide with Q), but the sets of normal and tangent unit vectors are different. The integrals involving $\ell_2(w)$ and $\ell_2^v(w)$ are taken along the curve $\partial\Sigma$, in the trigonometric sense, starting from the point with normal and tangent vectors $(\vec{n}_1, \vec{s}_1)$ to the point with normal and tangent vectors $(\vec{n}_2, \vec{s}_2)$. Thus, they take the following forms:

$$\int_{\partial\Sigma} \ell_2(w)\partial_s \operatorname{Tr}\delta w \,\mathrm{d}\bar{s} = \int_{Q_1}^{Q_2} \ell_2(w)\partial_s \operatorname{Tr}\delta w \,\mathrm{d}\bar{s}$$

$$= -\int_{\partial\Sigma} \partial_s \ell_2(w) \operatorname{Tr}\delta w \,\mathrm{d}\bar{s} + \big[\ell_2(w)(Q_2) - \ell_2(w)(Q_1)\big] \operatorname{Tr}\delta w(Q)$$

$$\int_{\partial\Sigma} \ell_2^v(w)\partial_s \operatorname{Tr}\delta w \,\mathrm{d}\bar{s} = \int_{Q_1}^{Q_2} \ell_2^v(w)\partial_s \operatorname{Tr}\delta w \,\mathrm{d}\bar{s}$$

$$= -\int_{\partial\Sigma} \partial_{\bar{s}} \ell_2^v(w) \operatorname{Tr}\delta w \,\mathrm{d}\bar{s} + \big[\ell_2^v(w)(Q_2) - \ell_2^v(w)(Q_1)\big] \operatorname{Tr}\delta w(Q)$$

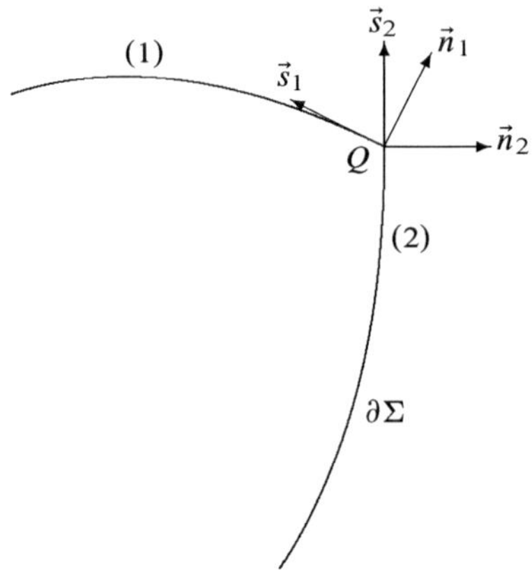

Figure 1.2. *The two sets of normal and tangent unit vectors at an angular point*

This result does not change the "clamped" and "simply supported" boundary conditions, but the "free boundary" condition must be modified as follows:

◇ constant thickness:

$$\begin{aligned} \mathrm{Tr}\,\Delta w - (1-\nu)\ \mathrm{Tr}\,\partial_{s^2} w &= 0 \\ \mathrm{Tr}\,\partial_n \Delta w + (1-\nu)\ \partial_{\bar{s}} \mathrm{Tr}\,\partial_n \partial_s w &= 0 \\ \mathrm{Tr}\,\partial_n \partial_s w(Q_1) &= \ \mathrm{Tr}\,\partial_n \partial_s w(Q_2) \end{aligned}$$

◇ variable thickness:

$$\begin{aligned} &\mathrm{Tr}\left(D\Delta w\right) - (1-\nu)\ \mathrm{Tr}\left(D\partial_{s^2} w\right) = 0 \\ &\mathrm{Tr}\,\partial_n\left(D\Delta w\right) + (1-\nu)\big[\ \mathrm{Tr}\,\partial_n\left(D\partial_{s^2} w\right) \\ &\qquad - \mathrm{Tr}\,\partial_s\left(D\partial_n\partial_s w\right) + \partial_{\bar{s}}\mathrm{Tr}\left(D\partial_n\partial_s w\right)\big] = 0 \\ &\mathrm{Tr}\left(D\partial_n\partial_s w\right)(Q_1) = \ \mathrm{Tr}\left(D\partial_n\partial_s w\right)(Q_2) \end{aligned}$$

The additional condition expresses that the twisting moments (rotation around the normal direction) must remain continuous even if the normal direction has a jump.

1.3. Beams

Let Ω be a space domain defined by: $\Omega = [-L/2 \leq x_1 \leq L/2, (x_2, x_3) \in \sigma(x_1)]$, the dimensions of the cross-section σ being small compared to L. An elastic solid occupying such a "long" and "linear" domain is called a beam. The aim of this section is to establish the equations governing the pure bending of a beam with a constant rectangular cross-section, that is $\sigma \equiv (0 < |x_2| < d/2, 0 < |x_3| < h/2)$. By *pure bending*, we mean that all the points of the solid which, at rest, are in a given plane parallel to the (x_1, x_2)-plane remain in this plane. Furthermore, it is assumed that the strains are small. Nevertheless, the displacement of a point can be rather large compared to d and h.

Beams with less simple geometries and submitted to more complicated deformations are commonly studied in many classical text-books (see, for example [LAN 67]): circular, elliptical cross-sections, combination of bending in two different directions, twisting around the axis, etc. The interest of the beam here is purely mathematical: indeed, the beam resonance modes provide a good basis for the computation of the resonance modes for an *in vacuo* or fluid-loaded rectangular plate. It is thus sufficient to pay attention to beams with a constant rectangular cross-section, made of a homogenous, isotropic and purely elastic material.

Considering a beam as a narrow rectangular plate, its displacement satisfies equation (1.5′). The hypothesis of pure flexion implies that the displacement is independent of the variable x_2 and an integration with respect to this variable leads to:

$$\int_{-L/2}^{L/2} \left[\frac{Eh^3d}{12(1-\nu^2)} w,_{11}\delta w,_{11} + \mu_s h d\,\ddot{w}\delta w\right] \mathrm{d}x_1 = \int_{-L/2}^{L/2} F\delta w \tag{1.13}$$

$$\text{with} \quad F = fd = \text{force per unit length}$$

After integrations by parts, we obtain:

$$\int_{-L/2}^{L/2} \left[\frac{Eh^3d}{12(1-\nu^2)} w,_{1111} + \mu_s h d\,\ddot{w}\right] \delta w \,\mathrm{d}x_1 + \frac{Eh^3d}{12(1-\nu^2)} \Big[w,_{11}\delta w,_1(L/2) - w,_{11}\delta w,_1(-L/2) - w,_{111}\delta w(L/2) + w,_{111}\delta w(-L/2) \Big] = \int_{-L/2}^{L/2} F\delta w \tag{1.14}$$

Classically, the factor ν^2, which is always small (less than 0.12) is neglected. Thus, the partial differential equation which governs the beam flexion is:

$$EI\frac{\partial^4 w}{\partial x_1^4} + \mu_s hd\frac{\partial^2 w}{\partial t_2} = F$$
$$\text{with} \qquad I = \frac{h^3 d}{12} \tag{1.15}$$

The quantity I is called the *inertia momentum* of the cross-section with respect to a line parallel to $x_3 = 0$ and $D = EI$ is called the *flexural rigidity* of the beam. This equation remains valid for any constant cross-section which is symmetric with respect to the plane x_3 and the inertia moment is defined by:

$$I = \int_\sigma x_3^2 \, \mathrm{d}x_1 \, \mathrm{d}x_2$$

Three boundary conditions at $x_1 = L/2$ and $x_1 = -L/2$ are commonly used:

◇ clamped boundaries: $w = 0 \quad , \quad \partial w/\partial x_1 = 0;$

◇ free boundaries: $\partial^2 w/\partial x_1^2 = 0 \quad , \quad \partial^3 w/\partial x_1^3 = 0;$

◇ simply supported boundaries: $w = 0 \quad , \quad \partial^2 w/\partial x_1^2 = 0;$

Final Comment

To establish the plate equation, it is generally assumed that there exists a *neutral surface*, that is, a surface along which the distance between two points remains constant: this is of course the case for a constant thickness plate and for a plate with a thickness which varies symmetrically with respect to the x_3-plane. The method used here does not require such a hypothesis.

Nevertheless, the result of the integration over variable x_3 which is performed to obtain equations (1.6) or (1.11) depends on the location of the plane $x_3 = 0$. Thus, the more accurate approximation is certainly obtained if this plane coincides with the neutral surface. The same remark applies equally to the beam equation.

1.4. Circular Cylindrical Shells

Let us consider a circular cylindrical surface Σ and a three-dimensional domain Ω defined in cylindrical coordinates by

$$\Sigma \equiv \{\rho = R \ , \ 0 \le \varphi < 2\pi \ , \ -L/2 < z < +L/2\}$$
$$\Omega \equiv \{\rho = R + r \text{ with } -h/2 < r < +h/2 \ , \ 0 \le \varphi < 2\pi \ , \ -L/2 < z < +L/2\}$$

where h, the thickness of Ω, is small compared to both R and $L/2$ (see Figure 1.3). An elastic, homogenous and isotropic solid – characterized by a density μ_s, a Young's modulus E and a Poisson's ratio ν – occupies Ω. The two boundaries $r = -h/2$ and $r = +h/2$ of the elastic solid are free (zero or negligible forces). The displacement of a point of the solid is denoted by $U\vec{e}_z + V\vec{e}_\varphi + W\vec{e}_r$.

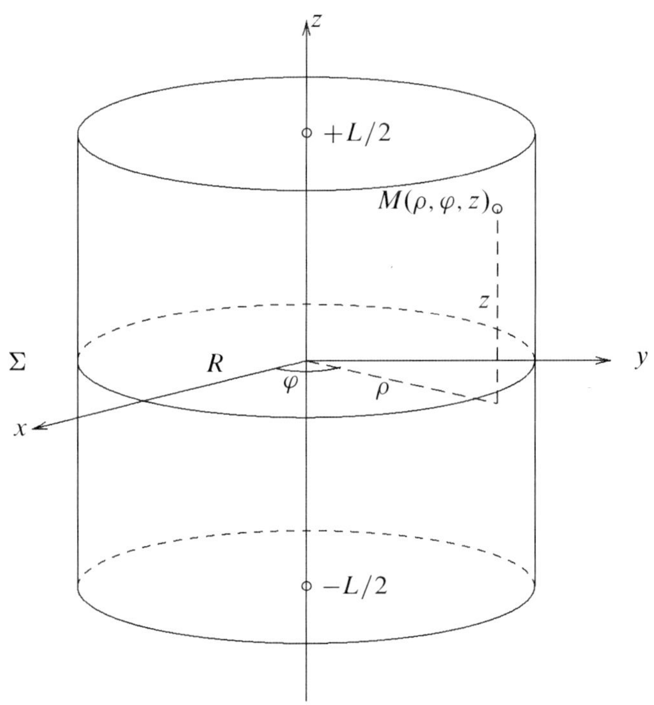

Figure 1.3. *The coordinate system of the cylindrical shell*

The strain tensor $\mathcal{D}_{ij}$ is thus:

$$\begin{aligned}
\mathcal{D}_{zz} &= U,_z & \mathcal{D}_{z\varphi} &= \frac{1}{2}\left[\frac{U,_\varphi}{R+r} + V,_z\right] \\
\mathcal{D}_{\varphi\varphi} &= \frac{V,_\varphi}{R+r} + \frac{W}{R+r} & \mathcal{D}_{zr} &= \frac{1}{2}\left[W,_z + U,_r\right] \\
\mathcal{D}_{rr} &= W,_r & \mathcal{D}_{\varphi r} &= \frac{1}{2}\left[V,_r - \frac{V}{R+r} + \frac{W,_\varphi}{R+r}\right]
\end{aligned} \tag{1.16}$$

Hooke's law relates this tensor to the stress tensor $\mathcal{S}_{ij}$ by:

$$\begin{aligned}
\mathcal{S}_{zz} &= \frac{E}{(1+\nu)(1-2\nu)}\left[(1-\nu)\mathcal{D}_{zz} + \nu(\mathcal{D}_{\varphi\varphi} + \mathcal{D}_{rr})\right] \\
\mathcal{S}_{z\varphi} &= \frac{E}{1+\nu}\mathcal{D}_{z\varphi} \\
\mathcal{S}_{\varphi\varphi} &= \frac{E}{(1+\nu)(1-2\nu)}\left[(1-\nu)\mathcal{D}_{\varphi\varphi} + \nu(\mathcal{D}_{rr} + \mathcal{D}_{zz})\right] \\
\mathcal{S}_{\varphi r} &= \frac{E}{1+\nu}\mathcal{D}_{\varphi r} \\
\mathcal{S}_{rr} &= \frac{E}{(1+\nu)(1-2\nu)}\left[(1-\nu)\mathcal{D}_{rr} + \nu(\mathcal{D}_{zz} + \mathcal{D}_{\varphi\varphi})\right] \\
\mathcal{S}_{rz} &= \frac{E}{1+\nu}\mathcal{D}_{rz}
\end{aligned} \tag{1.17}$$

The hypothesis that h is small prompts us to look for truncated Taylor series for the shell displacement and stress tensor:

$$\begin{aligned}
U &= U^0(z,\varphi) + rU^1(z,\varphi) + \mathcal{O}(r^2) \\
V &= V^0(z,\varphi) + rV^1(z,\varphi) + \mathcal{O}(r^2) \\
W &= W^0(z,\varphi) + rW^1(z,\varphi) + \mathcal{O}(r^2) \\
\mathcal{S}_{ij} &= \mathcal{S}^0_{ij}(z,\varphi) + r\mathcal{S}^1_{ij}(z,\varphi) + \mathcal{O}(r^2) \quad \text{with } i,j = r,\varphi,z
\end{aligned}$$

The "free boundary" condition for $r = \pm h/2$, together with the approximation $1/(R+r) \simeq R^{-1}(1-r/R)$, implies

$$(1-\nu)\mathcal{D}_{rr} + \nu(\mathcal{D}_{zz} + \mathcal{D}_{\varphi\varphi}) = \mathcal{O}(r^2)$$

$$W^0,_z + U^1 \pm \frac{h}{2}W^1,_z = 0 \quad , \quad \forall(\varphi,z)$$

$$V^1 - \frac{V^0 - W^0,_\varphi}{R} \mp \frac{h/2}{R}\left(V^1 - \frac{V^0 - W^0,_\varphi}{R} - W^1,_\varphi\right) = 0 \quad , \quad \forall(\varphi,z)$$

In the last two equations, the terms independent of h and the coefficients of h must be independently zero. As a first consequence, W^1 is independent of the two variables φ and z: making it equal to zero implies that the variation of the shell thickness is a second order phenomenon. The second consequence is that U^1 and V^1 are expressed in terms of V^0 and W^0 only:

$$U^1 = -W^0_{,z} \quad , \quad V^1 = \frac{V^0 - W^0_{,\varphi}}{R}$$

The only unknown functions are thus:

$$u = U^0, \quad v = V^0, \quad w = W^0$$

In what follows, we denote by d_{ij} and σ_{ij} the approximations of the strain and stress tensors which are defined by:

$$\begin{aligned}
d_{zz} &= u_{,z} - r w_{,zz} & d_{z\varphi} &= \frac{1}{2}\left(\frac{u_{,\varphi}}{R} + v_{,z} - \frac{2r}{R} w_{,z\varphi}\right) \\
d_{\varphi\varphi} &= \frac{1}{R}\left(v_{,\varphi} + w - \frac{r}{R} w_{,\varphi\varphi}\right) & d_{zr} &= 0 \\
d_{rr} &= -\frac{\nu}{1-\nu}\left(d_{zz} + d_{\varphi\varphi}\right) & d_{\varphi r} &= 0
\end{aligned} \tag{1.18}$$

$$\begin{aligned}
\sigma_{zz} &= \frac{E}{1-\nu^2}\left(d_{zz} + \nu d_{\varphi\varphi}\right) & \sigma_{z\varphi} &= \frac{E}{1+\nu} d_{z\varphi} \\
\sigma_{\varphi\varphi} &= \frac{E}{1-\nu^2}\left(d_{\varphi\varphi} + \nu d_{zz}\right) & \sigma_{zr} &= 0 \\
\sigma_{rr} &= 0 & \sigma_{\varphi r} &= 0
\end{aligned} \tag{1.18'}$$

The corresponding approximation of the potential energy density is expressed as:

$$\begin{aligned}
\mathrm{d}\mathcal{E}_p = \frac{E}{2(1-\nu^2)} \Bigg[& \left(u_{,z}^2 + \left(\frac{v_{,\varphi} + w}{R}\right)^2 + 2\nu \frac{u_{,z}\left(v_{,\varphi} + w\right)}{R} \right. \\
& \left. + \frac{1-\nu}{2}\left(\frac{u_{,\varphi}}{R} + v_{,z}\right)^2 \right) + r^2 \left(w_{,zz}^2 + \frac{w_{,\varphi\varphi}^2}{R^4} + \frac{2\nu}{R^2} w_{,zz}\, w_{,\varphi\varphi} \right. \\
& \left. + (1-\nu)\left(\frac{w_{,z\varphi}}{R}\right)^2 + (1-\nu)\left(\frac{w_{,\varphi z}}{R}\right)^2 \right) + r(\cdots) \Bigg]
\end{aligned}$$

In this expression, the tensor components $d_{z\varphi}$ (resp. $\sigma_{z\varphi}$) and $d_{\varphi z}$ (resp. $\sigma_{\varphi z}$) have been distinguished though their expressions are the same: indeed, they represent different physical quantities which, after the forthcoming integrations by parts, give different boundary terms.

Because the vector (u, v, w) do not depend on r, this expression can be integrated analytically with respect to this variable; the only terms to be accounted for are those which involve an even power (0 and 2) in r; the odd powers give an integral which is zero. Thus, the approximation of the potential energy of the shell is given by:

$$\mathcal{E}_p = \int_0^{2\pi} R\,\mathrm{d}\varphi \int_{-L/2}^{L/2} \mathrm{d}z \frac{Eh}{2(1-\nu^2)} \left\{ \nu\left(u_{,z} + \frac{v_{,\varphi} + w}{R}\right)^2 \right.$$

$$+ (1-\nu)\left[u_{,z}^2 + \left(\frac{v_{,\varphi} + w}{R}\right)^2\right] + \frac{1-\nu}{2}\left(\frac{u_{,\varphi}}{R} + v_{,z}\right)^2$$

$$\left. + \frac{h^2}{12}\left[\nu\left(w_{,zz} + \frac{w_{,\varphi\varphi}}{R^2}\right)^2 + (1-\nu)\left(w_{,zz}^2 + \frac{w_{,z\varphi}^2}{R^2} + \frac{w_{,\varphi z}^2}{R^2} + \frac{w_{,\varphi\varphi}^2}{R^4}\right)\right]\right\} \quad (1.19)$$

which is a positive quantity. The kinetic energy is

$$\mathcal{C} = \frac{\mu_s h}{2} \int_0^{2\pi} R d\varphi \int_{-L/2}^{L/2} \left(\dot{u}^2 + \dot{v}^2 + \dot{w}^2\right) \mathrm{d}z$$

Let f be the density of a force exerted on the shell with components (f_z, f_φ, f_r). The virtual works theorem leads to the variational equation of the thin shell:

$$\int_0^{2\pi} R d\varphi \int_{-L/2}^{L/2} dz \left\{ \frac{Eh}{1-\nu^2}\left[\nu\left(u_{,z} + \frac{v_{,\varphi} + w}{R}\right)\left(\delta u_{,z} + \frac{\delta v_{,\varphi} + \delta w}{R}\right)\right.\right.$$

$$+ (1-\nu)\left(u_{,z}\,\delta u_{,z} + \frac{v_{,\varphi} + w}{R}\frac{\delta v_{,\varphi} + \delta w}{R}\right) + \frac{1-\nu}{2}\left(\frac{u_{,\varphi}}{R} + v_{,z}\right)\left(\frac{\delta u_{,\varphi}}{R} + \delta v_{,z}\right)$$

$$+ \frac{h^2}{12}\left[\nu\left(w_{,zz} + \frac{w_{,\varphi\varphi}}{R^2}\right)\left(\delta w_{,zz} + \frac{\delta w_{,\varphi\varphi}}{R^2}\right) + (1-\nu)\left(w_{,zz}\,\delta w_{,zz} + \frac{w_{,z\varphi}\,\delta w_{,z\varphi}}{R^2}\right.\right.$$

$$\left.\left.\left.\left. + \frac{w_{,\varphi z}\,\delta w_{,\varphi z}}{R^2} + \frac{w_{,\varphi\varphi}\,\delta w_{,\varphi\varphi}}{R^4}\right)\right]\right] + \mu_s h(\ddot{u}\delta u + \ddot{v}\delta v + \ddot{w}\delta w)\right\}$$

$$= \int_0^{2\pi} R\,\mathrm{d}\varphi \int_{-L/2}^{L/2} \mathrm{d}z\left(f_z\delta u + f_\varphi\delta v + f_r\delta w\right) \quad (1.20)$$

Gathering the different terms, we obtain an equivalent expression:

$$\int_0^{2\pi} R\,\mathrm{d}\varphi \int_{-L/2}^{L/2} \mathrm{d}z \left\{ \frac{Eh}{1-\nu^2}\left[u_{,z}\,\delta u_{,z} + \frac{1}{R^2}(v_{,\varphi} + w)(\delta v_{,\varphi} + \delta w)\right.\right.$$

$$+ \frac{\nu}{R}(v_{,\varphi} + w)\delta u_{,z} + \frac{\nu}{R}u_{,z}\,(\delta v_{,\varphi} + \delta w) + \frac{1-\nu}{2}\left(\frac{u_{,\varphi}}{R} + v_{,z}\right)\left(\frac{\delta u_{,\varphi}}{R} + \delta v_{,z}\right)$$

$$+ \frac{h^2}{12}\left(w_{,zz}\,\delta w_{,zz} + \frac{1}{R^4}w_{,\varphi\varphi}\,\delta w_{,\varphi\varphi} + \frac{\nu}{R^2}w_{,\varphi\varphi}\,\delta w_{,zz} + \frac{\nu}{R^2}w_{,zz}\,\delta w_{,\varphi\varphi}\right.$$

$$+\frac{1-\nu}{R^2}w_{,z\varphi}\,\delta w_{,z\varphi}+\frac{1-\nu}{R^2}w_{,\varphi z}\,\delta w_{,\varphi z}\Big)\Big]+\mu_s h(\ddot{u}\delta u+\ddot{v}\delta v+\ddot{w}\delta w)\Big\}$$

$$=\int_0^{2\pi}R\,\mathrm{d}\varphi\int_{-L/2}^{L/2}\mathrm{d}z\big(f_z\delta u+f_\varphi\delta v+f_r\delta w\big) \tag{1.20'}$$

Integrations by parts are now performed and boundary integrals along the two circles $z=-L/2$ and $z=+L/2$ appear:

$$\int_0^{2\pi}Rd\varphi\int_{-L/2}^{L/2}dz\Big\{\frac{Eh}{1-\nu^2}\Big\{-\delta u\Big(u_{,zz}+\frac{1-\nu}{2R^2}u_{,\varphi\varphi}+\frac{1+\nu}{2R}v_{,\varphi z}+\frac{\nu}{R}w_{,z}\Big)$$

$$-\delta v\Big(\frac{1+\nu}{2R}u_{,z\varphi}+\frac{v_{,\varphi\varphi}}{R^2}+\frac{1-\nu}{2}v_{,zz}+\frac{1}{R^2}w_{,\varphi}\Big)+\delta w\Big[\frac{\nu}{R}u_{,z}+\frac{v_{,\varphi}}{R^2}+\frac{w}{R^2}$$

$$+\frac{h^2}{12}\Big(w_{,zzzz}+\frac{2}{R^2}w_{,zz\varphi\varphi}+\frac{1}{R^4}w_{,\varphi\varphi\varphi\varphi}\Big)\Big]\Big\}+\mu_s h(\ddot{u}\delta u+\ddot{v}\delta v+\ddot{w}\delta w)\Big\}$$

$$+\frac{Eh}{1-\nu^2}\int_0^{2\pi}Rd\varphi\Big\{\delta u\Big[u_{,z}+\frac{\nu}{R}(v_{,\varphi}+w)\Big]+\delta v\frac{1-\nu}{2}\Big(\frac{u_{,\varphi}}{R}+v_{,z}\Big)$$

$$+\frac{h^2}{12}\Big[-\delta w\Big(w_{,zzz}+\frac{w_{,\varphi\varphi z}}{R^2}\Big)+\frac{1-\nu}{R^2}\delta w_{,\varphi}\,w_{,\varphi z}$$

$$+\delta w_{,z}\Big(w_{,zz}+\frac{\nu}{R^2}w_{,\varphi\varphi}\Big)\Big]\Big\}_{z=-L/2}^{z=+L/2}$$

$$=\int_0^{2\pi}Rd\varphi\int_{-L/2}^{L/2}dz\big(f_z\delta u+f_\varphi\delta v+f_r\delta w\big) \tag{1.21}$$

This integral relationship must be satisfied for any virtual displacement vector $(\delta u,\delta v,\delta w)$: this can be achieved if the surface integrals and the boundary integrals cancel independently.

The surface integrals cancel for any virtual displacement vector if the following system of partial differential equations is satisfied:

$$\frac{Eh}{1-\nu^2}\mathcal{M}^c\begin{pmatrix}u\\v\\w\end{pmatrix}+\mu_s h\begin{pmatrix}\ddot{u}\\\ddot{v}\\\ddot{w}\end{pmatrix}=\begin{pmatrix}f_z\\f_\varphi\\f_r\end{pmatrix} \tag{1.22}$$

where the matrix operator $\mathcal{M}^c$ is:

$$\begin{pmatrix} -\frac{\partial^2}{\partial z^2}-\frac{1-\nu}{2R^2}\frac{\partial^2}{\partial\varphi^2} & -\frac{1+\nu}{2R}\frac{\partial^2}{\partial z\partial\varphi} & -\frac{\nu}{R}\frac{\partial}{\partial z} \\ -\frac{1+\nu}{2R}\frac{\partial^2}{\partial z\partial\varphi} & -\frac{1-\nu}{2}\frac{\partial^2}{\partial z^2}-\frac{1}{R^2}\frac{\partial^2}{\partial\varphi^2} & -\frac{1}{R^2}\frac{\partial}{\partial\varphi} \\ \frac{\nu}{R}\frac{\partial}{\partial z} & \frac{1}{R^2}\frac{\partial}{\partial\varphi} & \frac{1}{R^2}+\frac{h^2}{12}\left[\frac{\partial^4}{\partial z^4} + \frac{2}{R^2}\frac{\partial^4}{\partial z^2\partial\varphi^2}+\frac{1}{R^4}\frac{\partial^4}{\partial\varphi^4}\right] \end{pmatrix} \tag{1.22'}$$

This is the simplest approximation that can be established for a circular cylindrical shell; it is known as the *Donnell and Mushtari shell approximation*. Instead of matrix (1.22′), most authors prefer to use a symmetric matrix obtained by changing the sign of the last equation (1.22).

To cancel the boundary integrals, it is necessary to impose four boundary conditions to the shell displacement. The most well-known "natural" boundary conditions are:

◇ Clamped boundary:

$$\mathrm{Tr}\,u = 0 \qquad \mathrm{Tr}\,v = 0$$
$$\mathrm{Tr}\,w = 0 \qquad \mathrm{Tr}\,w,_z = 0$$

◇ Free boundary:

$$\mathrm{Tr}\,u,_z + \frac{\nu}{R}\,\mathrm{Tr}\left[v,_\varphi + w\right] = 0 \qquad \frac{\mathrm{Tr}\,u,_\varphi}{R} + \mathrm{Tr}\,v,_z = 0$$
$$\mathrm{Tr}\,w,_{zz} + \frac{\nu}{R^2}\,\mathrm{Tr}\,w,_{\varphi\varphi} = 0$$
$$\mathrm{Tr}\,w,_{zzz} + \frac{\mathrm{Tr}\,w,_{z\varphi\varphi}}{R^2} + \frac{1-\nu}{R^2}(\,\mathrm{Tr}\,w,_{z\varphi}\,),_\varphi = 0$$

◇ Shear diaphragm (other terminologies: simply or freely supported boundary):

$$\mathrm{Tr}\,v = 0 \qquad \mathrm{Tr}\,w = 0$$
$$\mathrm{Tr}\,u,_z + \frac{\nu}{R}\,\mathrm{Tr}\left[v,_\varphi + w\right] = 0 \qquad \mathrm{Tr}\,w,_{zz} + \frac{\nu}{R^2}\,\mathrm{Tr}\,w,_{\varphi\varphi} = 0$$

Physically, this condition corresponds to a shell closed by a very thin plate: due to its high in-plane rigidity, components v and w of the displacement of the shell boundaries are close to zero; because of its negligible flexural rigidity, it exerts negligible longitudinal force and bending moment.

When any of these boundary conditions are satisfied, no energy is lost across the shell boundaries. They are thus called *conservative boundary conditions.*

The names "free boundary" and "shear diaphragm" are justified by the physical interpretation of the boundary terms. The coefficients of the displacement components δu, δv and δw are force line densities:

$$\begin{aligned} F_{zz} &= \frac{Eh}{1-\nu^2}\left[u,_z + \frac{\nu}{R}\left(v,_\varphi + w\right)\right] \\ F_{z\varphi} &= \frac{Eh}{1-\nu^2}\frac{1-\nu}{2}\left(\frac{u,_\varphi}{R} + v,_z\right) \\ F_{zr} &= -\frac{Eh^3}{12(1-\nu^2)}\left(w,_{zzz} + \frac{w,_{\varphi\varphi z}}{R^2}\right) \end{aligned} \tag{1.22-a}$$

The force F_{zz} is tangential to the shell in the z-direction (compression-dilatation force) and the force $F_{z\varphi}$ is tangential to the shell in the φ-direction (tangential shearing force). The third force, F_{zr}, is a shearing force, normal to the shell. It must be noted that it is of order h^3, while the other forces are of order h: this is in accordance with the hypothesis that the components $\mathcal{S}_{rz}$, $\mathcal{S}_{r\varphi}$ and $\mathcal{S}_{rr}$ of the stress tensor are negligible along the shell boundaries $r = \pm h/2$.

The term proportional to $\delta w,_\varphi / R$, the derivative of the virtual normal displacement in the φ-direction, is a twisting moment line density:

$$M_{z\varphi} = \frac{Eh^3}{12(1-\nu^2)}\frac{1-\nu}{R} w,_{\varphi z} \tag{1.22-b}$$

Finally, the term proportional to $\delta w,_z$, the z-derivative of the virtual normal displacement, is a bending moment line density:

$$M_{zz} = \frac{Eh^3}{12(1-\nu^2)}\left(w,_{zz} + \frac{\nu}{R^2} w,_{\varphi\varphi}\right) \tag{1.22-c}$$

The force line densities in the tangential directions and the line moment densities can be deduced from the approximations of the stress tensor components by an integration over the thickness of the shell. Because of the hypothesis that the two faces of the shell are free, the shearing force density cannot be deduced from the approximation of the r-components of the stress tensor.

More accurate cylindrical shell equations have been developed (in particular by W. FLÜGGE [FLU 90]) which are valid under less restrictive assumptions: thicker structures, damped material, etc.

REMARK.– We have restricted this study to a cylindrical shell limited by two circles. It is, of course, possible to consider cylindrical shells limited by any contour: the integration by parts of equation (1.20′) results in boundary integrals which are less simple than those which appear in expression (1.21), but the quantities involved are of the same physical nature.

1.5. Spherical Shells

Let Σ be a spherical surface and Ω be a three-dimensional domain defined in spherical coordinates by (see Figure 1.4):

$$\Sigma \equiv \{\rho = R, 0 \le \varphi < 2\pi, \theta_1 < \theta < \theta_2\}$$
$$\Omega \equiv \{R - h/2 < \rho = R + r < R + h/2, 0 \le \varphi < 2\pi, \theta_1 < \theta < \theta_2\}$$

where h, the thickness of Ω, is small compared to R. An elastic, homogenous

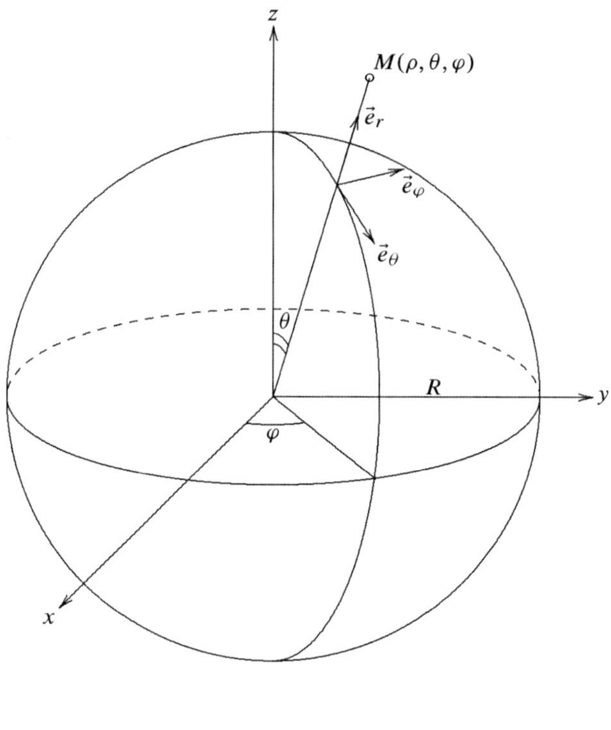

Figure 1.4. *The coordinate system for the spherical shell*

and isotropic solid – characterized by a density μ_s, a Young's modulus E and a Poisson's ratio ν – occupies Ω; the boundaries $r = -h/2$ and $r = h/2$ are free.

The displacement of a point of the solid is denoted by $U\vec{e}_\theta + V\vec{e}_\varphi + W\vec{e}_r$. The strain tensor $\mathcal{D}_{ij}$ is given by:

$$
\begin{aligned}
\mathcal{D}_{\theta\theta} &= \frac{U_{,\theta}}{R+r} + \frac{W}{R+r} \quad , \quad \mathcal{D}_{\theta\varphi} = \frac{1}{2}\frac{1}{R+r}\left[\frac{U_{,\varphi}}{\sin\theta} + V_{,\theta} - V \operatorname{cotg}\theta\right] \\
\mathcal{D}_{\varphi\varphi} &= \frac{U \operatorname{cotg}\theta}{R+r} + \frac{1}{R+r}\frac{V_{,\varphi}}{\sin\theta} + \frac{W}{R+r} \\
\mathcal{D}_{\theta r} &= \frac{1}{2}\left[\frac{W_{,\theta}}{R+r} + U_{,r} - \frac{U}{R+r}\right] \\
\mathcal{D}_{rr} &= W_{,r} \quad , \quad \mathcal{D}_{\varphi r} = \frac{1}{2}\left[\frac{1}{R+r}\frac{W_{,\varphi}}{\sin\theta} + V_{,r} - \frac{V}{R+r}\right]
\end{aligned}
\tag{1.23}
$$

Using Hooke's law, the stress tensor is written as:

$$
\begin{aligned}
\mathcal{S}_{\theta\theta} &= \frac{E}{(1+\nu)(1-2\nu)}\left[(1-\nu)\mathcal{D}_{\theta\theta} + \nu(\mathcal{D}_{\varphi\varphi} + \mathcal{D}_{rr})\right] \quad , \quad \mathcal{S}_{\theta\varphi} = \frac{E}{1+\nu}\mathcal{D}_{\theta\varphi} \\
\mathcal{S}_{\varphi\varphi} &= \frac{E}{(1+\nu)(1-2\nu)}\left[(1-\nu)\mathcal{D}_{\varphi\varphi} + \nu(\mathcal{D}_{rr} + \mathcal{D}_{\theta\theta})\right] \quad , \quad \mathcal{S}_{\varphi r} = \frac{E}{1+\nu}\mathcal{D}_{\varphi r} \\
\mathcal{S}_{rr} &= \frac{E}{(1+\nu)(1-2\nu)}\left[(1-\nu)\mathcal{D}_{rr} + \nu(\mathcal{D}_{\theta\theta} + \mathcal{D}_{\varphi\varphi})\right] \quad , \quad \mathcal{S}_{r\theta} = \frac{E}{1+\nu}\mathcal{D}_{r\theta}
\end{aligned}
\tag{1.24}
$$

1.5.1. *Approximation of the Strain and Stress Tensors and Application of the Virtual Works Theorem*

As for the cylindrical shell, the displacement and the stress tensor components are sought as truncated Taylor series:

$$
\begin{aligned}
U &= U^0(\theta,\varphi) + rU^1(\theta,\varphi) + \mathcal{O}(r^2) \\
V &= V^0(\theta,\varphi) + rV^1(\theta,\varphi) + \mathcal{O}(r^2) \\
W &= W^0(\theta,\varphi) + rW^1(\theta,\varphi) + \mathcal{O}(r^2) \\
\mathcal{S}_{ij} &= \mathcal{S}^0_{ij}(\theta,\varphi) + r\mathcal{S}^1_{ij}(\theta,\varphi) + \mathcal{O}(r^2) \ \text{ with } \ i,j = r,\theta,\varphi
\end{aligned}
$$

Using the approximation:

$$
\frac{1}{R+r} = \frac{1}{R}\left(1 - \frac{r}{R}\right) + \mathcal{O}(r^2)
$$

the components $\mathcal{D}_{\theta r}$ and $\mathcal{D}_{\varphi r}$ of the strain tensor take the form:

$$
\begin{aligned}
\mathcal{D}_{\theta r} &= \frac{1}{2}\left[-\frac{U^0}{R} + U^1 + \frac{W^0_{,\theta}}{R}\right] - \frac{r}{2R}\left[-\frac{U^0}{R} + U^1 + \frac{W^0_{,\theta}}{R} - W^1_{,\theta}\right] + \mathcal{O}(r^2) \\
\mathcal{D}_{\varphi r} &= \frac{1}{2}\left[-\frac{V^0}{R} + V^1 + \frac{W^0_{,\varphi}}{R\sin\theta}\right] - \frac{r}{2R}\left[-\frac{V^0}{R} + V^1 + \frac{W^0_{,\varphi}}{R\sin\theta} - \frac{W^1_{,\varphi}}{\sin\theta}\right] + \mathcal{O}(r^2)
\end{aligned}
$$

The free boundary condition for $r = \pm h/2$ leads to:

$$\mathcal{S}^0_{ir}(\theta,\varphi) = \mathcal{S}^1_{ir}(\theta,\varphi) = 0 \ , \ i = r,\theta,\varphi$$

which implies that:

$$\begin{aligned}
\mathcal{D}_{rr} &= -\frac{\nu}{1-\nu}\left(\mathcal{D}_{\theta\theta} + \mathcal{D}_{\varphi\varphi}\right) \\
-\frac{U^0}{R} + U^1 + \frac{W^0_{,\theta}}{R} &= -\frac{V^0}{R} + V^1 + \frac{W^0_{,\varphi}}{R\sin\theta} = 0 \\
-\frac{U^0}{R} + U^1 + \frac{W^0_{,\theta}}{R} - W^1_{,\theta} &= -\frac{V^0}{R} + V^1 + \frac{W^0_{,\varphi}}{R\sin\theta} - \frac{W^1_{,\varphi}}{\sin\theta} = 0
\end{aligned} \qquad (1.25)$$

The first result is that W^1 is independent of both variables θ and φ: by taking it equal to zero, the thickness variations of the shell are made negligible. The components U^1 and V^1 are expressed in terms of $u = U^0$, $v = V^0$ and $w = W^0$ as:

$$U^1 = \frac{u - w_{,\theta}}{R} \quad , \quad V^1 = \frac{1}{R}\left(v - \frac{w_{,\varphi}}{\sin\theta}\right)$$

In the following, we denote by d_{ij} and σ_{ij} the approximations of the strain and stress tensors components. We have:

$$\begin{aligned}
d_{\theta\theta} &= \frac{1}{R}\Big[u_{,\theta} + w - \frac{r}{R}w_{,\theta\theta}\Big] \quad , \quad d_{\varphi r} = 0 \\
d_{\varphi\varphi} &= \frac{1}{R}\Big[u\,\mathrm{cotg}\,\theta + \frac{v_{,\varphi}}{\sin\theta} + w - \frac{r}{R\sin\theta}\Big(w_{,\theta}\cos\theta + \frac{w_{,\varphi\varphi}}{\sin\theta}\Big)\Big] \\
d_{\theta\varphi} &= \frac{1}{2R}\Big[\frac{u_{,\varphi}}{\sin\theta} - v\,\mathrm{cotg}\,\theta + v_{,\theta} + \frac{2r}{R\sin\theta}\left(w_{,\varphi}\,\mathrm{cotg}\,\theta - w_{,\theta\varphi}\right)\Big] \\
d_{rr} &= -\frac{\nu}{1-\nu}\left[d_{\theta\theta} + d_{\varphi\varphi}\right] \quad , \quad d_{\theta r} = 0
\end{aligned} \qquad (1.26)$$

$$\begin{aligned}
\sigma_{\theta\theta} &= \frac{E}{1-\nu^2}\left(d_{\theta\theta} + \nu d_{\varphi\varphi}\right) & \sigma_{\theta\varphi} &= \frac{E}{1-\nu^2}(1-\nu)d_{\theta\varphi} \\
\sigma_{\varphi\varphi} &= \frac{E}{1-\nu^2}\left(d_{\varphi\varphi} + \nu d_{\theta\theta}\right) & \sigma_{\theta r} &= 0 \\
\sigma_{rr} &= 0 & \sigma_{\varphi r} &= 0
\end{aligned} \qquad (1.26')$$

With these approximations, the potential energy density writes:

$$\begin{aligned}
\mathrm{d}\mathcal{E} &= \frac{1}{2}\left[\sigma_{\theta\theta}d_{\theta\theta} + \sigma_{\varphi\varphi}d_{\varphi\varphi} + \sigma_{\theta\varphi}d_{\theta\varphi} + \sigma_{\varphi\theta}d_{\varphi\theta}\right] \\
&= \frac{E}{2(1-\nu^2)}\left[(d_{\theta\theta} + \nu d_{\varphi\varphi})d_{\theta\theta} + \nu\left(d_{\varphi\varphi} + \nu d_{\theta\theta}\right)d_{\varphi\varphi} + (1-\nu)(d^2_{\theta\varphi} + d^2_{\varphi\theta})\right]
\end{aligned}$$

The total potential energy is obtained by integrating this expression over the domain Ω. To be consistent with the approximations already made, the integration element is approximated by:

$$R^2(1 + r/R)^2 \sin\theta \ \mathrm{d}\theta \ \mathrm{d}\varphi \ \mathrm{d}r \simeq R^2 \sin\theta \ \mathrm{d}\theta \ \mathrm{d}\varphi \ \mathrm{d}r$$

This finally gives:

$$\begin{aligned}
\mathcal{E} = \frac{Eh}{2(1-\nu^2)} \int_0^{2\pi} \mathrm{d}\varphi \int_{\theta_1}^{\theta_2} \sin\theta \, \mathrm{d}\theta \Big\{ \nu\big(u \operatorname{cotg}\theta + u_{,\theta} + \frac{v_{,\varphi}}{\sin\theta} + 2w\big)^2 \\
+ (1-\nu)\Big[\big(u \operatorname{cotg}\theta + \frac{v_{,\varphi}}{\sin\theta} + w\big)^2 + \big(u_{,\theta} + w\big)^2 + \frac{1}{2}\big(\frac{u_{,\varphi}}{\sin\theta} - v \operatorname{cotg}\theta + v_{,\theta}\big)^2\Big] \\
+ \frac{h^2}{12R^2}\Big[\nu\big(w_{,\theta} \operatorname{cotg}\theta + \frac{w_{,\varphi\varphi}}{\sin^2\theta} + w_{,\theta\theta}\big)^2 + (1-\nu)\Big(\big(w_{,\theta} \operatorname{cotg}\theta + \frac{w_{,\varphi\varphi}}{\sin^2\theta}\big)^2 + w_{,\theta\theta}^2\Big) \\
+ \frac{(1-\nu)}{\sin^2\theta}\big(w_{,\varphi} \operatorname{cotg}\theta - w_{,\theta\varphi}\big)^2 + \frac{(1-\nu)}{\sin^2\theta}\big(w_{,\varphi} \operatorname{cotg}\theta - w_{,\varphi\theta}\big)^2\Big]\Big\}
\end{aligned}$$

which is a positive quantity. Assuming that the shell is excited by a force with density $(f_\theta, f_\varphi, f_r)$, the virtual works theorem leads to the variational form of the spherical shell equation:

$$\begin{aligned}
\frac{Eh}{(1-\nu^2)} \int_0^{2\pi} \mathrm{d}\varphi \int_{\theta_1}^{\theta_2} \sin\theta \, \mathrm{d}\theta \\
\Big\{ \nu\big(u \operatorname{cotg}\theta + u_{,\theta} + \frac{v_{,\varphi}}{\sin\theta} + 2w\big)\big(\delta u \operatorname{cotg}\theta + \delta u_{,\theta} + \frac{\delta v_{,\varphi}}{\sin\theta} + 2\delta w\big) \\
+ (1-\nu)\Big[\big(u \operatorname{cotg}\theta + \frac{v_{,\varphi}}{\sin\theta} + w\big)\big(\delta u \operatorname{cotg}\theta + \frac{\delta v_{,\varphi}}{\sin\theta} + \delta w\big) + \big(u_{,\theta} + w\big)\big(\delta u_{,\theta} + \delta w\big) \\
+ \frac{1}{2}\big(\frac{u_{,\varphi}}{\sin\theta} - v \operatorname{cotg}\theta + v_{,\theta}\big)\big(\frac{\delta u_{,\varphi}}{\sin\theta} - \delta v \operatorname{cotg}\theta + \delta v_{,\theta}\big)\Big] \\
+ \frac{h^2}{12R^2}\Big[\nu\big(w_{,\theta} \operatorname{cotg}\theta + \frac{w_{,\varphi\varphi}}{\sin^2\theta} + w_{,\theta\theta}\big)\big(\delta w_{,\theta} \operatorname{cotg}\theta + \frac{\delta w_{,\varphi\varphi}}{\sin^2\theta} + \delta w_{,\theta\theta}\big) \\
+ (1-\nu)\Big(\big(w_{,\theta} \operatorname{cotg}\theta + \frac{w_{,\varphi\varphi}}{\sin^2\theta}\big)\big(\delta w_{,\theta} \operatorname{cotg}\theta + \frac{\delta w_{,\varphi\varphi}}{\sin^2\theta}\big) + w_{,\theta\theta}\, \delta w_{,\theta\theta}\Big) \\
+ \frac{(1-\nu)}{\sin^2\theta}\big(w_{,\varphi} \operatorname{cotg}\theta - w_{,\theta\varphi}\big)\big(\delta w_{,\varphi} \operatorname{cotg}\theta - \delta w_{,\theta\varphi}\big) \\
+ \frac{(1-\nu)}{\sin^2\theta}\big(w_{,\varphi} \operatorname{cotg}\theta - w_{,\varphi\theta}\big)\big(\delta w_{,\varphi} \operatorname{cotg}\theta - \delta w_{,\varphi\theta}\big)\Big]\Big\}
\end{aligned}$$

$$+\mu_s h\int_0^{2\pi}\mathrm{d}\varphi\int_{\theta_1}^{\theta_2}\sin\theta\,\mathrm{d}\theta\big(\ddot{u}\,\delta u+\ddot{v}\,\delta v+\ddot{w}\,\delta w\big)$$
$$=\int_0^{2\pi}\mathrm{d}\varphi\int_{\theta_1}^{\theta_2}\sin\theta\,\mathrm{d}\theta\big(f_\theta\,\delta u+f_\varphi\,\delta v+f_r\,\delta w\big)\quad(1.27)$$

This expression can be rewritten as:

$$\frac{Eh}{1-\nu^2}\int_0^{2\pi}\mathrm{d}\varphi\int_{\theta_1}^{\theta_2}\sin\theta\,\mathrm{d}\theta\Big\{\delta u\,\mathrm{cotg}\,\theta\Big[u\,\mathrm{cotg}\,\theta+\nu u_{,\theta}+\frac{v_{,\varphi}}{\sin\theta}+(1+\nu)w\Big]$$
$$-\delta v\frac{1-\nu}{2}\,\mathrm{cotg}\,\theta\Big[\frac{u_{,\varphi}}{\sin\theta}-v\,\mathrm{cotg}\,\theta+v_{,\theta}\Big]$$
$$+(1+\nu)\delta w\Big[u\,\mathrm{cotg}\,\theta+u_{,\theta}+\frac{v_{,\varphi}}{\sin\theta}+2w\Big]$$
$$+\delta u_{,\varphi}\frac{1-\nu}{2\sin\theta}\Big[\frac{u_{,\varphi}}{\sin\theta}-v\,\mathrm{cotg}\,\theta+v_{,\theta}\Big]$$
$$+\frac{\delta v_{,\varphi}}{\sin\theta}\Big[u\,\mathrm{cotg}\,\theta+\nu u_{,\theta}+\frac{v_{,\varphi}}{\sin\theta}+(1+\nu)w\Big]$$
$$+\delta w_{,\varphi}\frac{h^2}{12R^2}2(1-\nu)\frac{\mathrm{cotg}\,\theta}{\sin^2\theta}\Big[w_{,\varphi}\,\mathrm{cotg}\,\theta-w_{,\varphi\theta}\Big]$$
$$+\delta u_{,\theta}\Big[\nu u\,\mathrm{cotg}\,\theta+u_{,\theta}+\nu\frac{v_{,\varphi}}{\sin\theta}+(1+\nu)w\Big]$$
$$+\frac{1-\nu}{2}\delta v_{,\theta}\Big[\frac{u_{,\varphi}}{\sin\theta}-v\,\mathrm{cotg}\,\theta+v_{,\theta}\Big]$$
$$+\delta w_{,\theta}\frac{h^2}{12R^2}\,\mathrm{cotg}\,\theta\Big[w_{,\theta}\,\mathrm{cotg}\,\theta+\nu w_{,\theta\theta}+\frac{w_{,\varphi\varphi}}{\sin^2\theta}\Big]$$
$$+\delta w_{,\varphi\varphi}\frac{h^2}{12R^2}\frac{1}{\sin^2\theta}\Big[w_{,\theta}\,\mathrm{cotg}\,\theta+\nu w_{,\theta\theta}+\frac{w_{,\varphi\varphi}}{\sin^2\theta}\Big]$$
$$+\delta w_{,\theta\theta}\frac{h^2}{12R^2}\Big[\nu w_{,\theta}\,\mathrm{cotg}\,\theta+w_{,\theta\theta}+\nu\frac{w_{,\varphi\varphi}}{\sin^2\theta}\Big]$$
$$-\delta w_{,\varphi\theta}\frac{h^2}{12R^2}\frac{(1-\nu)}{\sin^2\theta}\Big[w_{,\varphi}\,\mathrm{cotg}\,\theta-w_{,\varphi\theta}\Big]$$
$$-\delta w_{,\theta\varphi}\frac{h^2}{12R^2}\frac{(1-\nu)}{\sin^2\theta}\Big[w_{,\varphi}\,\mathrm{cotg}\,\theta-w_{,\theta\varphi}\Big]\Big\}$$
$$+\mu_s h\int_0^{2\pi}\mathrm{d}\varphi\int_{\theta_1}^{\theta_2}\sin\theta\,\mathrm{d}\theta\big[\ddot{u}\,\delta u+\ddot{v}\,\delta v+\ddot{w}\,\delta w\big]$$
$$=\int_0^{2\pi}\mathrm{d}\varphi\int_{\theta_1}^{\theta_2}\sin\theta\,\mathrm{d}\theta\big[f_\theta\,\delta u+f_\varphi\,\delta v+f_r\,\delta w\big]\qquad(1.27')$$

Integrations by parts are then performed and boundary integrals appear:

$$\frac{Eh}{(1-\nu^2)}\int_0^{2\pi} R\,\mathrm{d}\varphi \int_{\theta_1}^{\theta_2} R\sin\theta\,\mathrm{d}\theta \Big\{\delta u\Big[\mathcal{M}^s_{\theta\theta}u + \mathcal{M}^s_{\theta\varphi}v + \mathcal{M}^s_{\theta r}w\Big]$$

$$+\,\delta v\Big[\mathcal{M}^s_{\varphi\theta}u + \mathcal{M}^s_{\varphi\varphi}v + \mathcal{M}^s_{\varphi r}w\Big] + \delta w\big[\mathcal{M}^s_{r\theta}u + \mathcal{M}^s_{r\varphi}v + \mathcal{M}^s_{rr}w\big]\Big\}$$

$$+\int_0^{2\pi} R\,\mathrm{d}\varphi\left\{\sin\theta\left[\delta u F_{\theta\theta} + \delta v F_{\theta\varphi} + \delta w F_{\theta r} + \frac{\delta w_{,\varphi}}{R\sin\theta}M_{\theta\varphi} + \frac{\delta w_{,\theta}}{R}M_{\theta\theta}\right]\right\}_{\theta=\theta_1}^{\theta=\theta_2}$$

$$+\,\mu_s h\int_0^{2\pi}\mathrm{d}\varphi\int_{\theta_1}^{\theta_2}\sin\theta\,\mathrm{d}\theta\big[\ddot u\,\delta u + \ddot v\,\delta v + \ddot w\,\delta w\big]$$

$$= \int_0^{2\pi}\mathrm{d}\varphi\int_{\theta_1}^{\theta_2}\sin\theta\,\mathrm{d}\theta\big[f_\theta\,\delta u + f_\varphi\,\delta v + f_r\,\delta w\big] \quad (1.28)$$

In this expression, the components $\mathcal{M}^s_{ij}$ of the matrix operator $\mathcal{M}^s$ are given by:

$$\mathcal{M}^s_{\theta\theta} = -\frac{1}{R^2}\left[\frac{\partial^2}{\partial\theta^2} + \frac{1-\nu}{2\sin^2\theta}\frac{\partial^2}{\partial\varphi^2} + \operatorname{cotg}\theta\frac{\partial}{\partial\theta} - (\operatorname{cotg}^2\theta + \nu)\right]$$

$$\mathcal{M}^s_{\theta\varphi} = -\frac{1}{R^2}\left[\frac{1+\nu}{2\sin\theta}\frac{\partial^2}{\partial\theta\partial\varphi} - \frac{3-\nu}{2}\frac{\operatorname{cotg}\theta}{\sin\theta}\frac{\partial}{\partial\varphi}\right] \quad (1.29)$$

$$\mathcal{M}^s_{\theta r} = -\frac{1+\nu}{R^2}\frac{\partial}{\partial\theta}$$

$$\mathcal{M}^s_{\varphi\theta} = -\frac{1}{R^2}\Big[\frac{1+\nu}{2\sin\theta}\frac{\partial^2}{\partial\theta\partial\varphi} + \frac{3-\nu}{2}\frac{\operatorname{cotg}\theta}{\sin\theta}\frac{\partial}{\partial\varphi}\Big]$$

$$\mathcal{M}^s_{\varphi\varphi} = -\frac{1}{R^2}\Big[\frac{1-\nu}{2}\frac{\partial^2}{\partial\theta^2} + \frac{1}{\sin^2\theta}\frac{\partial^2}{\partial\varphi^2}$$

$$+\frac{1-\nu}{2}\operatorname{cotg}\theta\frac{\partial}{\partial\theta} - \frac{1-\nu}{2}(\operatorname{cotg}^2\theta - 1)\Big] \quad (1.29')$$

$$\mathcal{M}^s_{\varphi r} = -\frac{1+\nu}{R^2\sin\theta}\frac{\partial}{\partial\varphi}$$

$$\mathcal{M}^s_{r\theta} = \frac{1+\nu}{R^2}\left(\frac{\partial}{\partial\theta} + \operatorname{cotg}\theta\right) \quad , \quad \mathcal{M}^s_{r\varphi} = \frac{1+\nu}{R^2 \sin\theta}\frac{\partial}{\partial\varphi}$$

$$\begin{aligned}\mathcal{M}^s_{rr} = \; & \frac{2(1+\nu)}{R^2} + \frac{h^2}{12R^4}\left(\frac{\partial^4}{\partial\theta^4} + \frac{2}{\sin^2\theta}\frac{\partial^4}{\partial\theta^2\partial\varphi^2} + \frac{1}{\sin^4\theta}\frac{\partial^4}{\partial\varphi^4}\right.\\ & +2\operatorname{cotg}\theta\frac{\partial^3}{\partial\theta^3} - \frac{2\operatorname{cotg}\theta}{\sin^2\theta}\frac{\partial^3}{\partial\theta\partial\varphi^2} - (1+\nu+\operatorname{cotg}^2\theta)\frac{\partial^2}{\partial\theta^2}\\ & \left. + \frac{3-\nu+4\operatorname{cotg}^2\theta}{\sin^2\theta}\frac{\partial^2}{\partial\varphi^2} + (2-\nu+\operatorname{cotg}^2\theta)\operatorname{cotg}\theta\frac{\partial}{\partial\theta}\right)\end{aligned} \tag{1.29''}$$

It must be remarked that, in equation (1.28), the term $\delta w_{,\varphi} M_{\theta\varphi}$ can be integrated by parts and thus replaced by $-\delta w (M_{\theta\varphi})_{,\varphi}$.

The terms involved in the boundary integrals have the following expressions:

$$\begin{aligned} F_{\theta\theta} &= \frac{Eh}{(1-\nu^2)R}\left(u_{,\theta} + \nu u \operatorname{cotg}\theta + \nu\frac{v_{,\varphi}}{\sin\theta} + (1+\nu)w\right)\\ F_{\theta\varphi} &= \frac{Eh}{(1-\nu^2)R}\frac{1-\nu}{2}\left(\frac{u_{,\varphi}}{\sin\theta} + v_{,\theta} - v\operatorname{cotg}\theta\right)\\ F_{\theta r} &= \frac{Eh^3}{12(1-\nu^2)R^3}\left\{\operatorname{cotg}\theta\left(\frac{w_{,\varphi\varphi}}{\sin^2\theta} + \nu w_{,\theta\theta} + w_{,\theta}\operatorname{cotg}\theta\right)\right.\\ & \quad - \frac{1}{\sin\theta}\left[\sin\theta\left(\nu\frac{w_{,\varphi\varphi}}{\sin^2\theta} + w_{,\theta\theta} + \nu w_{,\theta}\operatorname{cotg}\theta\right)\right]_{,\theta}\\ & \quad \left. - \frac{1-\nu}{\sin^2\theta}\left[w_{,\theta\varphi} - w_{,\varphi}\operatorname{cotg}\theta\right]_{,\varphi}\right\}\\ M_{\theta\theta} &= \frac{Eh^3}{12(1-\nu^2)R^2}\left(w_{,\theta\theta} + \nu\frac{w_{,\varphi\varphi}}{\sin^2\theta} + \nu w_{,\theta}\operatorname{cotg}\theta\right)\\ M_{\theta\varphi} &= \frac{Eh^3}{12(1-\nu^2)R^2}\frac{1-\nu}{\sin\theta}\left(w_{,\theta\varphi} - w_{,\varphi}\operatorname{cotg}\theta\right)\end{aligned} \tag{1.30}$$

(the symbol Tr in front of each expression has been omitted).

The first three terms represent line force densities: $F_{\theta\theta}$ is tangential to the shell surface and normal to the circle θ = constant (tangential compression-dilatation force); $F_{\theta\varphi}$ is tangent to the shell surface and to the circle θ = constant (tangential shearing force); F_{rr} is normal to the shell surface (normal shearing force). The other two terms are moment line densities: $M_{\theta\theta}$ is a bending moment around the tangent to the curve θ = constant; $M_{\theta\varphi}$ is a twisting moment around the direction normal to this circle and tangent to the shell surface.

Equation (1.28) must be satisfied for any virtual displacement: thus, the surface integrals and the boundary integrals must cancel separately. This leads first to the spherical shell equation – known as the *Donnell and Mushtari approximation* – which is written as:

$$\frac{Eh}{1-\nu^2}\mathcal{M}^s\begin{pmatrix} u \\ v \\ w \end{pmatrix} + \mu_s h \begin{pmatrix} \ddot{u} \\ \ddot{v} \\ \ddot{w} \end{pmatrix} = \begin{pmatrix} f_\theta \\ f_\varphi \\ f_r \end{pmatrix} \tag{1.31}$$

The cancellation of the boundary integrals is obtained by adding four boundary conditions. Let us give the most commonly used:

- the simplest condition describes a clamped boundary:

$$\mathrm{Tr}\, u = 0 \qquad \mathrm{Tr}\, v = 0$$
$$\mathrm{Tr}\, w = 0 \qquad \mathrm{Tr}\, w,_\theta = 0$$

- using the interpretation given above of the boundary terms, a free boundary is described by the following conditions:

$$\mathrm{Tr}\, u,_\theta + \nu\, \mathrm{Tr}\, u \operatorname{cotg}\theta + \nu \frac{\mathrm{Tr}\, v,_\varphi}{\sin\theta} + (1+\nu)\, \mathrm{Tr}\, w = 0$$

$$\frac{\mathrm{Tr}\, u,_\varphi}{\sin\theta} + \mathrm{Tr}\, v,_\theta - \mathrm{Tr}\, v \operatorname{cotg}\theta = 0$$

$$\nu\, \mathrm{Tr}\, w,_{\theta\theta} + \frac{\mathrm{Tr}\, w,_{\varphi\varphi}}{\sin^2\theta} + \mathrm{Tr}\, w,_\theta \operatorname{cotg}\theta = 0$$

$$\begin{aligned}\mathrm{Tr}\Bigg\{ &\operatorname{cotg}\theta \left(\frac{w,_{\varphi\varphi}}{\sin^2\theta} + \nu w,_{\theta\theta} + w,_\theta \operatorname{cotg}\theta\right) \\ &- \frac{1}{\sin\theta}\left[\sin\theta\left(\nu\frac{w,_{\varphi\varphi}}{\sin^2\theta} + w,_{\theta\theta} + \nu w,_\theta \operatorname{cotg}\theta\right)\right],_\theta \\ &\qquad + \frac{1-\nu}{\sin^2\theta}\Big[w,_{\theta\varphi} - w,_\varphi \operatorname{cotg}\theta\Big],_\varphi \Bigg\} \\ &+ \mathrm{Tr}\left[\frac{1-\nu}{\sin\theta}\left(w,_{\theta\varphi} - w,_\varphi \operatorname{cotg}\theta\right)\right],_\varphi = 0\end{aligned}$$

Again, these are conservative boundary conditions.

More accurate spherical shell equations have been developed (in particular by W. FLÜGGE [FLU 90]).

REMARK.– Here we have considered a spherical shell limited by two parallel circles. If the shell is limited by different curves, the boundary integral terms have different expressions but their physical meaning is, of course, exactly the same.

1.5.2. *Regularity Conditions at the Apexes*

Assume now that the shell extends from $\theta_1 = 0$ to any value $\theta_2 < \pi$. It is necessary to impose boundary conditions at the apex point $\theta_1 = 0$, but they cannot be arbitrarily chosen: they must express the fact that this point behaves like any ordinary point of the shell, in particular all the efforts in its neighborhood are continuous and finite. This result is due to [MAU 01].

More precisely, let us consider a circle $\theta = \theta_1 \neq 0$. The virtual work done by the forces and moments which are exerted on it must tend to 0 when $\theta_1 \to 0$.

Normal shearing force.– The cancellation of the virtual work due to the normal shearing force:

$$F_{\theta r} = \frac{Eh^3}{12(1-\nu^2)R^3}\left\{w,_{\theta\theta\theta} - \frac{w,_{\varphi\varphi\theta}}{\sin^2\theta} + w,_{\theta\theta}\,\operatorname{cotg}\theta \right.$$
$$\left. +2\nu w,_{\varphi\varphi}\frac{\operatorname{cotg}\theta}{\sin^2\theta} + \frac{1+(1-\nu)\cos^2\theta}{\sin^2\theta}w,_\theta\right\}$$

is obtained if $F_{\theta_1 r}\sin\theta_1 \longrightarrow 0$ for $\theta_1 \to 0$. Let us seek w of the form:

$$w = w_0(\theta) + \theta\bar{w}(\theta,\varphi) \quad , \quad \int_0^{2\pi} \bar{w}(\theta,\varphi)\,\mathrm{d}\varphi = 0$$

where w_0 and $\bar{w}$ are assumed to have Taylor series around $\theta = 0$. The term $w,_{\theta\theta}\,\operatorname{cotg}\theta$ has zero contribution if:

$$w_0,_{\theta\theta}(0) + 2\bar{w},_\theta(0,\varphi) = 0\ \forall\varphi \quad \Longrightarrow \quad w_0,_{\theta\theta}(0) = \bar{w},_\theta(0,\varphi) = 0$$

The term $w,_{\varphi\varphi}\,\operatorname{cotg}\theta/\sin^2\theta$ has no contribution if $\bar{w},_{\varphi\varphi}(0,\varphi) = 0$. The virtual work of the term:

$$-\frac{w,_{\varphi\varphi\theta}}{\sin^2\theta} + \frac{1+(1-\nu)\cos^2\theta}{\sin^2\theta}w,_\theta$$

is zero if the following condition is fulfilled:

$$\lim_{\theta\to 0}\left\{\frac{-\bar{w},_{\varphi\varphi} + (2-\nu)(w_0,_\theta + \bar{w})}{\theta} - \bar{w},_{\varphi\varphi\theta} + (2-\nu)\bar{w},_\theta\right\} = 0$$

As a consequence of the former conditions, we have:

$$w_0,_\theta(0) = \bar{w}(0,\varphi) = 0$$

This implies that $w,_{\varphi\varphi}$ is zero at $\theta = 0$.

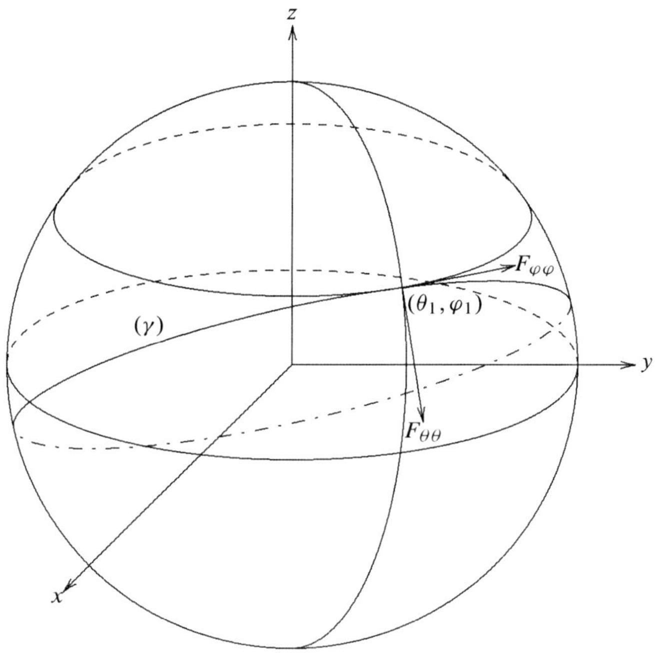

Figure 1.5. *Regularity problems at the apex: relationship between the tangential forces $F_{\theta\theta}$ and $F_{\varphi\varphi}$ at a point with coordinates $\theta_1 \to 0$*

Tangent forces.– The forces $F_{\theta\theta}$ and $F_{\theta\varphi}$ produce virtual works which tend to zero for $\theta_1 \to 0$ if:

$$\lim_{\theta_1 \to 0} \left\{\theta u,_\theta + \nu(u + v,_\varphi) + \theta(1+\nu)w\right\} = 0 \quad , \quad \lim_{\theta_1 \to 0} \left\{u,_\varphi + \theta v,_\theta - v\right\} = 0$$

This results in two conditions:

$$u(0,\varphi) + v,_\varphi(0,\varphi) = 0 \qquad u,_\varphi(0,\varphi) - v(0,\varphi) = 0$$

These two conditions are not sufficient. Indeed, let us remark that the shell displacement, being a periodic function of φ, can be expanded into a Fourier series with respect to this variable. Four conditions at the apex are required for each Fourier component: the former two equalities reduce to only one for the components -1 and +1. It is thus necessary to find an additional condition.

Relationship between the tangential forces $F_{\theta\theta}$ and $F_{\varphi\varphi}$.– The force $F_{\theta\theta}(\theta_1, \varphi_1)$ is exerted on a length element of the circle $\theta = \theta_1$ around the point $\varphi = \varphi_1$, in a direction normal to this circle. However, it can also be considered as the force exerted in the same direction on a length element of the main circle γ (see Figure 1.5). It is thus of the same nature as the force $F_{\varphi\varphi}(\theta_1, \varphi_1)$ which is exerted on the length element of the circle $\varphi = \varphi_1$ around the point $\theta = \theta_1$. This property remains true when $\theta_1 \to 0$. The angle between the two circles γ and $\varphi = \varphi_1$ being equal to $\pi/2$, we must have:

$$F_{\theta\theta}(0, \varphi) = F_{\varphi\varphi}(0, \varphi - \pi/2)$$

(at the other apex, we have $F_{\theta\theta}(0, \varphi) = F_{\varphi\varphi}(0, \varphi + \pi/2)$.) The force $F_{\varphi\varphi}$ has the following expression:

$$F_{\varphi\varphi} = \frac{Eh}{(1-\nu^2)R}\left(\nu u_{,\theta} + u \operatorname{cotg}\theta + \frac{v_{,\varphi}}{\sin\theta} + (1+\nu)w\right)$$

Let us seek u and v of the form:

$$\begin{aligned} u(\theta, \varphi) &= u_1(\theta)e^{\imath\varphi} + u_{-1}(\theta)e^{-\imath\varphi} + \theta\bar{u}(\theta, \varphi) \\ v(\theta, \varphi) &= v_0(\theta) + v_1(\theta)e^{\imath\varphi} + v_{-1}(\theta)e^{-\imath\varphi} + \theta\bar{v}(\theta, \varphi) \end{aligned}$$

where the function $\bar{u}(\theta, \varphi)$ is orthogonal to $e^{\imath\varphi}$ and $e^{-\imath\varphi}$, and $\bar{v}(\theta, \varphi)$ is orthogonal to 1, $e^{\imath\varphi}$ and $e^{-\imath\varphi}$. We immediately have:

$$v_0(0) = 0 \qquad u_1(0) + \imath\, v_1(0) = 0 \qquad u_{-1}(0) - \imath\, v_{-1}(0) = 0$$

The relationship between the two tangential forces is expressed by:

$$\begin{aligned} & u_{1,\theta}(0)(1+\imath\nu)e^{\imath\varphi} + u_{-1,\theta}(0)(1-\imath\nu)e^{-\imath\varphi} \\ & \qquad + (1+\nu)\left[\bar{u}(0,\varphi) - \bar{u}(0, \varphi - \pi/2)\right] = 0 \ \forall\varphi \\ & \lim_{\theta_1 \to 0} \frac{1}{\theta_1}\left(\nu\bar{v}_{,\varphi}(0,\varphi) - \bar{v}_{,\varphi}(0, \varphi - \pi/2)\right) = 0 \ \forall\varphi \end{aligned}$$

This provides the missing equalities:

$$u_{1,\theta}(0) = 0 \qquad u_{-1,\theta}(0) = 0 \qquad \bar{u}(0,\varphi) = 0 \qquad \bar{v}_{,\varphi}(0,\varphi) = 0$$

The boundary conditions at the apex.– We can now gather all the results. First, the shell displacement has the following form:

$$
\begin{aligned}
u(\theta,\varphi) &= u_1(\theta)\mathrm{e}^{\imath\varphi} + u_{-1}(\theta)\mathrm{e}^{-\imath\varphi} + \theta\bar{u}(\theta,\varphi) \\
v(\theta,\varphi) &= v_0(\theta) + v_1(\theta)\mathrm{e}^{\imath\varphi} + v_{-1}(\theta)\mathrm{e}^{-\imath\varphi} + \theta\bar{v}(\theta,\varphi) \\
w(\theta,\varphi) &= w_0(\theta) + \theta\bar{w}(\theta,\varphi)
\end{aligned}
$$

with:

$$
\begin{aligned}
\int_0^{2\pi} \bar{w}(\theta,\varphi)\,\mathrm{d}\varphi &= \int_0^{2\pi} \bar{v}(\theta,\varphi)\,\mathrm{d}\varphi \\
&= \int_0^{2\pi} \bar{u}(\theta,\varphi)\mathrm{e}^{\pm\imath\varphi}\,\mathrm{d}\varphi = \int_0^{2\pi} \bar{v}(\theta,\varphi)\mathrm{e}^{\pm\imath\varphi}\,\mathrm{d}\varphi = 0
\end{aligned}
\tag{1.32}
$$

The various functions which appear in these formulae satisfy the following boundary conditions:

$$
\begin{gathered}
v_0(0) = 0 \\
u_1(0) + \imath\, v_1(0) = 0 \quad , \quad u_{1},_{\theta}(0) = 0 \\
u_{-1}(0) - \imath\, v_{-1}(0) = 0 \quad , \quad u_{-1},_{\theta}(0) = 0 \\
\bar{u}(0,\varphi) = \bar{v},_{\varphi}(0,\varphi) = 0 \\
w_0,_{\theta}(0) = w_0,_{\theta\theta}(0) = \bar{w}(0,\varphi) = \bar{w},_{\theta}(0,\varphi) = 0
\end{gathered}
\tag{1.32$'$}
$$

If $\theta_1 > 0$ but $\theta_2 = \pi$, the boundary conditions which are required at this last point are easily deduced from the former conditions. If the two points $\theta_1 = 0$ and $\theta_2 = \pi$ belong to the shell, the coefficient θ in expressions (1.32) is replaced by $\theta(\pi - \theta)$; the boundary conditions in θ_1 remain unchanged; in θ_2, the second and fourth conditions are replaced by $u_1(\pi) - \imath\, v_1(\pi) = 0$ and $u_{-1}(\pi) + \imath\, v_{-1}(\pi) = 0$ respectively.

1.6. Variational Form of the Equations Governing Harmonic Vibrations of Plates and Shells

In most numerical methods, the plate or shell displacement is sought as a truncated series of basis functions. The coefficients of the expansion are obtained by minimizing a bilinear form. The variational principle involved is nothing else but the mathematical translation of the virtual work theorem that we have used to obtain the partial differential equations for plates and shells. This section is restricted to the case of vibrations which depend harmonically on the time variable (the factor $\mathrm{e}^{-\imath\omega t}$ is omitted).

1.6.1. *Variational Form of the Plate Equation*

Let us consider a plate which occupies a domain Σ, with boundary $\partial\Sigma$. The plate displacement w satisfies any set of natural boundary conditions along $\partial\Sigma$. Equation (1.5) requires that w is a twice differentiable function (almost everywhere), that it is square integrable and that its first and second order derivatives are square integrable too. These properties, together with the boundary conditions, define a functional space (a Sobolev space) that we denote by $\mathbf{H}(\Sigma)$. Let $\hat{w}$ be any function in $\mathbf{H}(\Sigma)$ and $\hat{w}^*$ its complex conjugate. The variational form of the plate equation is:

$$\int_\Sigma \frac{Eh^3}{12(1-\nu^2)}\Big[\nu\big(w_{,11}+w_{,22}\big)\big(\hat{w}^*_{,11}+\hat{w}^*_{,22}\big)+(1-\nu)\big(w_{,11}\,\hat{w}^*_{,11}+w_{,22}\,\hat{w}^*_{,22}$$

$$+\,w_{,12}\,\hat{w}^*_{,12}+w_{,21}\,\hat{w}^*_{,21}\big)\Big]\,\mathrm{d}\Sigma-\mu_s h\omega^2\int_\Sigma w\,\hat{w}^*\,\mathrm{d}\Sigma=\int_\Sigma f\,\hat{w}^*\,\mathrm{d}\Sigma \quad (1.33)$$

This equation is identical to (1.5) in which δw has been replaced by $\hat{w}^*$ and a harmonic time dependence has been assumed. Let us recall that the left hand side integrals are positive for $\hat{w}=w$. This property will be used in the next chapters. Another form of the variational form of the plate equation is deduced from expression (1.5′):

$$\int_\Sigma \frac{Eh^3}{12(1-\nu^2)}\Big[\big(w_{,11}+w_{,22}\big)\big(\hat{w}^*_{,11}+\hat{w}^*_{,22}\big)+(1-\nu)\big(2w_{,12}\,\hat{w}^*_{,12}$$

$$-w_{,11}\,\hat{w}^*_{,22}-w_{,22}\,\hat{w}^*_{,11}\big)\Big]\,\mathrm{d}\Sigma-\mu_s h\omega^2\int_\Sigma w\,\hat{w}^*\,\mathrm{d}\Sigma=\int_\Sigma f\,\hat{w}^*\,\mathrm{d}\Sigma \quad (1.33')$$

Many numerical methods are based on the following scheme. Let $v^{(j)}(j=1,2,\ldots,\infty)$ be a basis of $\mathbf{H}(\Sigma)$. The solution w is expressed as a series expansion of these basis functions

$$w=\sum_{j=1}^{\infty}\alpha_j v^{(j)}$$

This expansion is introduced into equation (1.33). This equation is satisfied for any $\hat{w}$ belonging to $\mathbf{H}(\Sigma)$ if it is satisfied for each of the basis functions. Thus, equation (1.33) is written for each $v^{(k)}(k=1,2,\ldots,\infty)$, and an infinite system of linear algebraic equations is obtained which determines the coefficients α_j. In practice, w is approximated by series truncated at a finite order, say n, and the equation, written for the first n basis functions, leads to a linear system of finite order which gives an approximation of the first n coefficients of the expansion.

1.6.2. *Variational Form of the Shells Equations*

Let us consider a shell – cylindrical or spherical – occupying a domain Σ with boundaries $\partial\Sigma$. Let (u, v, w) be the components of the shell displacement which satisfy any set of natural boundary conditions. Equations (1.20) and (1.27) require that the functions u and v are differentiable (almost everywhere) and square integrable as well as their first derivatives; the function w must be twice differentiable and square integrable as well as its derivatives up to order 2. These properties, together with the boundary conditions, define a functional space that, for simplicity, we denote again by $\mathbf{H}(\Sigma)$ and which is the product of three functional spaces, one for each component of the shell displacement. Let $(\hat{u}, \hat{v}, \hat{w})$ be any vector belonging to $\mathbf{H}(\Sigma)$. If in equation (1.20) or (1.27) a harmonic dependence is assumed and $(\delta u, \delta v, \delta w)$ is replaced by $(\hat{u}^*, \hat{v}^*, \hat{w}^*)$, we obtain an equation of the form:

$$\mathcal{A}(u, v, w; \hat{u}, \hat{v}, \hat{w}) - \mu_s h\omega^2 \int_\Sigma (u\hat{u}^* + v\hat{v}^* + w\hat{w}^*)\,\mathrm{d}\Sigma = \int_\Sigma (f_1\hat{u}^* + f_2\hat{v}^* + f_3\hat{w}^*)\,\mathrm{d}\Sigma \quad (1.34)$$

where (f_1, f_2, f_3) are the components of the excitation force. This is the variational form of the shell equations. For the cylindrical shell, the bilinear form $\mathcal{A}(u, v, w; \hat{u}, \hat{v}, \hat{w})$ is defined by:

$$\begin{aligned}
\mathcal{A}^c(u, v, w; \hat{u}, \hat{v}, \hat{w}) = {} & \\
\int_0^{2\pi} R\,\mathrm{d}\varphi \int_{-L}^{L} \mathrm{d}z \frac{Eh}{1-\nu^2} \Bigg[& \nu\left(u_{,z} + \frac{v_{,\varphi} + w}{R}\right)\left(\hat{u}_{,z}^* + \frac{\hat{v}_{,\varphi}^* + \hat{w}^*}{R}\right) \\
& + (1-\nu)\left(u_{,z}\,\hat{u}_{,z}^* + \frac{v_{,\varphi} + w}{R}\,\frac{\hat{v}_{,\varphi}^* + \hat{w}^*}{R}\right) + \frac{1-\nu}{2}\left(\frac{u_{,\varphi}}{R} + v_{,z}\right)\left(\frac{\hat{u}_{,\varphi}^*}{R} + \hat{v}_{,z}^*\right) \\
& + \frac{h^2}{12}\Bigg[\nu\left(w_{,zz} + \frac{w_{,\varphi\varphi}}{R^2}\right)\left(\hat{w}_{,zz}^* + \frac{\hat{w}_{,\varphi\varphi}^*}{R^2}\right) + (1-\nu)\Bigg(w_{,zz}\,\hat{w}_{,zz}^* + \frac{w_{,z\varphi}\,\hat{w}_{,z\varphi}^*}{R^2} \\
& + \frac{w_{,\varphi z}\,\hat{w}_{,\varphi z}^*}{R^2} + \frac{w_{,\varphi\varphi}\,\hat{w}_{,\varphi\varphi}^*}{R^4}\Bigg)\Bigg]\Bigg]
\end{aligned} \quad (1.34')$$

For the spherical shell, it is defined by:

$$\mathcal{A}^s(u, v, w; \hat{u}, \hat{v}, \hat{w}) = \frac{Eh}{(1-\nu^2)} \int_0^{2\pi} \mathrm{d}\varphi \int_{\theta_1}^{\theta_2} \sin\theta\,\mathrm{d}\theta$$

$$\Bigg\{ \nu\left(u \operatorname{cotg}\theta + u_{,\theta} + \frac{v_{,\varphi}}{\sin\theta} + 2w\right)\left(\hat{u}^* \operatorname{cotg}\theta + \hat{u}_{,\theta}^* + \frac{\hat{v}_{,\varphi}^*}{\sin\theta} + 2\hat{w}^*\right)$$

$$
\begin{aligned}
&+(1-\nu)\Big[\Big(u\operatorname{cotg}\theta+\frac{v,_\varphi}{\sin\theta}+w\Big)\Big(\hat{u}^*\operatorname{cotg}\theta+\frac{\hat{v},^*_\varphi}{\sin\theta}+\hat{w}^*\Big)+\big(u,_\theta+w\big)\big(\hat{u},^*_\theta+\hat{w}^*\big)\\
&\qquad+\frac{1}{2}\Big(\frac{u,_\varphi}{\sin\theta}-v\operatorname{cotg}\theta+v,_\theta\Big)\Big(\frac{\hat{u},^*_\varphi}{\sin\theta}-\hat{v}^*\operatorname{cotg}\theta+\hat{v},^*_\theta\Big)\Big]\\
&+\frac{h^2}{12R^2}\Big[\nu\big(w,_\theta\operatorname{cotg}\theta+\frac{w,_{\varphi\varphi}}{\sin^2\theta}+w,_{\theta\theta}\big)\big(\hat{w},^*_\theta\operatorname{cotg}\theta+\frac{\hat{w},^*_{\varphi\varphi}}{\sin^2\theta}+\hat{w},^*_{\theta\theta}\big)\\
&\quad+(1-\nu)\Big(\Big(w,_\theta\operatorname{cotg}\theta+\frac{w,_{\varphi\varphi}}{\sin^2\theta}\Big)\Big(\hat{w},^*_\theta\operatorname{cotg}\theta+\frac{\hat{w},^*_{\varphi\varphi}}{\sin^2\theta}\Big)+w,_{\theta\theta}\,\hat{w},^*_{\theta\theta}\Big)\\
&\qquad+\frac{(1-\nu)}{\sin^2\theta}\big(w,_\varphi\operatorname{cotg}\theta-w,_{\theta\varphi}\big)\big(\hat{w},^*_\varphi\operatorname{cotg}\theta-\hat{w},^*_{\theta\varphi}\big)\\
&\qquad\quad+\frac{(1-\nu)}{\sin^2\theta}\big(w,_\varphi\operatorname{cotg}\theta-w,_{\varphi\theta}\big)\big(\hat{w},^*_\varphi\operatorname{cotg}\theta-\hat{w},^*_{\varphi\theta}\big)\Big]\Big\}
\end{aligned}
\qquad (1.34'')
$$

It is immediately seen that $\mathcal{A}^c(u,v,w;u,v,w)$ and $\mathcal{A}^s(u,v,w;u,v,w)$, being sums of positive terms, are positive. This property will be used in the next chapter. As for the plate equation, the variational form of shells equations is commonly used in numerical methods.

REMARK.– Similar variational equations are established for any time dependence of the excitation. More complex boundary conditions or excitation of plates and shells by forces or momentums applied at the boundaries can equally be introduced.

1.7. Exercises

1.Complete the proof of equation (1.6).

2.Prove equation (1.10).

3.Establish the equations for (a) a beam with varying thickness $h(x_1)$; (b) a beam with varying width $d(x_1)$.

4.Prove equation (1.21).

5.Consider a circular cylindrical shell with boundaries $z = \pm L/2$ and $\theta = \pm\theta_0$. Establish the equivalent of (1.21). Give the physical meaning of the quantities which appear in the integrals along the boundaries $\theta = \pm\theta_0$.

6.Prove equation (1.28).

Chapter 2

Vibratory Response of Thin Structures *in vacuo*: Resonance Modes, Forced Harmonic Regime, Transient Regime

2.1. Introduction

This chapter deals with the classical methods commonly used for solving problems related to the vibrations of thin structures.

The first problem solved is the calculation of the resonance frequencies and modes of a structure, because of their practical importance: it is known that, if a structure is excited at a frequency close enough to one of its resonance frequencies, the vibration amplitude can reach such high levels that the structure might be damaged or even collapse. The second problem is the calculation of the response of a structure to a harmonic or a transient excitation.

The first structure considered is a beam with a constant cross-section and infinite or finite length. This simple structure plays a fundamental role in the understanding of *in vacuo* vibrations. Indeed, the behavior of any other structure is quite similar to that of a beam in the sense that:

1. any structure has resonance modes, that is any structure can present, for a zero amplitude excitation, non-zero amplitude oscillations (called "free oscillations");

2. the free oscillations occur for a countable infinite number of frequencies called "resonance frequencies" or "eigenfrequencies" to each of which corresponds a particular displacement field of the structure called "resonance mode" or "eigenmode";

3. the set of resonance frequencies and resonance modes is characteristic of the structure: shape, boundary conditions, material;

4. the response of the structure to any excitation, when it exists, can be expanded into a convergent series of the resonance modes;

5. the theorems defining the conditions of existence and uniqueness of the response of a structure to an excitation are identical for any kind of structure.

The resonance frequencies and modes of a beam are defined mathematically and, as far as possible, we give their analytical expressions. When no analytical expression is known, analytical approximations are proposed. Then, it is shown how the resonance modes enable us to express the response of a beam to a harmonic excitation.

The second family of structures is the thin plate with constant or variable thickness. We start with the vibrations of an infinite plate and look first for possible free oscillations. The plate response to a point unit force which satisfies the physics of the phenomenon (energy conservation) is then defined and its analytical expression is established. This response is called *Green's function of the infinite plate* and it is a *Green's kernel* of the plate equation.

Next, the response of a plate with finite dimensions to a harmonic excitation is considered. After some general considerations and theorems, we develop the more common methods used to calculate the resonance frequencies and modes of circular and rectangular plates. Here again, analytical methods, when they exist, are presented. We also propose classical analytical approximation methods. We then develop the *Boundary Element Method*, which applies to plates with non-simple shapes or boundary conditions. The section goes on with a short section on plates with variable thickness. All along, examples of results, possibly compared to experiments, are presented.

By the end, the response of an infinite plate to an impact point force is established. It is shown that the plate equation is not of the hyperbolic type, as the equation governing the vibrations of an elastic medium is, but is of the parabolic type: this implies that the solution corresponds to waves propagating with an infinite velocity. This is, of course, one of the limitations of the plate equation.

A third section is devoted to cylindrical shells. We follow the same scheme as for plates. First, the free oscillations of an infinite cylindrical shell are examined. The analytical solution obtained depends on wavenumbers which are roots of a fourth order polynomial: such roots have an analytical expression, but it is not very easy to use. Thus, the free oscillations of a cylindrical shell are not, properly speaking, known analytically.

Then, the response of an infinite cylindrical shell to a three component point force (one component parallel to each coordinate unit vector), which satisfies the energy conservation principle, is defined. This implies that this response is a tensor of order 3; it is called *Green's tensor* of the shell equation.

After a section devoted to general theorems on the existence and uniqueness of the response of a finite length shell to a harmonic excitation, the resonance modes of finite length shells are studied for two simple boundary conditions. Examples of eigenmode shapes are given.

We end this section with methods to compute the response of a finite length shell to a harmonic excitation. The first method proposed is the expansion of the solution into a series of resonance modes. Then, the general aspects of the Boundary Element Method are proposed.

The final section is devoted to general considerations on spherical shells.

2.2. Vibrations of Constant Cross-Section Beams

We first establish the solutions for the homogenous beam equation which describe how free oscillations can take place and the response of an infinite beam to a time harmonic point force. We then use these solutions to find the free oscillations of finite length beams which are known as *resonance modes*. In the final section, the resonance modes are used to describe the response of a beam to a harmonic excitation.

2.2.1. *Independent Solutions for the Homogenous Beam Equation*

Let us consider an infinite beam of rectangular cross-section $\sigma = \{-d/2 < y < +d/2, -h/2 < z < +h/2\}$, made of a homogenous and isotropic material characterized by a Young's modulus E and a density μ_s. We consider pure bending in the z direction. The displacement $\widetilde{w}(x,t)$ satisfies equation (1.15):

$$EI\frac{\partial^4\widetilde{w}(x,t)}{\partial x^4}+\bar{\mu}\frac{\partial^2\widetilde{w}(x,t)}{\partial t^2}=0 \qquad \text{with} \quad I=\frac{h^3d}{12} \quad , \quad \bar{\mu}=\mu_s hd$$

Looking for free oscillations of the form $\widetilde{w}(x,t)=w(x)\mathrm{e}^{-\imath\omega t}$ (harmonic oscillations at frequency $f=\omega/2\pi$), the function w satisfies the following differential equation:

$$\frac{\mathrm{d}^4 w(x)}{\mathrm{d}x^4}-\lambda^4 w(x)=0 \qquad \text{with} \quad \lambda^4=\frac{\bar{\mu}\omega^2}{EI} \tag{2.1}$$

Let us first notice that, for $\omega=0$ (or, equivalently, $\lambda=0$), any polynomial of degree up to 3 is a solution. For any non-zero frequency, a classical method consists of seeking w as an integral similar to a Laplace integral, i.e.:

$$w(x)=\int_C \widehat{w}(p)\mathrm{e}^{px}\,\mathrm{d}p \tag{2.2}$$

where C is a closed contour in the complex plane. To ensure that w is not zero, $\widehat{w}$ must be a meromorphic function inside C (presence of one pole, at least). Using (2.2), equation (2.1) becomes:

$$\int_C (p^4-\lambda^4)\widehat{w}(p)\mathrm{e}^{px}\,\mathrm{d}p=0 \tag{2.3}$$

This equation is satisfied if the function to be integrated is holomorphic. The polynomial $p^4-\lambda^4$ has four roots: $\pm\imath\lambda$ and $\pm\lambda$. Let us choose:

$$\widehat{w}(p)=\frac{A}{p-\imath\lambda}$$

Equation (2.3) is satisfied and we obtain (integration by the residue method):

$$w(x)=\int_C \frac{A}{p-\imath\lambda}\mathrm{e}^{px}\,\mathrm{d}p=2\imath\pi A\mathrm{e}^{\imath\lambda x}$$

For reasons which will become clear in the next section, we choose $A=1/8\imath\pi\lambda^3$. Doing the same with the three other roots, we finally find four independent solutions of (2.1):

$$\begin{aligned} w_1(x)&=\frac{\mathrm{e}^{\imath\lambda x}}{4\lambda^3} \quad , \quad & w_2(x)&=\frac{\mathrm{e}^{-\imath\lambda x}}{4\lambda^3} \\ w_3(x)&=-\imath\frac{\mathrm{e}^{\lambda x}}{4\lambda^3} \quad , \quad & w_4(x)&=-\imath\frac{\mathrm{e}^{-\lambda x}}{4\lambda^3} \end{aligned} \tag{2.4}$$

Any solution $w(x)$ is a linear combination of these four independent solutions.

These four functions are often called *free waves*: the first two are called *propagating free waves*, while the latter two are called *damped free waves*. The associated parameter λ is called *the wave number*. The reason for such a denomination is that w_1 and w_2 have the same form as waves propagating along a vibrating string in the $x > 0$ direction and in the $x < 0$ direction respectively. However, the corresponding propagation velocity is not constant like the propagation velocity in a vibrating string. It is frequency-dependent:

$$c(\omega) = \left(\frac{EI}{\bar{\mu}}\right)^{1/4} \sqrt{\omega}$$

This shows that w_1 and w_2 are not physically propagating waves. This is not a surprising result. Indeed, in an elastic solid, there exist only two kinds of propagating waves – compression waves and shearing waves – which both have a given velocity. In the present case, it would be better to use the name of *pseudo-waves*, but this is not the custom.

Here, an important remark must be made. To establish the beam equation, we started from the three-dimensional elasticity equations which imply that time dependent perturbations propagate with finite velocities. From a mathematical point of view, the equations are of the hyperbolic type. As a consequence of the approximations made, the beam equation is of the parabolic type (same type as the heat equation) and the propagation velocity of a perturbation is infinite. So, the beam equation cannot describe its transient response correctly if the length of the beam is larger than a few percent of the wavelengths which will appear in the three-dimension elastic solid. Nevertheless, this is not really a constraining restriction because the velocity of waves in a solid is a few thousand meters per second and, thus, the wavelengths are always large compared to the beam length. The same remark applies to plate and shell equations.

2.2.2. *Response of an Infinite Beam to a Point Harmonic Force*

Let us consider an infinite beam excited by a point harmonic force, $EI\delta \mathrm{e}^{-\imath\omega t}$, located at $x = 0$ (δ is the Dirac distribution at $x = 0$). The beam displacement has, of course, the same time dependence: $\tilde{\Gamma}(x,t) = \Gamma(x)\mathrm{e}^{-\imath\omega t}$. The function $\Gamma(x)$ satisfies the following equation:

$$\frac{\mathrm{d}^4\Gamma(x)}{\mathrm{d}x^4} - \lambda^4\Gamma(x) = \delta \tag{2.5}$$

The uniqueness of the solution can be ensured by mathematical conditions deduced from physical considerations, in particular that the energy conservation principle must be fulfilled. The simplest method is to apply the *limit absorption principle*, which is deduced from the physical observation that a purely elastic solid is an idealization of a weakly absorbing material. It can be stated as follows:

THEOREM 2.1 (LIMIT ABSORPTION PRINCIPLE).– *Let ε be a small positive parameter. Let us consider the equation:*

$$\frac{\mathrm{d}^4 \Gamma_\varepsilon(x)}{\mathrm{d}x^4} - \left[\lambda(1+\imath\varepsilon)\right]^4 \Gamma_\varepsilon(x) = \delta \tag{2.5'}$$

It has a unique bounded solution $\Gamma_\varepsilon(x)$ which has a unique limit $\Gamma(x)$ for $\varepsilon \to 0$. This limit is the unique solution of (2.5) which describes the physical phenomenon.

The easiest method to find the solution of (2.5′) is to make use of the space Fourier transform. Let $\widehat{w}(\xi)$ be the Fourier transform of $w(x)$ defined by:

$$\widehat{w}(\xi) = \int_{-\infty}^{+\infty} w(x)\mathrm{e}^{-2\imath\pi x\xi}\,\mathrm{d}x \quad , \quad w(x) = \int_{-\infty}^{+\infty} \widehat{w}(\xi)\mathrm{e}^{+2\imath\pi x\xi}\,\mathrm{d}\xi$$

Applying this Fourier transform to equation (2.5′) leads to:

$$\widehat{\Gamma}_\varepsilon(\xi) = \frac{1}{16\pi^4\xi^4 - \lambda^4(1+\imath\varepsilon)^4} \Longrightarrow \Gamma_\varepsilon(x) = \frac{1}{EI}\int_{-\infty}^{+\infty} \frac{\mathrm{e}^{+2\imath\pi x\xi}}{16\pi^4\xi^4 - \lambda^4(1+\imath\varepsilon)^4}\,\mathrm{d}\xi$$

The residue theorem is applied as follows:

- ◇ for $x > 0$ the integration contour includes the real ξ-axis closed by the half-circle of radius $R \to +\infty$ in the upper complex half-plane ($\Im\xi > 0$); this contour contains the two poles $\lambda(1+\imath\varepsilon)/2\pi$ and $\imath\lambda(1+\imath\varepsilon)/2\pi$ and, thus:

$$\Gamma_\varepsilon(x) = \frac{1}{4\lambda^3}\left[\mathrm{e}^{\imath\lambda(1+\imath\varepsilon)x} - \imath\mathrm{e}^{-\lambda(1+\imath\varepsilon)x}\right]$$

- ◇ for $x < 0$ the integration contour includes the real ξ-axis closed by the half-circle of radius $R \to +\infty$ in the lower complex half-plane ($\Im\xi < 0$); this contour contains the two poles $-\lambda(1+\imath\varepsilon)/2\pi$ and $-\imath\lambda(1+\imath\varepsilon)/2\pi$ and, thus:

$$\Gamma_\varepsilon(x) = \frac{1}{4\lambda^3}\left[\mathrm{e}^{-\imath\lambda(1+\imath\varepsilon)x} - \imath\mathrm{e}^{+\lambda(1+\imath\varepsilon)x}\right]$$

Taking the limits for $\varepsilon \to 0$, these expressions lead to:

$$\Gamma(x) = \frac{1}{4\lambda^3}\left[e^{\imath\lambda|x|} - \imath e^{-\lambda|x|}\right] \quad \text{for} \quad -\infty < x < +\infty \tag{2.6}$$

For a point force located at x_0, the beam displacement $\Gamma(x, x_0)$ is given by:

$$\Gamma(x, x_0) = \frac{1}{4\lambda^3}\left[e^{\imath\lambda|x-x_0|} - \imath e^{-\lambda|x-x_0|}\right] \tag{2.6'}$$

$\Gamma(x, x_0)$ is the *Green's kernel* of the equation governing the oscillations of an infinite beam, which satisfies the energy conservation principle. It is commonly called *Green's function of the beam equation.* This expression of the Green's function justifies the choice of the basic functions (2.4).

2.2.3. *Resonance Modes of Finite Length Beams*

1. *General definition and theorems*

Let us consider a beam extending from $x = -L/2$ to $x = L/2$, with mass per unit length $\bar{\mu}$, Young's modulus E and inertia momentum I. At each end, $x = -L/2$ and $x = +L/2$, the displacement satisfies a couple of boundary conditions which can be chosen from the simple ones which were defined in section 1.3 (clamped, free, or simply supported end) or which corresponds to less simple situations.

DEFINITION 2.1 (RESONANCE MODES OF A BEAM).– *A resonance mode is a function $w(x)$ which satisfies the homogenous time harmonic beam equation*

$$w^{(IV)} - \lambda^4 w = 0$$

and a given set of boundary conditions ($w^{(IV)}$ is the fourth order derivative of w with respect to x). The corresponding value of λ is called the resonance wave number *and the corresponding frequency is called* the resonance frequency.

The following theorem can be proved.

THEOREM 2.2 (EXISTENCE OF RESONANCE MODES).– *A purely elastic beam, submitted to real boundary conditions, has a countable set of resonance modes and real resonance frequencies. The resonance modes are orthogonal to each other. The set of resonance modes is a basis on which any solution of the non-homogenous harmonic beam equation (when it exists) can be expanded.*

By *real boundary conditions* we mean boundary conditions with real coefficients. From a physical point of view, this means that there is no energy loss through the ends of the beam. If there is some energy loss (by material damping or through the beam ends), the resonance frequencies are complex.

The orthogonality of the resonance modes is expressed by the following relationship:

$$\int_{-L/2}^{+L/2} w_m(x) w_n(x)\,\mathrm{d}x = 0 \quad \text{for} \quad m \neq n$$

Let us prove it.

PROOFOF THE ORTHOGONALITY OF THE BEAM RESONANCE MODES.– Let $w_m(x)$ and $w_n(x)$ be two different modes and consider the integral

$$I = \int_{-L/2}^{+L/2} w''_m(x) w''_n(x)\,\mathrm{d}x$$

A first double integration by parts gives:

$$\begin{aligned}I &= \int_{-L/2}^{+L/2} w_m^{IV}(x) w_n(x)\,\mathrm{d}x + w''_m w'_n\Big|_{x=-L/2}^{x=+L/2} - w'''_m w_n\Big|_{x=-L/2}^{x=+L/2}\\ &= -\lambda_m^4 \int_{-L/2}^{+L/2} w_m(x) w_n(x)\,\mathrm{d}x + w''_m w'_n\Big|_{x=-L/2}^{x=+L/2} - w'''_m w_n\Big|_{x=-L/2}^{x=+L/2}\end{aligned}$$

In the same way, we have:

$$I = -\lambda_n^4 \int_{-L/2}^{+L/2} w_m(x) w_n(x)\,\mathrm{d}x + w''_n w'_m\Big|_{x=-L/2}^{x=+L/2} - w'''_n w_m\Big|_{x=-L/2}^{x=+L/2}$$

The difference between these two equalities leads to:

$$\begin{aligned}0 = \left(\lambda_m^4 - \lambda_n^4\right) \int_{-L/2}^{+L/2} w_m(x) w_n(x)\,\mathrm{d}x + w''_m w'_n\Big|_{x=-L/2}^{x=+L/2} - w'''_m w_n\Big|_{x=-L/2}^{x=+L/2}\\ - w''_n w'_m\Big|_{x=-L/2}^{x=+L/2} + w'''_n w_m\Big|_{x=-L/2}^{x=+L/2}\end{aligned}$$

Because the two modes satisfy the same boundary conditions, the integrated terms cancel out and we can conclude that

$$\int_{-L/2}^{+L/2} w_m(x)w_n(x)\,\mathrm{d}x = 0 \quad \text{for} \quad m \neq n$$

□

The resonance modes of the beam are sought as a linear combination of basic solutions (2.4) of the beam equation

$$w(x) = A\frac{\mathrm{e}^{\imath\lambda x}}{4\lambda^3} + B\frac{\mathrm{e}^{-\imath\lambda x}}{4\lambda^3} + C\imath\frac{\mathrm{e}^{\lambda x}}{4\lambda^3} + D\imath\frac{\mathrm{e}^{-\lambda x}}{4\lambda^3}$$

The four boundary conditions lead to a homogenous system of four linear equations. This system has a non-zero solution if its determinant is zero: this occurs for a countable sequence of values λ_n of λ. To each λ_n, there corresponds a beam shape $w_n(x)$ which is determined up to a multiplicative constant. It is generally better to choose this constant such that the integral of $|w_n(x)|^2$ along the beam length ($\mathbf{L}^2$-norm) is equal to 1.

2. *The reduced beam equation*

We have to solve equation (2.1) for a beam with finite length L. Using the beam length as the length unit, this equation becomes:

$$\frac{\mathrm{d}^4 W(X)}{\mathrm{d}X^4} - \Lambda^4 W(X) = 0 \text{ for } -\frac{1}{2} < X < \frac{1}{2}$$
$$\text{with } X = \frac{x}{L}\,,\ \Lambda^4 = L^4\lambda^4\,,\ \omega = \sqrt{\frac{EI}{\tilde{\mu}}\frac{\Lambda^2}{L^2}}\,,\ W(X) = \alpha w(LX) \qquad (2.7)$$

where α is a constant determined by the choice that both w and W have a norm equal to unity. This gives:

$$w(x) = \frac{W(x/L)}{\sqrt{L}} \qquad (2.7')$$

Equation (2.7) is often called the *non-dimensional beam equation*. This terminology is somewhat confusing: indeed, the physical nature of the variables which appear in an equation – here, displacement and wavenumber – do not depend on the choice of units that measure them. We thus prefer to call (2.7) the *reduced beam equation*. This equation has the advantage of being valid for a class of beams. Indeed, once $W(X)$ is obtained (as done right below), $w(x)$ is easily deduced for any length L from equation (2.7′).

REMARK.– It is important to notice that variables occurring in an equation which describes a physical phenomenon are open to two interpretations. The first one is that they represent physical quantities and, thus, they have a dimension. When numerical results are looked for, though the equation has exactly the same form, the variables involved are all dimensionless: indeed, each one represents the ratio of a physical quantity to the quantity taken as measure unit (at least if a coherent system of units is adopted).

In the following paragraphs, devoted to determining the resonance modes of beams with different boundary conditions, the reduced beam equation will be used. The beam will satisfy the same boundary conditions at both ends. Thus, it is simpler to look for either even or odd modes, that is, for modes having one of the two following forms:

$$\begin{aligned} W^e(X) &= A\cos(\Lambda X) + B\cosh(\Lambda X) \\ W^o(X) &= A\sin(\Lambda X) + B\sinh(\Lambda X) \end{aligned}$$

3. *Resonances of a beam simply supported at both ends*

The beam displacement W satisfies the following boundary conditions:

$$W(-1/2) = W(+1/2) = W''(-1/2) = W''(+1/2) = 0 \tag{2.8}$$

Even resonance modes.– At $X = -1/2$ and $X = +1/2$, the boundary conditions provide two independent equations:

$$\begin{cases} A\cos(\Lambda/2) + B\cosh(\Lambda/2) = 0 \\ -A\cos(\Lambda/2) + B\cosh(\Lambda/2) = 0 \end{cases}$$

$$\Rightarrow \qquad A\cos(\Lambda/2) = 0 \quad , \quad B\cosh(\Lambda/2) = 0$$

Because the function $\cosh(\Lambda/2)$ is always larger than 1 for $\Lambda > 0$, the coefficient B must be zero. The coefficient A is non-zero if

$$\cos(\Lambda/2) = 0 \quad \Longrightarrow \quad \Lambda_n^e = (2n+1)\pi, \quad n = 0, 1, \ldots, \infty \tag{2.9}$$

and the modes have the form $W_n^e(X) = A\cos\big[(2n+1)\pi X\big]$. The value of the coefficient A for each mode is chosen so that the $\mathbf{L}^2$-norm of the mode is equal to one, that is:

$$A^2 \int_{-1/2}^{+1/2} \cos\big[(2n+1)\pi X\big]^2 \, \mathrm{d}X = 1$$

The even resonance modes are thus given by:

$$W_n^e(X) = \sqrt{2}\cos\bigl[(2n+1)\pi X\bigr], \quad n = 0, 1, \dots, \infty \tag{2.9'}$$

Odd resonance modes.– At $X = -1/2$ and $X = +1/2$, the boundary conditions provide the following two independent equations:

$$\begin{cases} A\sin(\Lambda/2) + B\sinh(\Lambda/2) = 0 \\ A\sin(\Lambda/2) - B\sinh(\Lambda/2) = 0 \end{cases}$$

$$\Rightarrow \qquad A\sin(\Lambda/2) = 0 \quad , \quad B\sinh(\Lambda/2) = 0$$

An obvious solution is $\Lambda = 0$. But, in that case, the corresponding mode would be identically zero. Such a mode is meaningless. For $\Lambda > 0$, $\sinh(\Lambda/2)$ is positive: we can conclude that $B = 0$ for all modes. The coefficient A is not zero if:

$$\sin(\Lambda/2) = 0 \quad \Longrightarrow \quad \Lambda_n^o = 2n\pi, \quad n = 1, 2, \dots, \infty \tag{2.10}$$

The value of the coefficient A being chosen so that the $\mathbf{L}^2$-norm of the mode is equal to one, the even resonance modes are given by:

$$W_n^o(X) = \sqrt{2}\sin\bigl(2n\pi X\bigr), \quad n = 1, 2, \dots, \infty \tag{2.10'}$$

Let us remark that the modes of a simply supported beam satisfy a one-dimensional Helmholtz equation and are identical to those of a vibrating string with fixed ends.

4. *Resonances of a beam clamped at both ends*

The beam displacement W satisfies the following boundary conditions:

$$W(-1/2) = W(+1/2) = W'(-1/2) = W'(+1/2) = 0 \tag{2.11}$$

Even resonance modes.– The boundary conditions lead to:

$$A\cos(\Lambda/2) + B\cosh(\Lambda/2) = 0$$
$$-A\sin(\Lambda/2) + B\sinh(\Lambda/2) = 0$$

Coefficients A and B are not zero if Λ is a solution of the transcendental equation

$$\begin{aligned} &\tan(\Lambda/2) = -\tanh(\Lambda/2) \\ \text{or} \quad &(1+\mathrm{e}^{-\Lambda})\sin(\Lambda/2) + (1-\mathrm{e}^{-\Lambda})\cos(\Lambda/2) = 0 \end{aligned} \tag{2.12}$$

which has no explicit solution (the second form is more suitable for numerical computation). Nevertheless, it is possible to obtain a very accurate approximation of the solutions. Indeed, as far as Λ is not very small, $\tanh(\Lambda/2)$ is close to 1. Thus, we can conclude that the resonance wavenumbers are given by:

$$\Lambda_n^e = \frac{(4n-1)\pi}{2} + \varepsilon_n^e \quad n = 1, 2, \ldots, \infty \quad \text{with} \quad |\varepsilon_n^e| \ll \frac{(4n-1)\pi}{2}$$

To calculate the ε_n^e, the functions which appear in (2.12) (second form) are expanded into Taylor series of this variable around $(4n-1)\pi/2$. By keeping the first order terms only, we obtain the following result:

$$\Lambda_n^e \simeq \frac{(4n-1)\pi}{2} + \frac{2e^{-(4n-1)\pi/2}}{1 + 2e^{-(4n-1)\pi/2}} \tag{2.13}$$

The even resonance modes are given by:

$$W_n^e(X) = A_n^e \left[\cos(\Lambda_n^e X) + \frac{\sin(\Lambda_n^e/2)}{\sinh(\Lambda_n^e/2)} \cosh(\Lambda_n^e X) \right] \tag{2.13$'$}$$

mode index n	exact Λ_n^e	approximated Λ_n^e	relative error
1	4.730040744862658	4.730038461575137	$4.73004\ 10^{-7}$
2	10.99560783800166	10.99560783800166	$1.11022\ 10^{-16}$
3	17.27875959474386	17.27875965739948	$3.62616\ 10^{-9}$
4	23.56194490192345	23.56194490204046	$4.96599\ 10^{-12}$
5	29.84513020910303	29.84513020910325	$7.41968\ 10^{-15}$
6	36.12831551628262	36.12831551628262	$1.11022\ 10^{-16}$
7	42.41150082346221	42.41150082346221	$1.11022\ 10^{-16}$
8	48.69468613064179	48.69468613064180	$1.11022\ 10^{-16}$
9	54.97787143782138	54.97787143782138	$1.11022\ 10^{-16}$
10	61.26105674500096	61.26105674500097	$1.11022\ 10^{-16}$

Table 2.1. *Comparison between the exact and the approximated resonance wavenumbers of the clamped beam of unit length*

Table 2.1 compares the exact solution (computed by Mathematica with 16 digits) and the above approximation of the first 10 resonance wavenumbers. The maximum relative error, which occurs on the first value, is of order 10^{-7}: this is quite a sufficient accuracy for engineering applications.

The normalization factors A_n^e are given by the following expression:

$$A_n^e = \left\{ \int_{-1/2}^{+1/2} \left[\cos(\Lambda_n^e X) + \frac{\sin(\Lambda_n^e/2)}{\sinh(\Lambda_n^e/2)} \cosh(\Lambda_n^e X) \right]^2 \mathrm{d}X \right\}^{-1/2}$$

$$= 2\sqrt{\Lambda_n^e} \left[6 \sin \Lambda_n^e + \frac{\Lambda_n^e (\cosh \Lambda_n^e - \cos \Lambda_n^e) - 3 \sinh \Lambda_n^e \ (\cos \Lambda_n^e - 1)}{\sinh^2(\Lambda_n^e/2)} \right]^{-1/2} \tag{2.13''}$$

This formula requires us to integrate several exponential functions and to express the result obtained in terms of trigonometric and hyperbolic functions.

Odd resonance modes.– The boundary conditions lead to:

$$A \sin(\Lambda/2) + B \sinh(\Lambda/2) = 0$$
$$A \cos(\Lambda/2) + B \cosh(\Lambda/2) = 0$$

Coefficients A and B are not zero if Λ is a solution of the transcendental equation

$$\tan(\Lambda/2) = \tanh(\Lambda/2)$$
$$\text{or } (1 + \mathrm{e}^{-\Lambda}) \sin(\Lambda/2) - (1 - \mathrm{e}^{-\Lambda}) \cos(\Lambda/2) = 0 \tag{2.14}$$

Using the same approximation method, we have:

$$\Lambda_n^o \simeq \frac{(4n+1)\pi}{2} - \frac{2\mathrm{e}^{-(4n+1)\pi/2}}{1 - 2\mathrm{e}^{-(4n+1)\pi/2}} \tag{2.15}$$

$$W_n^o(X) = A_n^o \left[\sin(\Lambda_n^o X) - \frac{\sin(\Lambda_n^o/2)}{\sinh(\Lambda_n^o/2)} \sinh(\Lambda_n^o X) \right] \tag{2.15'}$$

$$A_n^o = 2\sqrt{\Lambda_n^o} \sinh(\Lambda_n^o/2) \Big[\Lambda_n^o (\cos \Lambda_n^o - 2) - \sin \Lambda_n^o$$
$$+ \cosh \Lambda_n^o (\Lambda_n^o + \sin \Lambda_n^o) + \sinh \Lambda_n^o (\cos \Lambda_n^o - 1) \Big]^{-1/2} \tag{2.15''}$$

Approximation (2.15) is as accurate as (2.13).

5. *Resonances of a beam free at both ends*

The beam displacement W satisfies the following boundary conditions:

$$W''(-1/2) = W''(+1/2) = W'''(-1/2) = W'''(+1/2) = 0 \tag{2.16}$$

There are two particular modes, which both correspond to $\Lambda = 0$:

$$W_0^e = 1 \quad , \quad W_0^o = X \tag{2.16'}$$

The even mode is a transverse rigid translation of the beam, while the odd one is a rotation with extension, the beam remaining rectilinear. As far as vibrations are concerned, these modes have no interest. However, to avoid such rigid displacements appearing in the mathematical solution, the function which describes the external excitation must be orthogonal to both W_0^e and W_0^o.

For the other modes, it is easy to see that the equations which give the resonance wavenumbers are identical to the equations for the clamped beam. Thus, apart from the modes W_0^e and W_0^o which correspond to a zero wavenumber, the free beam has the same resonance wavenumbers as the clamped beam, although the modes are different:

$$\begin{aligned} W_n^e(X) &= A_n^e\left[\cos(\Lambda_n^e X) - \frac{\sin(\Lambda_n^e/2)}{\sinh(\Lambda_n^e/2)}\cosh(\Lambda_n^e X)\right] \\ W_n^o(X) &= A_n^o\left[\sin(\Lambda_n^o X) + \frac{\sin(\Lambda_n^o/2)}{\sinh(\Lambda_n^o/2)}\sinh(\Lambda_n^o X)\right] \end{aligned} \tag{2.16''}$$

where the normalization coefficients are given by:

$$\begin{aligned} A_n^e &= 2\sqrt{\Lambda_n^e}\sinh(\Lambda_n^e/2)\Big[-\Lambda_n^e\cos\Lambda_n^e + \cosh\Lambda_n^e\big(\Lambda_n^e - \sin\Lambda_n^e\big) \\ &\qquad + \sin\Lambda_n^e + \sinh\Lambda_n^e\big(\cos\Lambda_n^e - 1\big)\Big]^{-1/2} \\ A_n^o &= 2\sqrt{\Lambda_n^o}\sinh(\Lambda_n^o/2)\Big[\Lambda_n^o\big(\cos\Lambda_n^o - 2\big) + \cosh\Lambda_n^o\big(\Lambda_n^o - 3\sin\Lambda_n^o\big) \\ &\qquad + 3\sin\Lambda_n^o - 3\sinh\Lambda_n^o\big(\cos\Lambda_n^o - 1\big)\Big]^{-1/2} \end{aligned} \tag{2.16'''}$$

2.2.4. *Response of a Finite Length Beam to a Harmonic Force*

Let us consider a beam with length L and excited by a harmonic force with density $F(x)$. Its displacement $w(x)$ satisfies the non-homogenous beam equation:

$$\frac{d^4 w(x)}{dx^4} - \lambda^4 w(x) = F(x) \qquad x \in]-L/2, +L/2[\tag{2.17}$$

and a given set of boundary conditions at $x = -L/2$ and $x = +L/2$. The general method which is proposed does not require us to explicit these conditions.

Let λ_n be the resonance wavenumbers and $w_n(x)$ the corresponding (normalized) resonance modes of the beam (that is, the set of solutions for the homogenous equation which satisfy the given boundary conditions). The beam displacement is sought as a series of the resonance modes:

$$w(x) = \sum_{n=1}^{\infty} \alpha_n w_n(x) \tag{2.18}$$

The most straightforward way to find the coefficients α_n is to introduce (2.18) into equation (2.17), multiply both sides by w_m and integrate over the beam domain:

$$\int_{-L/2}^{+L/2} w_m(x) \sum_{n=1}^{\infty} \alpha_n \Big[\frac{\mathrm{d}^4 w_n(x)}{\mathrm{d}x^4} - \lambda^4 w_n(x)\Big] \mathrm{d}x$$

$$= \int_{-L/2}^{+L/2} F(x) w_m(x)\,\mathrm{d}x \quad , \quad m = 1, 2, \ldots, \infty$$

Using the equation satisfied by the functions w_n and their orthogonality property leads to:

$$\alpha_m(\lambda_n^4 - \lambda^4) = F_m \quad \text{with} \quad F_m = \int_{-L/2}^{+L/2} F(x) w_m(x)\,\mathrm{d}x$$

We see that, if λ is different from all the resonance wave numbers, the α_m are uniquely determined and so is the displacement $w(x)$ which is given by:

$$w(x) = \sum_{n=1}^{\infty} \frac{F_n}{\lambda_n^4 - \lambda^4} w_n(x) \tag{2.19}$$

If λ is equal to one of the resonance wavenumbers, say λ_r, two cases must be considered:

- ◇ F_r is different from zero, and thus the coefficient α_r is infinite: there is no solution of finite amplitude;
- ◇ F_r is zero (that is, $F(x)$ is orthogonal to w_r): the coefficient α_r is arbitrary and the solution $w(x)$ is determined modulo any function proportional to w_r.

This result shows the importance of resonance phenomena in mechanical engineering. Indeed, if the excitation frequency is equal to a resonance frequency of a structure, the response is theoretically infinite.

Physical materials are never purely elastic but always have some damping. This implies that the resonance frequencies are not purely real but have an imaginary part, which accounts for energy dissipation. As a consequence, the response of a realistic structure to any harmonic excitation is always well defined. But many materials (like metals) which are of common use have a very small damping. So, when the excitation frequency of a structure made of a slightly damped material is equal to (or just close to) the real part of a resonance frequency, the response becomes extremely large and the stresses induced rapidly exceed the elastic limit of the material.

2.3. Vibrations of Plates

This section is mainly devoted to the harmonic vibrations of constant thickness plates. We first look at the solutions of the homogenous plate equation and at the response of an infinite plate to a point harmonic force. We then describe the analytical methods for computing the resonance frequencies and modes of simple shape plates: circular plates and rectangular plates. We show that the response of a finite dimension plate to a harmonic force can be expressed as a series of resonance modes. The study of harmonic vibrations of constant thickness plates is completed by a section devoted to the Boundary Element Method: the advantage of BEM is that it applies to any shape of plates, and any type of boundary conditions and provides a tool for the numerical computation of both the resonance modes and the forced regime response.

The last two sections are devoted to the resonance modes of a variable thickness plate and the transient response of infinite beams and plates.

Let us recall the notations adopted in section 1.2 for the characteristics of a plate: μ_s = density of the solid; E = Young's modulus; ν = Poisson's ratio; h = plate thickness; $D = Eh^3/12(1-\nu^2)$ = plate rigidity; $\bar{\mu} = \mu_s h$ mass per unit area. We generally consider purely elastic materials, that is, materials within which there is no energy loss.

2.3.1. *Free Vibrations of an Infinite Plate*

Let us consider the homogenous plate equation

$$\mathbf{P}(w) \equiv \Delta^2 w - \lambda^4 w = 0 \qquad \text{with} \qquad \lambda^4 = \frac{\bar{\mu}}{D} \tag{2.20}$$

It can be written as:

$$\mathbf{P}(w) \equiv \big(\Delta - \lambda^2\big)\big(\Delta w + \lambda^2 w\big) = 0$$

or (2.21)

$$\mathbf{P}(w) \equiv \big(\Delta + \lambda^2\big)\big(\Delta w - \lambda^2 w\big) = 0$$

Thus, any solution of any of the following two equations

$$\text{(a)} \quad \Delta w_1 + \lambda^2 w_1 = 0 \qquad \text{or} \qquad \text{(b)} \quad \Delta w_2 - \lambda^2 w_2 = 0 \tag{2.22}$$

is a solution of equation (2.20). Let us show that any solution w of (2.20) is a linear combination of solutions of equations (2.22). Let W_1 and W_2 be defined by:

$$W_1 = \Delta w - \lambda^2 w \qquad , \qquad W_2 = \Delta w + \lambda^2 w$$

We must have:

$$\begin{aligned} \mathbf{P}(w) &\equiv \big(\Delta + \lambda^2\big) W_1 = 0 \quad \Longrightarrow \quad W_1 = \alpha w_1 \\ \mathbf{P}(w) &\equiv \big(\Delta - \lambda^2\big) W_2 = 0 \quad \Longrightarrow \quad W_2 = \beta w_2 \end{aligned}$$

We deduce that:

$$2\lambda^2 w \equiv W_2 - W_1 = \beta w_2 - \alpha w_1$$

This shows that w is a linear combination of solutions of the two-dimension Helmholtz equations (2.22) (a) and (b), which, thus, provide all the independent solutions of the plate equation. The independent solutions of the homogenous Helmholtz equations are well known. We just recall them for two coordinate systems: rectangular coordinates and cylindrical coordinates.

1. *Rectangular coordinates*

Let (x_1, x_2) be the coordinates of a point M. There are four independent solutions of equations (2.22) depending on x_1 only, which are:

$$\begin{aligned} &\text{(a)} \quad u_1^+ = \mathrm{e}^{\imath \lambda x_1} \quad , \quad u_1^- = \mathrm{e}^{-\imath \lambda x_1} \\ &\text{(b)} \quad u_2^+ = \mathrm{e}^{\lambda x_1} \quad , \quad u_2^- = \mathrm{e}^{-\lambda x_1} \end{aligned}$$

All other solutions are obtained by rotating the axes. We finally get two families of independent solutions:

$$\text{(a)} \quad w_1 = \mathrm{e}^{\imath \lambda (x_1 \cos\theta + x_2 \sin\theta)} \qquad , \qquad \text{(b)} \quad w_2 = \mathrm{e}^{\lambda (x_1 \cos\theta + x_2 \sin\theta)}$$

where θ is any angle between 0 and 2π.

2. *Cylindrical coordinates*

Let (ρ, θ) be the coordinates of a point M. In cylindrical coordinates, the expression of Helmholtz equation (2.22-a) is:

$$\frac{1}{\rho}\frac{\partial}{\partial \rho}\left(\rho\frac{\partial w}{\partial \rho}\right) + \frac{1}{\rho^2}\frac{\partial^2 w}{\partial \theta^2} + \lambda^2 w = 0 \tag{2.23}$$

The solutions of this equation are sought as the product of a function of ρ by a function of θ, $w(\rho, \theta) = W_\rho(\rho)W_\theta(\theta)$. Furthermore, it is necessary to choose for W_θ a periodic function with period 2π, that is

$$W_\theta(\theta) = e^{\imath n\theta} \qquad \text{with} \qquad n = -\infty, \cdots, -2, -1, 0, 1, 2, \cdots, +\infty$$

With such a choice, the function W_ρ must satisfy:

$$\frac{1}{\rho}\frac{\partial}{\partial \rho}\left(\rho\frac{\partial W_\rho}{\partial \rho}\right) + (\lambda^2 - \frac{n^2}{\rho^2})W_\rho = 0$$

This is a Bessel equation, the regular solutions of which are the *Bessel functions of the first kind* $J_n(\lambda\rho)$. So, the family of solutions of equation (2.22-a) is:

$$w_1(\rho, \theta) = J_n(\lambda\rho)e^{\pm\imath n\theta} \qquad \text{with} \qquad n = 0, 1, 2, \cdots, +\infty \tag{2.24}$$

In the same way, the family of solutions of equation (2.22-b) is given by:

$$w_2(\rho, \theta) = J_n(\imath\lambda\rho)e^{\pm\imath n\theta} = e^{\imath n\pi/2} I_n(\lambda\rho)e^{\pm\imath n\theta} \; , \; n = 0, 1, 2, \cdots, +\infty \tag{2.24$'$}$$

where $I_n(z)$ is the modified Bessel function. For details concerning the Bessel equation and the Bessel functions the reader can refer to [MOR 53] and [ABR 65], or to any other book on classical mathematics or mathematical physics.

We can construct an equivalent family of real independent solutions which are either even or odd in θ:

$$\begin{array}{lll} w_1^e(\rho, \theta) = J_n(\lambda\rho)\cos n\theta \; , & w_2^e(\rho, \theta) = I_n(\lambda\rho)\cos n\theta \; , & n = 0, 1, \ldots, \infty \\ w_1^o(\rho, \theta) = J_n(\lambda\rho)\sin n\theta \; , & w_2^o(\rho, \theta) = I_n(\lambda\rho)\sin n\theta \; , & n = 1, \ldots, \infty \end{array} \tag{2.24$''$}$$

REMARK.– The Bessel equation has another family of solutions, known as *Bessel functions of the second kind* $Y_n(\lambda\rho)$, which tend to infinity for $\rho \to 0$. These functions satisfy the homogenous Bessel equation for strictly positive ρ. In the range $\rho \geq 0$, they satisfy a non-homogenous equation and, so, must be rejected.

2.3.2. *Green's Kernel and Green's function for the Time Harmonic Plate Equation and Response of an Infinite Plate to a Harmonic Excitation*

Here again, it is assumed that the time dependence is $e^{-\imath\omega t}$. A Green's kernel $\Gamma(M)$ of the time harmonic plate equation is a solution of the plate equation

$$\left(\Delta^2 - \lambda^4\right)\Gamma = \frac{\delta}{D} \tag{2.25}$$

where, in the second member, the term δ represents a point force located at the axes origin. The Green's function for the infinite plate equation is the Green's kernel which satisfies the energy conservation principle. It can be defined by imposing that it satisfies the limit absorption principle (Theorem 2.1). To obtain the correct solution, λ is replaced by $\lambda(1+\imath\varepsilon)$, where ε is a small positive parameter which will tend to 0. Equation (2.25) becomes:

$$\left(\Delta^2 - \lambda^4(1+\imath\varepsilon)^4\right)\Gamma_\varepsilon = \frac{\delta}{D} \tag{2.25$'$}$$

Let us introduce the space Fourier transform $\widehat{\Gamma}_\varepsilon$ defined by:

$$\widehat{\Gamma}_\varepsilon(\xi_1,\xi_2) = \int_{-\infty}^{+\infty}\int_{-\infty}^{+\infty} \Gamma_\varepsilon(x_1,x_2)e^{-2\imath\pi(x_1\xi_1+x_2\xi_2)}\,dx_1\,dx_2$$

By applying this Fourier transform to equation (2.25′), we obtain:

$$\left(16\pi^4\xi^4 - \lambda^4(1+\imath\varepsilon)^4\right)\widehat{\Gamma}_\varepsilon = \frac{1}{D}$$

$$\Longrightarrow \quad \widehat{\Gamma}_\varepsilon = \frac{1}{D}\frac{1}{16\pi^4\xi^4 - \lambda^4(1+\imath\varepsilon)^4} \quad \text{with} \quad \xi = \sqrt{\xi_1^2+\xi_2^2}$$

It appears that $\widehat{\Gamma}_\varepsilon(\xi_1,\xi_2)$ does not depend on the two Fourier variables ξ_1 and ξ_2 but on the radial distance ξ only. So, as shown in many treatises on mathematics and mathematical physics, $\Gamma_\varepsilon(M)$ depends on the distance ρ between M and the coordinate origin, and it can be denoted $\Gamma_\varepsilon(\rho)$. Thus, the inverse Fourier transform reduces to a Bessel transform

$$\Gamma_\varepsilon(\rho) = 2\pi\int_0^\infty \widehat{\Gamma}_\varepsilon(\xi)J_0(2\pi\xi\rho)\rho\,d\rho$$

This integral can be transformed into an integral extending along the whole real axis as follows. Using the following properties of the Hankel function of the first kind [ABR 65]

$$H_0^1(2\pi\xi\rho) = J_0(2\pi\xi\rho) + \imath Y_0(2\pi\xi\rho)$$
$$H_0^{(1)}(2\pi\xi\rho e^{\imath\pi}) = -H_0^{(2)}(2\pi\xi\rho) = -J_0(2\pi\xi\rho) + \imath Y_0(2\pi\xi\rho)$$

we obtain:

$$J_0(2\pi\xi\rho) = \frac{1}{2}\big(H_0^{(1)}(2\pi\xi\rho) - H_0^{(1)}(2\pi\xi\rho e^{\imath\pi})\big)$$

$$\Longrightarrow \Gamma_\varepsilon(\rho) = \pi \int_{\infty e^{\imath\pi}}^{\infty} \widehat{\Gamma}_\varepsilon(\xi) H_0^{(1)}(2\pi\xi\rho)\rho\,\mathrm{d}\rho$$

This integral form of $\Gamma_\varepsilon(\rho)$ can be evaluated by the residue method. The integration contour is the real axis closed by a half circle in the upper half of the complex plane, with radius increasing to infinity. To obtain the residues involved we have to determine the poles of $\widehat{\Gamma}_\varepsilon$. These are given by:

$$\begin{aligned} 2\pi\zeta^+ &= \lambda(1+\imath\varepsilon) \quad , \quad 2\pi\zeta^- = -\lambda(1+\imath\varepsilon) \\ 2\pi\eta^+ &= \imath\lambda(1+\imath\varepsilon) \quad , \quad 2\pi\eta^- = -\imath\lambda(1-\imath\varepsilon) \end{aligned}$$

The final result is:

$$\begin{aligned} \Gamma_\varepsilon(\rho) &= \frac{\imath}{8D\lambda^2(1+\imath\varepsilon)^2}\big[H_0^{(1)}(\lambda(1+\imath\varepsilon)\rho) - H_0^{(1)}(\imath\lambda(1+\imath\varepsilon)\rho)\big] \\ \Gamma(\rho) &= \lim_{\varepsilon\to 0}\Gamma_\varepsilon(\rho) = \frac{\imath}{8D\lambda^2}\big[H_0^{(1)}(\lambda\rho) - H_0^{(1)}(\imath\lambda\rho)\big] \end{aligned} \tag{2.26}$$

In what follows, we will use the notation $H_0(z)$ for $H_0^{(1)}(z)$ except where there is a possibility of ambiguity.

The plate equation Green's function (like any Green's kernel of the plate equation) is regular at origin, together with its first order derivative. Its second order derivative has a logarithmic singularity. The third order derivative has a singularity which is of order ρ^{-1}. For $\rho \ll 1$, we have the following expansions:

$$\begin{aligned} \Gamma(\rho) &= \frac{\imath}{8\lambda^2 D} + \frac{\big[2(\gamma-1) - \imath\pi + \ln(\imath\lambda/4) + \ln\lambda + 2\ln\rho\big]\rho^2}{16\pi D} + \mathcal{O}(\rho^3) \\ \Gamma'(\rho) &= \frac{\big[2(\gamma-1/2) - \imath\pi + \ln(\imath\lambda/4) + \ln\lambda + 2\ln\rho\big]\rho}{8\pi D} + \mathcal{O}(\rho^3) \\ \Gamma''(\rho) &= \frac{2(\gamma+1/2) - \imath\pi + \ln(\imath\lambda/4) + \ln\lambda + 2\ln\rho}{8\pi D} + \frac{3\imath\lambda^2\rho^2}{128D} + \mathcal{O}(\rho^3) \\ \Gamma'''(\rho) &= \frac{1}{4\pi D\rho} + \frac{3\imath\lambda^2\rho}{64\pi D} + \mathcal{O}(\rho^3) \end{aligned} \tag{2.27}$$

where γ is the Euler constant.

If the point force is located at a point with coordinates (x_0, y_0), the response of the plate at a point (x, y) is obtained by replacing ρ with $\sqrt{(x-x_0)^2 + (y-y_0)^2}$ (by translating the axes origin to (x_0, y_0), the result is obvious). Various other notations are commonly used:

$$\Gamma\left(\sqrt{(x-x_0)^2 + (y-y_0)^2}\right),\ \Gamma(x-x_0, y-y_0),\ \Gamma\left(\rho(M, M_0)\right),\ \Gamma(M, M_0)$$

Response of an infinite plate to a harmonic force.– Let $f(x_1, x_2)$ be a force applied on a domain σ of the plate (σ is of finite or infinite extent). We seek the solution of

$$\left(\Delta^2 - \lambda^4\right) w(x_1, x_2) = \frac{f(x_1, x_2)}{D} \tag{2.28}$$

Using the plate equation Green's function, the solution of equation (2.28) is given by:

$$w(x_1, x_2) = \Gamma * f(x_1, x_2) = \int_\sigma \Gamma(x_1 - x_1', x_2 - x_2') f(x_1', x_2')\, \mathrm{d}x_1'\, \mathrm{d}x_2' \tag{2.29}$$

where $\Gamma * f$ is the space convolution product of Γ and f. The integral representation of the solution is valid as far as f is a function (and, of course, that the product Γf is integrable), while the convolution product representation must be used if f is not a function but a distribution.

REMARK.– Forces are not the only way to excite plates. Point momentum excitations can occur: they exert a twisting effort on a point of the plate. Such an excitation is mathematically described by the derivative $\vec{d} \cdot \vec{\nabla}\delta$ of the Dirac measure δ in the direction of a unit vector $\vec{d}$. The response of the plate is given by the convolution product of the Green's function by the excitation

$$\Gamma * \vec{d}(\cdot \vec{\nabla}\delta) = \vec{d} \cdot \vec{\nabla}\Gamma$$

which is the derivative of the Green's function in the direction of the vector $\vec{d}$.

2.3.3. *Harmonic Vibrations of a Plate of Finite Dimensions: General Definition and Theorems*

We now consider a plate which occupies a bounded domain Σ, with boundary $\partial\Sigma$. In what follows, we generally consider conservative boundary conditions, that is, we assume that there is no energy loss through $\partial\Sigma$. The relationships between the plate displacement and its derivatives, which express the boundary conditions have real coefficients.

The definition of the resonance modes of a plate is identical to that given for a beam.

DEFINITION 2.2 (RESONANCE MODES OF A PLATE).– *A* resonance mode *is a function which satisfies the homogenous time harmonic plate equation*

$$\Delta^2 w(M) - \lambda^4 w(M) = 0 \qquad \textit{for} \qquad M \in \Sigma$$

and a given set of homogenous boundary conditions along $\partial\Sigma$*. The corresponding value of* λ *is called* resonance wavenumber *and the corresponding frequency is called* resonance frequency.

The following theorem can be proved:

THEOREM 2.3 (EXISTENCE OF RESONANCE MODES).– *A purely elastic plate, submitted to conservative boundary conditions, has a countable set of resonance modes and real resonance frequencies. The set of resonance modes is an orthogonal basis on which any solution of the non-homogenous harmonic plate equation (when it exists) can be expanded.*

The resonance modes and resonance frequencies of a purely elastic plate, submitted to conservative boundary conditions, are also called *eigenmodes* and *eigenfrequencies*. In such a case, both terminologies can equally be used.

If there is any energy loss (by material damping or through the boundaries), the resonance frequencies are complex.

The orthogonality of the eigenmodes is easily proved, the steps of the proof being as for the beam resonance modes.

PROOF OF THE ORTHOGONALITY OF THE PLATE EIGENMODES.– For simplicity, the plate is assumed to have a regular contour. Furthermore, we assume that the displacement satisfies conservative boundary conditions, so the eigenwavenumbers are real and eigenmodes can be taken as real. Let us consider the following integral

$$I = \int_\Sigma \frac{Eh^3}{12(1-\nu^2)} \Big[\nu \big(w_{p,11} + w_{p,22} \big) \big(w_{q,11} + w_{q,22} \big)$$
$$+ (1-\nu) \big(w_{p,11}\, w_{q,11} + w_{p,22}\, w_{q,22} + w_{p,12}\, w_{q,12} + w_{p,21}\, w_{q,21} \big) \Big] \, \mathrm{d}\Sigma$$

where w_p and w_q are two different eigenmodes corresponding to two different eigenwavenumbers λ_p and λ_q. A first set of integrations by parts gives:

$$
\begin{aligned}
I = &\int_{\Sigma} \frac{Eh^3}{12(1-\nu^2)} \Delta^2 w_p w_q \,\mathrm{d}\Sigma \\
&+ \frac{Eh^3}{12(1-\nu^2)} \int_{\partial\Sigma} \Big\{ \big[\operatorname{Tr} \Delta w_p - (1-\nu) \operatorname{Tr} \partial_{s^2} w_p \big] \operatorname{Tr} \partial_n w_q \\
&\qquad - \big[(1-\nu)\, \partial_{\bar{s}} \operatorname{Tr} \partial_n \partial_s w_p + \operatorname{Tr} \partial_n \Delta w_p \big] \operatorname{Tr} w_q \Big\} \,\mathrm{d}\bar{s} \\
= {}& \lambda_p^4 \frac{Eh^3}{12(1-\nu^2)} \int_{\Sigma} w_p w_q \,\mathrm{d}\Sigma \\
&+ \frac{Eh^3}{12(1-\nu^2)} \int_{\partial\Sigma} \Big\{ \big[\operatorname{Tr} \Delta w_p - (1-\nu) \operatorname{Tr} \partial_{s^2} w_p \big] \operatorname{Tr} \partial_n w_q \\
&\qquad - \big[(1-\nu)\, \partial_{\bar{s}} \operatorname{Tr} \partial_n \partial_s w_p + \operatorname{Tr} \partial_n \Delta w_p \big] \operatorname{Tr} w_q \Big\} \,\mathrm{d}\bar{s}
\end{aligned}
$$

A second set of integrations by parts leads to:

$$
\begin{aligned}
I = {}& \lambda_q^4 \frac{Eh^3}{12(1-\nu^2)} \int_{\Sigma} w_p w_q \,\mathrm{d}\Sigma \\
&+ \frac{Eh^3}{12(1-\nu^2)} \int_{\partial\Sigma} \Big\{ \operatorname{Tr} \partial_n w_p \big[\operatorname{Tr} \Delta w_q - (1-\nu) \operatorname{Tr} \partial_{s^2} w_q \big] \\
&\qquad - \operatorname{Tr} w_p \big[(1-\nu)\, \partial_{\bar{s}} \operatorname{Tr} \partial_n \partial_s w_q + \operatorname{Tr} \partial_n \Delta w_q \big] \Big\} \,\mathrm{d}\bar{s}
\end{aligned}
$$

By subtracting these two expressions of I and accounting for the fact that w_p and w_q satisfy the same boundary conditions, we obtain:

$$
\left(\lambda_p^4 - \lambda_q^4\right) \int_{\Sigma} w_p w_q \,\mathrm{d}\Sigma = 0 \quad \Rightarrow \quad \int_{\Sigma} w_p w_q \,\mathrm{d}\Sigma = 0 \quad \text{for} \quad p \neq q
$$

which expresses the orthogonality between the eigenmodes. □

The following theorem gives the conditions of existence and uniqueness of the response of a plate to a harmonic excitation.

THEOREM 2.4 (EXISTENCE AND UNIQUENESS OF THE SOLUTION).– *Let $f(M)$ be the harmonic excitation of a plate (force, momentum, ...) with angular frequency ω and let $\lambda = (\omega^2 \bar{\mu}/D)^{1/4}$ be the corresponding wavenumber.*

◇ *If λ is different from all the resonance wavenumbers, then the solution w of the non-homogenous plate equation*

$$
\Delta^2 w(M) - \lambda^4 w(M) = f(M) \qquad \text{for} \qquad M \in \Sigma
$$

exists and is unique for any excitation $f(M)$.

◇ *If λ is equal to one resonance wavenumber, say λ_n, the solution of the non-homogenous plate equation does not exist unless $f(M)$ is orthogonal to the corresponding resonance mode $w_n(M)$.*

◇ *For $\lambda = \lambda_n$, if $f(M)$ is orthogonal to $w_n(M)$, a non-unique solution of the non-homogenous plate equation exists, which can be decomposed into a uniquely determined component $w^{(1)}(M)$ orthogonal to $w_n(M)$, and a component $w^{(2)}(M) = \alpha w_n(M)$ with arbitrary amplitude α.*

The proof of these theorems can be found in classical books of mathematical analysis (see, for example, [COU 57, JOH 86]).

Reduced plate equation.– In what follows, we will generally deal with a reduced form of the plate equation which is obtained by using a characteristic length of the plate as unit of length. For a circular plate, its radius is a natural unit of length. For a rectangular plate, its smallest dimension is used here. The results so obtained are valid for a class of plates, and not uniquely for a particular physical plate.

2.3.4. *Resonance Modes and Resonance Frequencies of Circular Plates with Uniform Boundary Conditions*

Let Σ be the circular domain $r < R$, with center O, occupied by the plate, and let $\partial\Sigma$ be its boundary. It is natural to use cylindrical coordinates (r, θ) centered at O. As suggested before, it is better to introduce a reduced plate equation by introducing the variable change $\rho = r/R$. The equation governing the reduced resonance modes is:

$$\left(\Delta^2 - \Lambda^4\right) W(\rho, \theta) = 0 \quad \text{for} \quad 0 < \rho < 1 \quad , \quad 0 \leq \theta < 2\pi \tag{2.30}$$

The resonance wavenumbers of the original plate are related to those of (2.30) by $\lambda = R\Lambda$.

Two boundary conditions must be added along $\partial\Sigma$ (circle $\rho = 1$). We will restrict this study to the uniform standard boundary conditions, that is: a/ the boundary conditions are the same all along $\partial\Sigma$, b/ the plate boundaries are either clamped, simply supported, or free.

The resonance modes are the solutions of the homogenous plate equation, which are 2π-periodic in θ. They are linear combinations of the independent solutions defined in section 2.3.1. To satisfy two boundary conditions, it is necessary to use two independent solutions. The use of the solutions defined in (2.24″) leads to real mode shapes of the form:

$$W^e(\rho,\theta) = \big(AJ_n(\Lambda\rho) + BI_n(\Lambda\rho)\big)\cos n\theta \quad n = 0,1,2,\cdots,\infty$$
$$W^o(\rho,\theta) = \big(AJ_n(\Lambda\rho) + BI_n(\Lambda\rho)\big)\sin n\theta \quad n = 1,2,\cdots,\infty \tag{2.31}$$

By introducing these functions into the plate equation, it appears clearly that, for $n = 1,2,\ldots,\infty$, the resonance wavenumbers of the even modes and of the odd modes are identical: so, apart from resonance wavenumbers corresponding to modes independent of θ, the other resonance wavenumbers are of the order two: that is, two different mode shapes (one even in θ, and the other one odd) correspond to each resonance wavenumber. All the modes have n nodal diameters (diameters with zero displacement).

To be able to write boundary conditions, we must first express the boundary operators in cylindrical coordinates. In the original coordinate system (r,θ), the unit vectors are $\vec{e}_r$ and $\vec{e}_\theta$. They depend on θ, as can be seen by expressing them in rectangular coordinates:

$$\vec{e}_r = \vec{i}\cos\theta + \vec{j}\sin\theta \quad , \quad \vec{e}_\theta = -\vec{i}\sin\theta + \vec{j}\cos\theta$$

and thus:

$$\frac{\partial\vec{e}_r}{\partial\theta} = \vec{e}_\theta \quad , \quad \frac{\partial\vec{e}_\theta}{\partial\theta} = -\vec{e}_r$$

If P is a point on $\partial\Sigma$, the exterior normal vector $\vec{n}(P)$ is identical to $\vec{e}_r(P)$; the tangent vector $\vec{s}(P)$ is identical to $\vec{e}_\theta(P)$. The components of the gradient operator are $(\partial/\partial r, r^{-1}\partial/\partial\theta)$.

The expressions of the various boundary operators are obtained by using the definitions given in section 1.2. In what follows, M is a point inside Σ and P a point on $\partial\Sigma$.

1. Boundary operators of order one:

$$\begin{aligned}\mathrm{Tr}\,\partial_n w(P) &= \lim_{M\to P}\Big\{\vec{n}(P)\cdot\vec{\nabla}_M w(M)\Big\}\\ &= \lim_{M\to P}\Big\{\vec{n}(P)\cdot\Big[\vec{e}_r(M)w_{,r}(M) + \vec{e}_\theta(M)\frac{w_{,\theta}(M)}{r(M)}\Big]\Big\} = w_{,r}(P)\end{aligned}$$

because $\lim_{M\to P}\vec{e}_\theta(M) = \vec{s}(P)$ which is orthogonal to $\vec{n}(P)$. For a similar reason, we have:

$$\begin{aligned}\mathrm{Tr}\,\partial_s w(P) &= \lim_{M\to P}\Big\{\vec{s}(P)\cdot\vec{\nabla}_M w(M)\Big\}\\ &= \lim_{M\to P}\Big\{\vec{s}(P)\cdot\Big[\vec{e}_r(M)w_{,r}(M) + \vec{e}_\theta(M)\frac{w_{,\theta}(M)}{r(M)}\Big]\Big\} = \frac{w_{,\theta}(P)}{R}\end{aligned}$$

2. Boundary operators of order two:

$$\begin{aligned}\mathrm{Tr}\,\partial_{s^2} w(P) &= \lim_{M\to P}\left\{\vec{s}(P)\cdot\vec{\nabla}_M\left[\vec{s}(P)\cdot\vec{\nabla}_M w(M)\right]\right\}\\ &= \lim_{M\to P}\left\{\vec{s}(P)\cdot\vec{\nabla}_M\left[\vec{s}(P)\cdot\left(\vec{e}_r(M)w_{,r}(M)+\vec{e}_\theta(M)\frac{w_{,\theta}(M)}{r(M)}\right)\right]\right\}\end{aligned}$$

The radial component of $\vec{\nabla}_M\left[\vec{s}(P)\cdot\vec{\nabla}_M w(M)\right]$ will give no contribution to the limit. The angular component is:

$$\frac{1}{r}\left[\vec{n}(P)\cdot\vec{e}_r(M)\left(\frac{w_{,\theta\theta}(M)}{r(M)}+w_{,r}(M)\right)+\vec{s}(P)\cdot\vec{e}_r(M)\cdots\right]$$

This leads to the following result:

$$\mathrm{Tr}\,\partial_{s^2} w(P) = \frac{w_{,\theta\theta}(P)}{R^2}+\frac{w_{,r}(P)}{R}$$

In a very similar way, we have:

$$\begin{aligned}\mathrm{Tr}\,\partial_n\partial_s w(P) &= \lim_{M\to P}\left\{\vec{n}(P)\cdot\vec{\nabla}_M\left[\vec{s}(P)\cdot\vec{\nabla}_M w(M)\right]\right\}\\ &= \frac{w_{,r\theta}(P)}{R}-\frac{w_{,\theta}(P)}{R^2}\end{aligned}$$

3. Boundary operators of order three:

They are easy to obtain from the former results and we have

$$\mathrm{Tr}\,\partial_n\Delta w(P) = \partial_r\Delta w(P)$$

$$\partial_{\vec{s}}\,\mathrm{Tr}\,\partial_n\partial_s w(P) = \frac{w_{,r\theta\theta}(P)}{R^2}-\frac{w_{,\theta\theta}(P)}{R^3}$$

More detailed studies of circular plates can be found in several textbooks and, in particular, in [LEI 69] which contains many useful tables and figures. Here, we only present the most important aspects.

1. *Clamped boundary*

The boundary conditions along the circle $\rho = 1$

$$\mathrm{Tr}\,W(\rho,\theta) = \mathrm{Tr}\,\partial_\rho W(\rho,\theta) = 0$$

lead to the following system of two linear equations

$$AJ_n(\Lambda)+BI_n(\Lambda)=0 \quad , \quad AJ'_n(\Lambda)+BI'_n(\Lambda)=0$$

which has non-zero solutions for Λ solution of the equation

$$J_n(\Lambda)I_n'(\Lambda) - J_n'(\Lambda)I_n(\Lambda) = 0 \tag{2.32}$$

Using the recursion formulae of Bessel functions which relates $\mathcal{Z}_n'$ to $\mathcal{Z}_n$ and $\mathcal{Z}_{n+1}$, this equation takes a simpler form:

$$\Big[J_n(\Lambda)I_{n+1}(\Lambda) + J_{n+1}(\Lambda)I_n(\Lambda)\Big]\sqrt{2\pi\Lambda}\mathrm{e}^{-\Lambda} = 0 \tag{2.32$'$}$$

where the multiplicative coefficient $\sqrt{2\pi\Lambda}\mathrm{e}^{-\Lambda}$ is introduced to balance the asymptotic exponential growth of the modified Bessel functions. Thus, equation (2.32 $'$) is fairly easy to solve. Many mathematical softwares include powerful programs of Bessel functions and methods for solving non-linear equations. It is, in general, necessary to have a rough localization of the roots. For this purpose, a graphic approach is often very efficient.

It can be shown that equation (2.32$'$) has an infinite sequence of real solutions Λ_{ns}, numbered from $s = 0$ to ∞ (convention in use in reference [LEI 69]), each of which defining two resonance modes:

$$\begin{aligned}
W_{ns}^e(\rho,\theta) &= A_{ns}\left[\frac{J_n(\Lambda_{ns}\rho)}{J_n(\Lambda_{ns})} - \frac{I_n(\Lambda_{ns}\rho)}{I_n(\Lambda_{ns})}\right]\cos n\theta, \quad n = 0, 1, \cdots \quad s = 0, 1, \cdots \\
W_{ns}^o(\rho,\theta) &= A_{ns}\left[\frac{J_n(\Lambda_{ns}\rho)}{J_n(\Lambda_{ns})} - \frac{I_n(\Lambda_{ns}\rho)}{I_n(\Lambda_{ns})}\right]\sin n\theta, \quad n = 1, 2, \cdots \quad s = 0, 1, \cdots
\end{aligned} \tag{2.33}$$

Normalization factors A_{ns} are often chosen so that the modes have a unit $\mathbf{L}^2$-norm. The index s corresponds to the number of nodal circles, the clamped boundary not being included. Figure 2.1(a) represents the first member of (2.32$'$) for $n = 3$ and shows how to approximately localize the resonance wavenumbers. Figure 2.1(b) shows the absolute value of the even mode $W_{32}^e(\rho,\theta)$. It must be remarked that the odd mode $W_{32}^o(\rho,\theta)$ is deduced from the even one by a rotation of $\pi/2n$.

2. *Simply supported boundary*

The boundary conditions along the circle $\rho = 1$

$$\mathrm{Tr}\, W(\rho,\theta) = 0 \quad , \quad \mathrm{Tr}\, \Delta W(\rho,\theta) - (1-\nu)\,\mathrm{Tr}\,\partial_{s^2} W(\rho,\theta) = 0$$

lead to the system of equations:

$$\begin{aligned}
&AJ_n(\Lambda) + BI_n(\Lambda) = 0 \\
&A\Big[\Lambda^2 J_n''[\Lambda] + \Lambda J_n'(\Lambda) - n^2 J_n(\Lambda) - (1-\nu)\big(-n^2 J_n(\Lambda) + \Lambda J_n'(\Lambda)\big)\Big] \\
&\quad + B\Big[\Lambda^2 I_n''[\Lambda] + \Lambda I_n'(\Lambda) - n^2 I_n(\Lambda) - (1-\nu)\big(-n^2 I_n(\Lambda) + \Lambda J_n'(\Lambda)\big)\Big] = 0
\end{aligned}$$

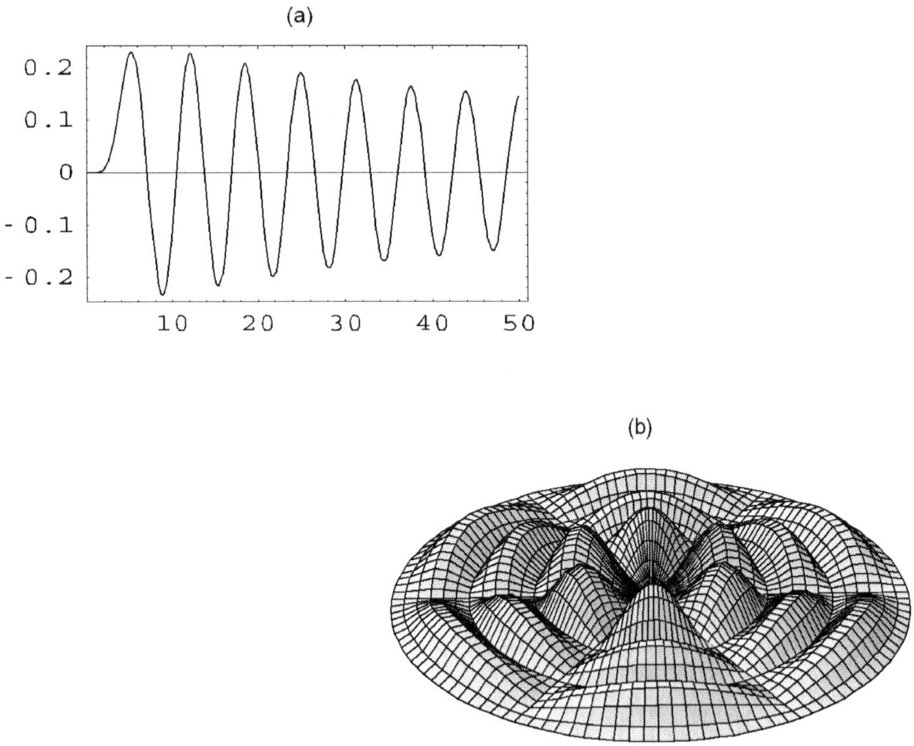

Figure 2.1. *Clamped plate: (a) – graphic solution of equation (2.32′) for $n = 3$; (b) – even mode $W^{e}_{3\,2}(\rho, \theta)$ corresponding to $\Lambda_{3\,2} = 13.795$*

which has a non-zero solution if its determinant is zero. Using the Bessel equations and the recursion formulae, the resonance wavenumber equation reduces to:

$$\left[J_{n+1}(\Lambda)I_n(\Lambda) + I_{n+1}(\Lambda)J_n(\Lambda) - \frac{2\Lambda}{1-\nu}J_n(\Lambda)I_n(\Lambda)\right]\sqrt{2\pi\Lambda}\mathrm{e}^{-\Lambda} = 0 \quad (2.34)$$

where a multiplicative factor has been introduced to balance the exponential growth due to the modified Bessel functions. The resonance modes have the expression given by (2.33), but with different resonance wavenumbers. It should be remarked that the resonance wavenumbers depend on the Poisson's ratio. Figure 2.2(a) represents the first member of (2.34) for $n = 3$. Figure 2.2(b) shows the absolute value of the odd mode $W^{o}_{3\,2}(\rho, \theta)$.

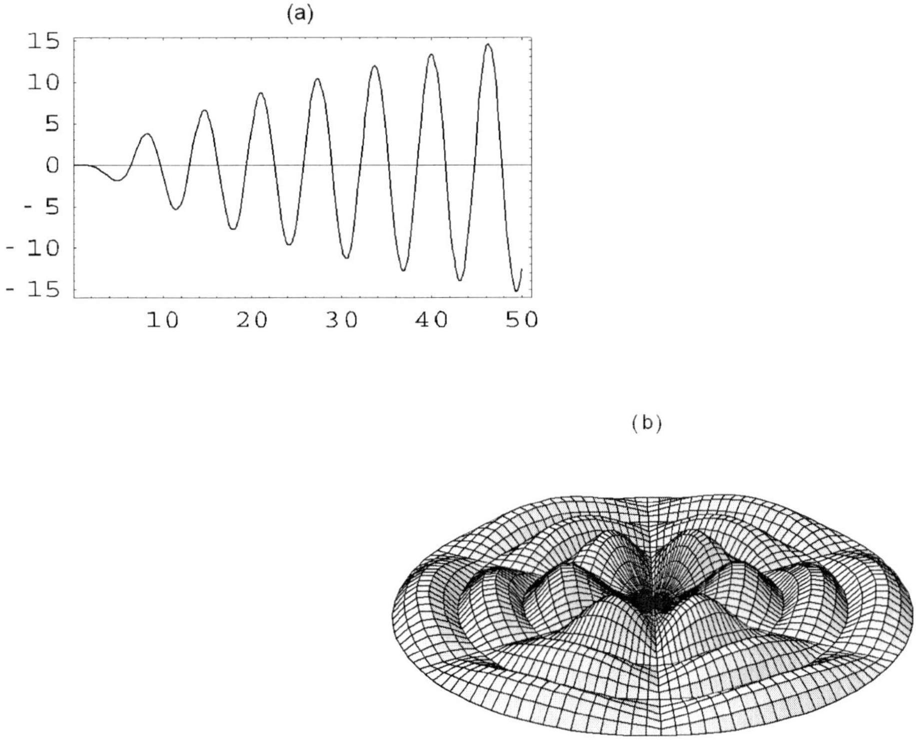

Figure 2.2. *Simply supported plate, with $\nu = 0.33$: (a) – graphic resolution of equation (2.34) for $n = 3$; (b) – Odd mode $W^o_{3\,2}(\rho, \theta)$ corresponding to $\Lambda_{3\,2} = 12.989$*

3. *Completely free boundary*

The boundary conditions along the circle $\rho = 1$ are:

$$\mathrm{Tr}\,\Delta W - (1-\nu)\,\mathrm{Tr}\,\partial_{s^2} W = 0 \quad , \quad \mathrm{Tr}\,\partial_n \Delta W + (1-\nu)\partial_s\,\mathrm{Tr}\,\partial_n \partial_s W = 0$$

As for the completely free beam, $\Lambda = 0$ is a resonance wavenumber which corresponds to three modes:

$$W_0^1 = 1 \quad , \quad W_0^2 = \rho\cos\theta \quad , \quad W_0^3 = \rho\sin\theta$$

This first resonance mode is a translation of the plate without deformation. The other two correspond to a rotation of the plate with an in-plane extension only. As far as transverse vibrations are concerned, these modes are of no interest.

For $\Lambda \neq 0$, the boundary conditions lead to:

$$AL_1(n,\Lambda) + BL_2(n,\Lambda) = 0 \quad , \quad AL_3(n,\Lambda) + BL_4(n,\Lambda) = 0$$

where:

$$\begin{aligned}
L_1(n,\Lambda) &= -\Lambda^2 J_n(\Lambda) - \frac{1-\nu}{2}\Big[\Lambda J_{n-1}(\Lambda) - 2n^2 J_n(\Lambda) - \Lambda J_{n+1}(\Lambda)\Big] \\
L_2(n,\Lambda) &= \Lambda^2 I_n(\Lambda) - \frac{1-\nu}{2}\Big[\Lambda I_{n-1}(\Lambda) - 2n^2 I_n(\Lambda) + \Lambda I_{n+1}(\Lambda)\Big] \\
L_3(n,\Lambda) &= \frac{1}{2}\Big\{-\Lambda^3\big[J_{n-1}(\Lambda) - J_{n+1}(\Lambda)\big] \\
&\qquad + (1-\nu)n^2\big[\Lambda J_{n-1}(\Lambda] - 2J_n(\Lambda) - \Lambda J_{n+1}(\Lambda)\big]\Big\} \\
L_4(n,\Lambda) &= \frac{1}{2}\Big\{\Lambda^3\big[I_{n-1}(\Lambda) + I_{n+1}(\Lambda)\big] \\
&\qquad + (1-\nu)n^2\big[-\Lambda I_{n-1}(\Lambda] + 2I_n(\Lambda) - \Lambda I_{n+1}(\Lambda)\big]\Big\}
\end{aligned} \tag{2.35}$$

The resonance wavenumber equation is:

$$\Big[L_1[n,\Lambda)L_4(n,\Lambda] - L_2(n,\Lambda)L_3(n,\Lambda)\Big]\mathrm{e}^{-\Lambda}\sqrt{2\pi\Lambda}/\Lambda^5 = 0 \tag{2.35$'$}$$

where a multiplicative factor has again been introduced to balance the exponential growth due to the modified Bessel functions.

The resonance modes have the following expression:

$$\begin{aligned}
W_{ns}^e &= A_{ns}\left[\frac{J_n(\Lambda_{ns}\rho)}{L_1(n,\Lambda_{ns})} - \frac{I_n(\Lambda_{ns}\rho)}{L_2(n,\Lambda_{ns})}\right]\cos n\theta, \quad n = 0,1,\cdots \quad s = 0,1,\cdots \\
W_{ns}^o &= A_{ns}\left[\frac{J_n(\Lambda_{ns}\rho)}{L_1(n,\Lambda_{ns})} - \frac{I_n(\Lambda_{ns}\rho)}{L_2(n,\Lambda_{ns})}\right]\sin n\theta, \quad n = 1,2,\cdots \quad s = 0,1,\cdots
\end{aligned} \tag{2.36}$$

Figure 2.3(a) represents the first member of (2.35 $'$) for $n = 3$. Figure 2.3(b) shows the absolute value of the even mode $W_{32}^e(\rho,\theta)$.

4. *Final comments*

We have mentioned that, to each resonance wavenumber, there corresponds an even and an odd resonance mode. This implies that any linear combination of these two independent modes is equally a resonance mode which differs from the original ones by the position of the nodal radii. However, being a linear combination of two independent solutions of the homogenous plate equation, it cannot be considered as a new mode.

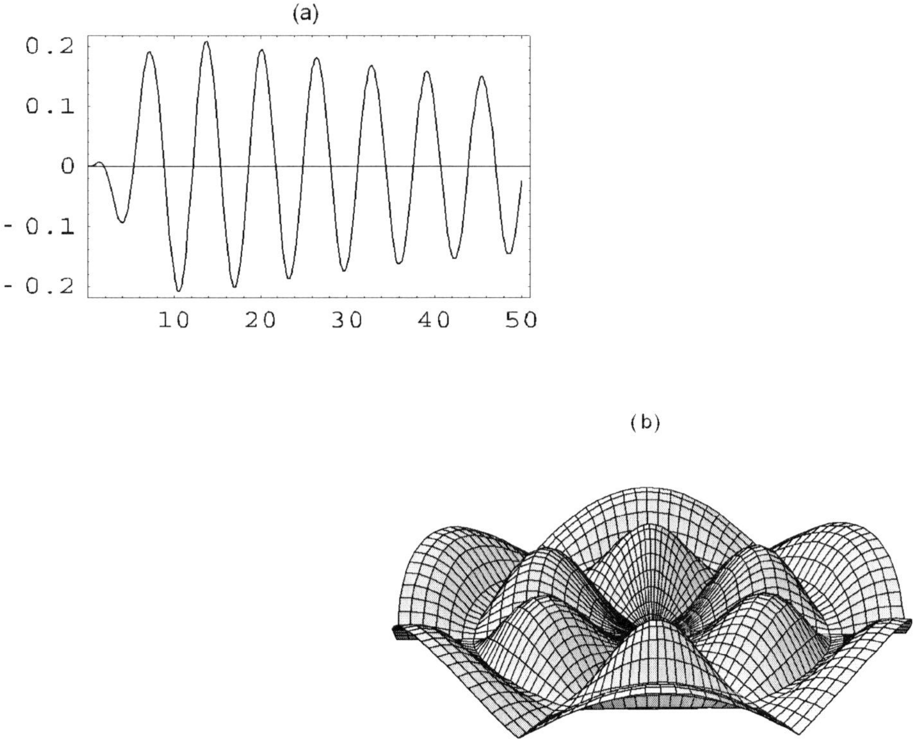

Figure 2.3. *Completely free plate, with* $\nu = 0.33$*: (a) – graphic solution of equation (2.35′) for* $n = 3$*; (b) – even mode* $W^e_{32}(\rho, \theta)$ *corresponding to* $\Lambda_{32} = 10.578$

Experimentally, it not very easy to obtain the Chladni patterns (nodal line patterns) of a circular plate, mainly if the material is quite perfectly isotropic: indeed, the angles origin is a mathematical concept and not a physical reality. It is necessary to force the plate to have a nodal line at a given position with an artificial anisotropy which can be obtained by imposing a zero displacement at some point, for example with the tip of a finger nail.

The former examples show that the lowest resonance wavenumber of the mode $(3, 2)$ corresponds to the completely free plate, while its highest value corresponds to the clamped plate. This is a general result which is true for every mode. The physical explanation is that a clamped plate is globally more rigid than a simply supported plate, which is itself globally more rigid than a completely free plate.

Another important aspect which has not been developed here is the calculation of the radii of nodal circular lines, that is the radii of the circles with zero displacement. Tables of such radii are often proposed together with tables of reduced wavenumbers (see, for example, [LEI 69]). They are not difficult to compute with any modern mathematical software, and this is a good exercise to become familiar with these academic, but nevertheless fundamental, problems.

2.3.5. *Resonance Modes and Resonance Frequencies of Rectangular Plates with Uniform Boundary Conditions*

Let us consider a rectangular plate with reduced dimensions 1 and $L \geq 1$. The reduced resonance wavenumbers Λ and resonance modes W are the solutions of the reduced homogenous plate equation

$$\left[\Delta^2 - \Lambda^4\right] W(x_1, x_2) = 0 \quad \text{with} \quad x_1 \in \left]-\frac{1}{2}, +\frac{1}{2}\right[\quad , \quad x_2 \in \left]-\frac{L}{2}, +\frac{L}{2}\right[\tag{2.37}$$

which satisfy a given set of boundary conditions.

Apart from the plate with simply supported boundaries, it is not possible to find an explicit form for the resonance modes, as we did for circular plates. The beam equation being very similar to the plate equation, it appears interesting to expand the plate modes as series of functions $V_n(x_1)V_m(x_2/L)$, where the functions V_n are beam modes which satisfy the same boundary conditions as the plate. We will see that such a choice of basis functions is particularly powerful.

We will limit this study to the three fundamental cases of uniform boundary conditions: simply supported plate, clamped plate and completely free plate. Other types of simple boundary conditions – as clamped along two opposite sides, and simply supported along the other two – lead to equations which can easily be solved (such problems are proposed as exercises).

In general, the resonance wavenumbers are of order 1. But there are particular cases (the square plate is the obvious example) where a resonance wavenumber is of order 2. In such a case, any linear combination of the two corresponding modes is, of course, a mode. But its nodal lines are completely different from the nodal lines of the original modes. A numerical result highlights this aspect.

1. *Simply supported boundary*

The boundary conditions take the form:

$$\begin{aligned} W = 0 \quad &, \quad W_{,11} + \nu W_{,22} = 0 \quad \text{for} \quad x_1 = \pm 1/2 \\ W = 0 \quad &, \quad W_{,22} + \nu W_{,11} = 0 \quad \text{for} \quad x_2 = \pm L/2 \end{aligned} \tag{2.38}$$

Condition $W = 0$ along the sides $x_1 = \pm 1/2$ implies that $W_{,22} = 0$, too. Similarly, $W_{,11} = 0$ along the sides $x_2 = \pm L/2$. The consequence is that the resonance wavenumbers do not depend on the Poisson's ratio, and the boundary conditions take a simpler form:

$$\begin{aligned} W &= 0 \quad , \quad W_{,11} = 0 \quad \text{for} \quad x_1 = \pm 1/2 \\ W &= 0 \quad , \quad W_{,22} = 0 \quad \text{for} \quad x_2 = \pm L/2 \end{aligned} \tag{2.38$'$}$$

Let $\{\bar{\lambda}_m, V_m(x)\}$ be the set of resonance wavenumbers and normalized resonance modes of a beam with unit length, simply supported at both ends – see equations (2.9-2.9′) and (2.10-2.10′) – and consider the function $W_{mn}(x_1, x_2) = V_m(x_1)V_n(x_2/L)L^{-1/2}$. Such a function satisfies boundary conditions (2.38) and the Helmholtz equation

$$\left[\Delta + (\bar{\lambda}_m^2 + \bar{\lambda}_n^2/L^2)\right]W_{mn} = 0$$

As said before (equation (2.21)), the plate operator can be written as $(\Delta - \Lambda^2)(\Delta + \Lambda^2)$. Thus, W_{mn} is solution of the plate equation:

$$\begin{aligned} &\left[\Delta^2 - (\bar{\lambda}_m^2 + \bar{\lambda}_n^2/L^2)^2\right]W_{mn} = \\ &\qquad \left[\Delta - (\bar{\lambda}_m^2 + \bar{\lambda}_n^2/L^2)\right]\left[\Delta + (\bar{\lambda}_m^2 + \bar{\lambda}_n^2/L^2)\right]W_{mn} = 0 \end{aligned}$$

We can conclude that the reduced orthonormalized resonance modes of a rectangular plate with simply supported boundaries is the set of functions $W_{mn}(x_1, x_2)$ defined above. The corresponding resonance wavenumbers are $(\bar{\lambda}_m^2 + \bar{\lambda}_n^2/L^2)^{1/2}$. More precisely, the resonance modes can be classified into 4 families:

1. "even-even" modes:

$$\begin{aligned} W_{2m\ 2n} &= \frac{2}{\sqrt{L}}\cos\left[(2m+1)\pi x_1\right]\cos\left[\frac{(2n+1)\pi x_2}{L}\right] \\ &\qquad m = 0, 1, \ldots \qquad n = 0, 1, \ldots \\ \Lambda_{2m\ 2n} &= \left[\left((2m+1)\pi\right)^2 + \left(\frac{(2n+1)\pi}{L}\right)^2\right]^{1/2} \end{aligned} \tag{2.39}$$

2. "even-odd" modes:

$$\begin{aligned} W_{2m\ 2n-1} &= \frac{2}{\sqrt{L}}\cos\left[(2m+1)\pi x_1\right]\sin\left[\frac{2n\pi x_2}{L}\right] \\ &\qquad m = 0, 1, \ldots \quad n = 1, 2, \ldots \\ \Lambda_{2m\ 2n-1} &= \left[\left((2m+1)\pi\right)^2 + \left(\frac{2n\pi}{L}\right)^2\right]^{1/2} \end{aligned} \tag{2.39$'$}$$

3. "odd-even" modes:

$$W_{2m-1\ 2n} = \frac{2}{\sqrt{L}} \sin\left[2m\pi x_1\right] \cos\left[\frac{(2n+1)\pi x_2}{L}\right]$$

$$m = 1, 2, \ldots \quad n = 0, 1, \ldots \qquad (2.39'')$$

$$\Lambda_{2m-1\ 2n} = \left[\left(2m\pi\right)^2 + \left(\frac{(2n+1)\pi}{L}\right)^2\right]^{1/2}$$

4. "odd-odd" modes:

$$W_{2m-1\ 2n-1} = \frac{2}{\sqrt{L}} \sin\left[2m\pi x_1\right] \sin\left[\frac{2n\pi x_2}{L}\right]$$

$$m = 1, 2, \ldots \quad n = 1, 3 \ldots \qquad (2.39''')$$

$$\Lambda_{2m-1\ 2n-1} = \left[\left(2m\pi\right)^2 + \left(\frac{2n\pi}{L}\right)^2\right]^{1/2}$$

n_1	0	0	1	0	1	2	0	1	2	3	...
n_2	0	1	0	2	1	0	3	2	1	0	...
m	1	2	3	4	5	6	7	8	9	10	...
number of nodal lines	0	1		2			3				...

Table 2.2. *Correspondence between the single index numbering of the modes $W_m(x_1, x_2)$ and the double index numbering $W_{n_1\ n_2}(x_1, x_2)$*

It is classical to number the resonance modes by the number of nodal lines in each direction, excluding the boundary. Here, the first index is the number of nodal lines parallel to the x_2-axis, and the second index is the number of nodal lines parallel to the x_1-axis. It can also be useful to number the resonance modes with a single index. To that end, we propose to class them into groups of modes having the same total number of nodal lines and to adopt the correspondence between the number of nodal lines in each direction and the single index of the mode which is presented in Table 2.2.

Figure 2.4 shows two representations of the mode (2,3) for a plate with reduced dimensions ($1, L = \pi/e$). Such a choice for L ensures that all resonances are of order one. As soon as L is a rational number, resonances of order two appear.

The simplest case is the square plate for which all resonances are of order two. Figure 2.5 shows two different aspects of the same resonance. The nodal

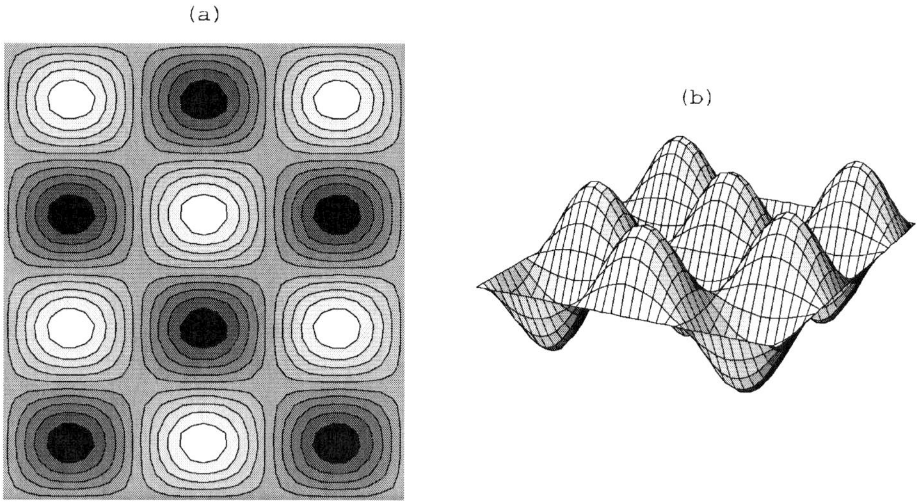

Figure 2.4. *Simply supported plate, ($L = \pi/\mathrm{e}$): (a) – contour representation of the mode $W_{23}(x_1, x_2)$ corresponding to $\Lambda_{23} = \sqrt{9 + 16\mathrm{e}^2/\pi^2}$; (b) – three-dimensional representation of the same mode*

lines of the standard mode (1,2) are straight lines; the mode (2,1) is deduced from it by a rotation of $\pi/2$. But a linear combination of these two modes, which corresponds to the same resonance wavenumber, presents a completely different pattern as shown in Figure 2.5(b).

For a rectangular plate with dimensions in a rational ratio N/M, the resonances of order two are not so many. For a ratio 3/2, the first resonance of order two of a simply supported plate appear for the modes (3,11) and (7,5). Figure 2.6 shows these two modes while Figure 2.7 presents a linear combination of these two modes, which is equally a resonance mode.

To obtain experimentally a given nodal pattern corresponding to a resonance frequency of order two, it is necessary to force the plate to have a node at a given position.

2. *Clamped boundary*

The boundary conditions have a simple form:

$$\begin{aligned} W &= 0\ ,\ W_{,2} = 0 \quad \text{on } x_1 = \pm 1/2 \\ W &= 0\ ,\ W_{,1} = 0 \quad \text{on } x_2 = \pm L/2 \end{aligned} \tag{2.40}$$

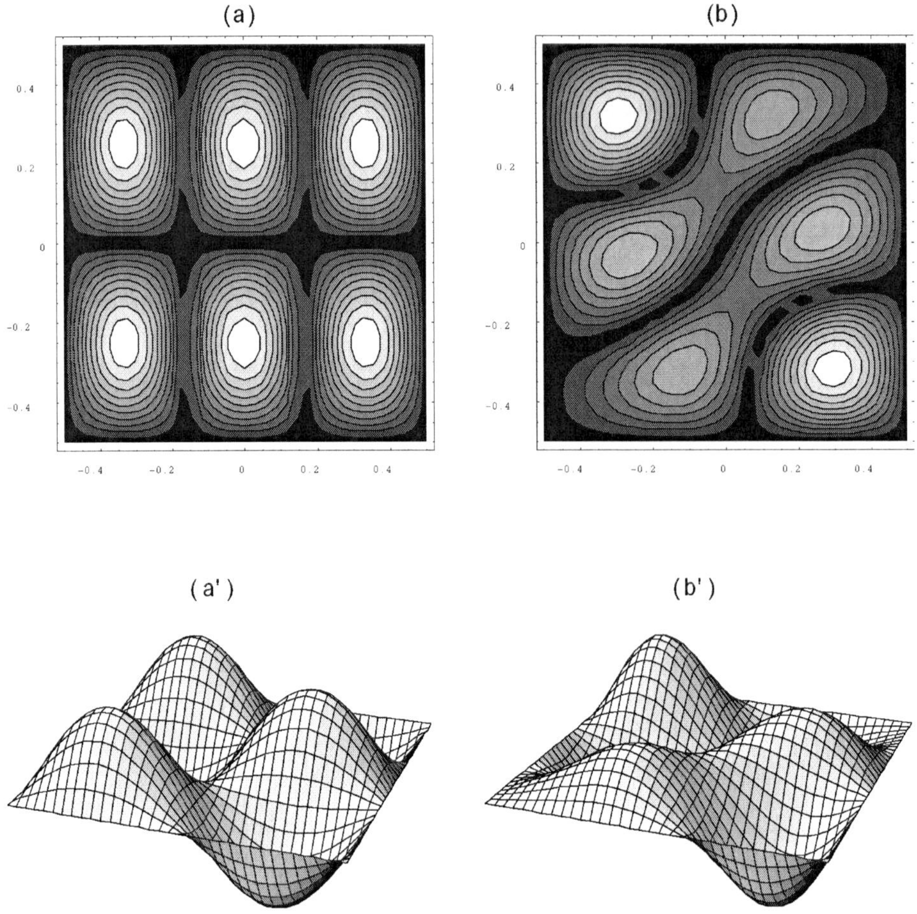

Figure 2.5. *Simply supported square plate: contour and 3D-representations of the mode $W_{12}(x_1, x_2)$ corresponding to $\Lambda_{12} = \sqrt{13}$ (a-a'); contour and 3D-representations of the linear combination $\cos 2\pi/3 W_{12}(x_1, x_2) + \sin 2\pi/3 W_{21}(x_1, x_2)$ of the modes $W_{12}(x_1, x_2)$ and $W_{21}(x_1, x_2)$ (b,b')*

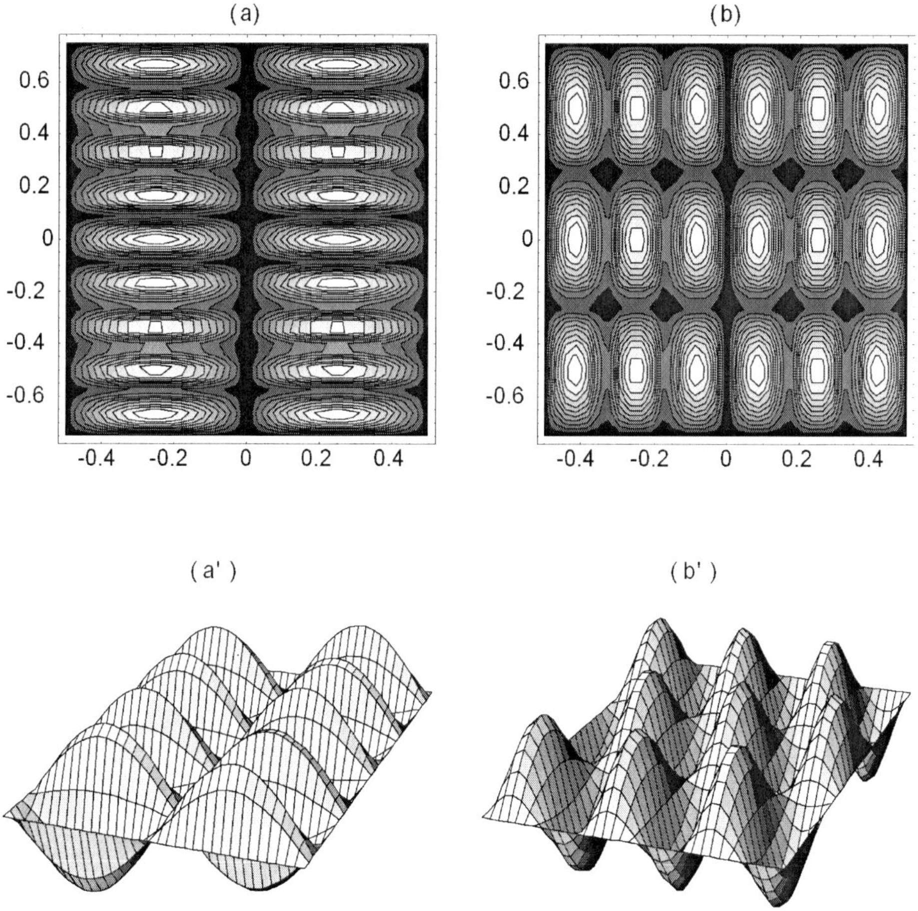

Figure 2.6. *Simply supported plate, ($L = 3/2$): contour representation and 3D-representation of two modes of a rectangular plate corresponding to the same resonance eigenvalue $\Lambda = 4\pi\sqrt{5}$: mode $W_{3\ 11}(x_1, x_2)$ (a-a') and $W_{7\ 5}(x_1, x_2)$ (b-b')*

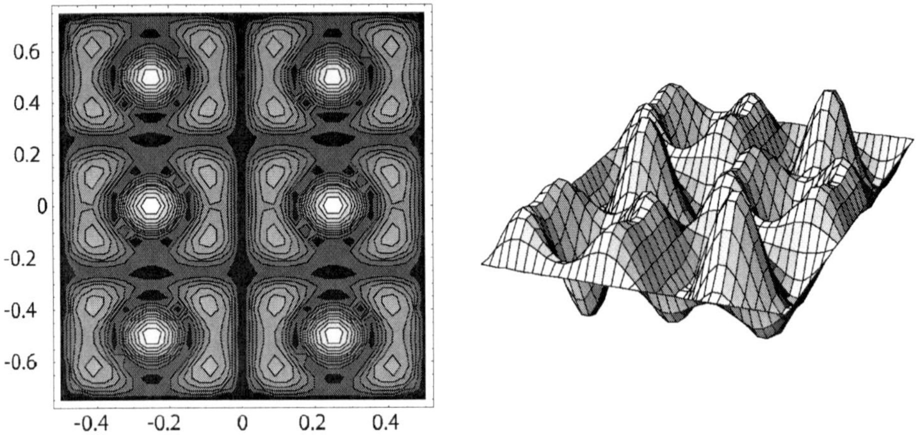

Figure 2.7. *Simply supported plate, ($L = 3/2$): contour representation and 3D-representation of the linear combination of two modes* $\sin(2\pi/3)(W_{3\ 11}x_1, x_2) + \cos(2\pi/3)W_{7\ 5}$

Let $\{\bar{\lambda}_m, V_m(x)\}$ be the set of resonance wavenumbers and normalized resonance modes of a beam with unit length, clamped at both ends, and consider the set of functions $U_{mn}(x_1, x_2) = V_m(x_1)V_n(x_2/L)L^{-1/2}$. Such functions satisfy boundary conditions (2.40), but they do not satisfy the plate equation. Nevertheless, this set of functions is an orthonormal basis for the functional space, which any solution of the clamped plate boundary value problem belongs to. Thus, we can seek the resonance modes of the clamped plate equation as series of U_{mn}. As we did for the simply supported plate, it is useful to consider four families of resonance modes. Denoting the even and odd beam modes by $V_m^e(x)$ and $V_m^o(x)$ respectively, the plate modes are sought as:

1. "even-even" modes:

$$W_{2r\ 2s} = \sum_{m=1}^{\infty}\sum_{n=1}^{\infty} w_{2r\ 2s}^{nm} V_m^e(x_1)V_n^e(x_2/L)L^{-1/2}$$

$$r \text{ and } s = 0, 1, \ldots \qquad (2.41)$$

2. "even-odd" modes:

$$W_{2r\ 2s+1} = \sum_{m=1}^{\infty}\sum_{n=1}^{\infty} w_{2r\ 2s+1}^{nm} V_m^e(x_1)V_n^o(x_2/L)L^{-1/2}$$

$$r \text{ and } s = 0, 1, \ldots \qquad (2.41')$$

3. "odd-even" modes:

$$W_{2r+1\,2s} = \sum_{m=1}^{\infty}\sum_{n=1}^{\infty} w_{2r+1\,2s}^{nm} V_m^o(x_1) V_n^e(x_2/L) L^{-1/2}$$

$$r \text{ and } s = 0, 1, \ldots \qquad (2.41'')$$

4. "odd-odd" modes:

$$W_{2r+1\,2s+1} = \sum_{m=1}^{\infty}\sum_{n=1}^{\infty} w_{2r+1\,2s+1}^{nm} V_m^o(x_1) V_n^o(x_2/L) L^{-1/2}$$

$$r \text{ and } s = 0, 1, \ldots \qquad (2.41''')$$

The resonance modes are numbered with two indices using the same convention as for the simply supported plate. They can also be numbered with a single index as proposed in the former section and, if so, these are denoted by $W_n(x_1, x_2)$.

To find the coefficients of the former expansions, use is made of the Ritz-Galerkin method. Let us recall the variational form of the homogenous reduced plate equation for a harmonic time dependence (formula (1.33′)):

$$\int_{-1/2}^{+1/2}\int_{-L/2}^{+L/2} \Big[\Delta W \Delta \delta W + (1-\nu)\big(W_{,12}\,\delta W_{,12} + W_{,21}\,\delta W_{,21} - W_{,11}\,\delta W_{,22} - W_{,22}\,\delta W_{,11}\big) - \Lambda^4 W \delta W\Big]\,dx_1\,dx_2 = 0$$

Let $W = \sum w^{mn} U_{mn}$ represent the expansion of any mode of a given family in terms of the corresponding basis functions U_{mn}. The variational plate equation must be satisfied for any function δW. This is achieved if the following sequence of equations, obtained by replacing successively δW with each function U_{pq}, are verified:

$$\sum_{mn} w^{mn} \int_{-1/2}^{+1/2}\int_{-L/2}^{+L/2} \Big[\Delta U_{mn} \Delta U_{pq} + (1-\nu)\big(U_{mn,12}\, U_{pq,12} + U_{mn,21}\, U_{pq,21} - U_{mn,11}\, U_{pq,22} - U_{mn,22}\, U_{pq,11}\big) - \Lambda^4 U_{mn} U_{pq}\Big]\,dx_1\,dx_2 = 0$$

$$\text{for all } (p, q) \qquad (2.42)$$

First of all, we remark that for a homogenous plate, the term factor of $(1-\nu)$ is zero. Indeed, because $U_{mn}(x_1, x_2) = V_m(x_1)V_n(x_2/L)$, we have

$$\int_{-1/2}^{+1/2}\int_{-L/2}^{+L/2} U_{mn,12}(x_1,x_2)U_{pq,12}(x_1,x_2)\,\mathrm{d}x_1\,\mathrm{d}x_2$$

$$= \frac{1}{L^3}\int_{-1/2}^{+1/2} V'_m(x_1)V'_p(x_1)\,\mathrm{d}x_1 \int_{-L/2}^{+L/2} V'_n(x_2/L)V'_q(x_2/L)\,\mathrm{d}x_2$$

$$= \frac{1}{L^3}\left[-\int_{-1/2}^{+1/2} V''_m(x_1)V_p(x_1)\,\mathrm{d}x_1 + V'_m(x_1)V_p(x_1)\Big|_{x_1=-1/2}^{x_1=+1/2}\right]$$

$$\times\left[-\int_{-L/2}^{+L/2} V_n(x_2/L)V''_q(x_2/L)\,\mathrm{d}x_2 + V'_n(x_2/L)V_q(x_2/L)\Big|_{x_2=-L/2}^{x_2=+L/2}\right]$$

$$= \int_{-1/2}^{+1/2}\int_{-L/2}^{+L/2} U_{mn,11}(x_1,x_2)U_{pq,22}(x_1,x_2)\,\mathrm{d}x_1\,\mathrm{d}x_2$$

the integrated terms being zero because of the boundary conditions satisfied by the beam modes. In the same way, we have

$$\int_{-1/2}^{+1/2}\int_{-L/2}^{+L/2} U_{mn,21}(x_1,x_2)U_{pq,21}(x_1,x_2)\,\mathrm{d}x_1\,\mathrm{d}x_2$$

$$= \int_{-1/2}^{+1/2}\int_{-L/2}^{+L/2} U_{mn,22}(x_1,x_2)U_{pq,11}(x_1,x_2)\,\mathrm{d}x_1\,\mathrm{d}x_2$$

Accounting for the equation satisfied by the beam resonance modes $V_m(x)$

$$\int_{-1/2}^{+1/2} \left[V''_m(x)\delta V''(x) - \bar{\lambda}_m^4 V_m(x)\delta V(x)\right]\mathrm{d}x = 0 \quad \text{for all} \quad \delta V(x)$$

and remembering that the functions U_{mn} are orthogonal and have a unit norm, equation (2.42) becomes:

$$\sum_{mn} w^{mn}\left[\left(\bar{\lambda}_p^4 + \frac{\bar{\lambda}_q^4}{L^4}\right)\delta_{mn}^{pq} + 2\int_{-1/2}^{+1/2}\int_{-L/2}^{+L/2} U_{mn,12}U_{pq,12}\,\mathrm{d}x_1\,\mathrm{d}x_2\right] = \Lambda^4 w^{pq}$$

$$\text{for all} \quad (p,q) \qquad (2.43)$$

where δ_{mn}^{pq} is the Kronecker symbol. We are left with an infinite system of linear homogenous equations which has a non-zero-solution if Λ^4 is an eigenvalue. The eigenvalues of an infinite system cannot be calculated analytically, approximations must be used.

mode	Reduced wavenumbers Λ_{mn} Ritz method	Warburton	Relative error
0 , 0	26.478	26.561	$0.312\ 10^{-2}$
1 , 0	65.677	65.916	$0.364\ 10^{-2}$
0 , 1	39.631	39.806	$0.442\ 10^{-2}$
2 , 0	124.879	125.244	$0.292\ 10^{-2}$
1 , 1	77.999	78.404	$0.519\ 10^{-2}$
0 , 2	62.060	62.353	$0.471\ 10^{-2}$
3 , 0	203.842	204.297	$0.223\ 10^{-2}$
2 , 1	136.977	137.601	$0.456\ 10^{-2}$
1 , 2	99.047	99.480	$0.437\ 10^{-2}$
0 , 3	93.212	93.662	$0.483\ 10^{-2}$
4 , 0	302.550	303.068	$0.171\ 10^{-2}$
3 , 1	215.920	216.733	$0.377\ 10^{-2}$
2 , 2	157.446	157.948	$0.319\ 10^{-2}$
1 , 3	128.916	129.496	$0.450\ 10^{-2}$
0 , 4	132.716	133.319	$0.455\ 10^{-2}$
5 , 0	420.998	421.562	$0.134\ 10^{-2}$
4 , 1	314.620	315.617	$0.317\ 10^{-2}$
3 , 2	236.110	236.856	$0.316\ 10^{-2}$
2 , 3	186.562	186.907	$0.185\ 10^{-2}$
1 , 4	167.394	168.222	$0.495\ 10^{-2}$
0 , 5	180.401	181.142	$0.411\ 10^{-2}$

Table 2.3. *Reduced resonance wavenumbers of a clamped rectangular plate, for $L = \pi/2$: comparison between the Ritz-Galerkin method results and Warburton's approximation*

The simplest approximation, known as *Warburton's approximation*, consists in neglecting the non-diagonal terms. This gives:

$$\Lambda_{mn}^4 \simeq \left(\bar{\lambda}_m^4 + \frac{\bar{\lambda}_n^4}{L^4}\right) + \int_{-1/2}^{+1/2}\int_{-L/2}^{+L/2} (U_{mn,12})^2 \, \mathrm{d}x_1 \, \mathrm{d}x_2 \tag{2.44}$$

and

$$W_{mn}(x_1, x_2) \simeq U_{mn}(x_1, x_2) = \frac{1}{\sqrt{L}} V_m(x_1) V_n(x_2/L)$$

This approximation is based on the assumption that the non-diagonal terms are never "too large" compared to the diagonal ones. As a consequence, the resonance modes of the plate are close to the functions U_{mn}: in particular, they have the same number of nodal lines.

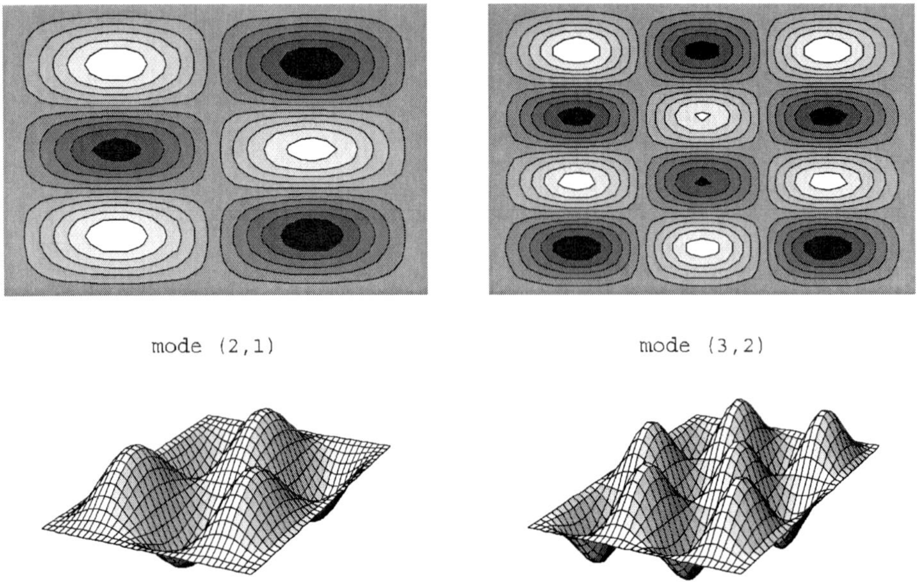

Figure 2.8. *Rectangular plate with clamped boundaries (sides ratio=$\pi/2$): three-dimensional and contour representations on modes (2,1) and (3,2)*

More accurate resonance wavenumbers and modes are obtained by truncation of system (2.43). The result depends strongly on the number of basis functions which have been accounted for. To avoid too large errors in the results, the resonance modes must be approximated with equivalent accuracies in each direction. This is correctly achieved if the highest order basis function has approximately the same density of nodal lines in both directions. For example, if $L = 1.5$, we account for all basis functions which have up to M nodal lines parallel to the x_2-axis and N nodal lines parallel to the x_1-axis with N as close as possible to $1.5 \times M$. This is an easy task when the functions U_{mn} are renumbered as in Table 2.2 for resonance modes. In Table 2.3 the reduced resonance wavenumbers of a plate with $L = \pi/2$ are presented for all modes having up to a total of 5 nodal lines, that is 21 modes. The computation has been conducted with 54 basis functions. This, in principle, enables us to have the same number of resonance modes, but, because the accuracy of the results decreases as the order of the mode increases, only around one half of the resonance wavenumbers have a good accuracy. Warburton's approximation is compared to the Ritz-Galerkin method result, which, of course, is much more accurate: however, it can be seen that Warburton's method provides resonance frequencies with an accuracy, which is quite sufficient for engineering purposes.

Figure 2.8 shows representations of modes (2,1) and (3,2). They look very similar to modes of a simply supported plate: the main difference is that the slope of the plate is zero along the boundary.

3. *Completely free plate*

This is the less simple case of a rectangular plate. We will give the main outlines of the methods used to solve it.

We first notice that, here again, $\Lambda = 0$ is a resonance wavenumber, with three resonance modes

$$W_0^1 = 1 \quad , \quad W_0^2 = x_1 \quad , \quad W_0^3 = x_2$$

The first mode is a translation of the plate in the normal direction. The other two are rotations with an in-plane extension of the plate. So, these modes are of no interest as far as vibration motions are concerned. To avoid them, the excitation functions considered must be orthogonal to these modes. From a physical point of view, this means that the excitations considered do not produce any global displacement of the plate: this is the case for acoustical excitations.

The main difficulty of this problem is the choice of a basis of functions which satisfy the boundary conditions. Another possibility is to expand any resonance mode on a basis of the space of functions which are square integrable on the plate domain (the classical Hilbert space $\mathbf{L}^2(\Sigma)$) and to use the Ritz-Galerkin method to determine the expansion coefficients. As we did before, we can use products of modes of the completely free beam (in the literature, and especially in reference [LEI 69], other functions are proposed). As in the case of the clamped plate, such functions partly satisfy the plate equation and, as a consequence, are close to the resonance modes shape. But, they do not satisfy the boundary conditions. Thus, an additional term appears in the resonance mode equation.

Let $(\bar{\lambda}_m, V_m)$ be the reduced resonance wavenumbers and modes of the unit length beam, with free ends. The set of functions $U_{mn}(x_1, x_2) = V_m(x_1)$ $V_n(x_2/L)L^{-1/2}$ is not a basis of the space which the completely free plate resonance modes belong to. Indeed, they do not satisfy the boundary conditions, which are:

$$\begin{array}{c} \mathrm{Tr}\,\Delta W - (1-\nu)\,\mathrm{Tr}\,\partial_{s^2} W = 0\ ,\ \ \mathrm{Tr}\,\partial_n \Delta W + (1-\nu)\partial_s\,\mathrm{Tr}\,\partial_n \partial_s W = 0 \\ \text{along} \quad (x_1 = \pm 1/2, -L/2 < x_2 < L/2) \\ \text{and} \quad (-1/ < x_1 < +1/2, x_2 = \pm L/2) \\ W_{,12} = 0 \quad \text{at the points} \\ (x_1 = 1/2, x_2 = L/2)\ ,\ (x_1 = 1/2, x_2 = -L/2)\ , \\ (x_1 = -1/2, x_2 = -L/2) \text{ and } (x_1 = -1/2, x_2 = L/2) \end{array} \qquad (2.45)$$

where the tangential and normal derivatives are identical to $\pm\partial_{x_1}$ or $\pm\partial_{x_2}$, depending on the side of the plate which is involved. Nevertheless, they constitute a basis of orthonormalized functions on which the resonance modes of the plate can be expanded. Let $W = \sum w^{mn} U_{mn}$ be this expansion, and consider the following integral

$$I = \int_{-1/2}^{+1/2}\int_{-L/2}^{+L/2}\Big[\Delta W \Delta U_{pq} + (1-\nu)\big(2W_{,12}\,U_{pq,12} - W_{11}U_{pq,22} - W_{,22}\,U_{pq,11}\big)\Big]\,\mathrm{d}x_1\,\mathrm{d}x_2$$

A first double integration by parts, which accounts for the partial differential equation and the boundary conditions satisfied by W, shows that

$$I = \Lambda^4 \int_{-1/2}^{+1/2}\int_{-L/2}^{+L/2} W U_{pq}\,\mathrm{d}x_1\,\mathrm{d}x_2$$

A second double integration by parts, which accounts for the differential equation satisfied by the beam modes, leads to:

$$I = \left[\bar{\lambda}_p^4 + \frac{\bar{\lambda}_q^4}{L^4}\right]\int_{-1/2}^{+1/2}\int_{-L/2}^{+L/2} W U_{pq}\,\mathrm{d}x_1\,\mathrm{d}x_2 + 2\int_{-1/2}^{+1/2}\int_{-L/2}^{+L/2} W_{,12}\,U_{pq,12}\,\mathrm{d}x_1\,\mathrm{d}x_2$$
$$+ \int_{\partial\widetilde{\Sigma}}\Big[\ell_1(W)\,\mathrm{Tr}\,\partial_n U_{pq} - \mathrm{Tr}\,\partial_n \Delta W\,\mathrm{Tr}\,U_{pq} + \ell_2(W)\,\mathrm{Tr}\,\partial_s U_{pq}\Big]\,\mathrm{d}\bar{s}$$

$\partial\widetilde{\Sigma}$ is the plate boundary in reduced coordinates; ℓ_1 and ℓ_2 are the boundary operators defined in section 1.2 and which are not difficult to calculate for a rectangular boundary. These two expressions of the integral I, together with the series expansion of W leads to a system of homogenous equations that are very similar to (2.43):

$$\sum_{mn} w^{mn}\Bigg[\left(\bar{\lambda}_p^4 + \frac{\bar{\lambda}_q^4}{L^4}\right)\delta_{mn}^{pq} + 2\int_{-1/2}^{+1/2}\int_{-L/2}^{+L/2} U_{mn,12}\,U_{pq,12}\,\mathrm{d}x_1\,\mathrm{d}x_2$$
$$+ \int_{\partial\widetilde{\Sigma}}\Big[\ell_1(U_{mn})\,\mathrm{Tr}\,\partial_n U_{pq} - \mathrm{Tr}\,\partial_n \Delta U_{mn}\,\mathrm{Tr}\,U_{pq} + \ell_2(U_{mn})\,\mathrm{Tr}\,\partial_s U_{pq}\Big]\,\mathrm{d}\bar{s}\Bigg] = \Lambda^4 w^{pq}$$

for all (p, q) (2.46)

Because the functions U_{pq} have a simple analytic form and the integration domains – the plate domain and its boundary – are geometrically simple, all integrals involved in equations (2.46) can be calculated analytically. As done for the clamped plate, the resonance problem is solved by truncation of this infinite system.

2.3.6. *Response of a Plate to a Harmonic Excitation: Resonance Modes Series Representation*

Let Σ be the domain occupied by the plate, $\partial\Sigma$ its boundary with $\vec{n}$ and $\vec{s}$ as normal and tangent unit vectors. The plate is excited by a harmonic force $\Re\big(f(M)\mathrm{e}^{-\imath\omega t}\big)$, where $f(M)$ is a force density. The plate displacement has the same time dependence and can be written as $\Re\big(w(M)\mathrm{e}^{-\imath\omega t}\big)$. The displacement amplitude $w(M)$ is the unique solution – if it exists – of the harmonic plate equation:

$$\big[\Delta^2 - \lambda^4\big]w(M) = \frac{f(M)}{D} \quad M \in \Sigma \quad , \quad \text{with} \quad \lambda^4 = \bar{\mu}\omega^2/D \tag{2.47}$$

which satisfies a given set of boundary conditions (clamped, simply supported, completely free, etc.).

1. *Straightforward determination of the resonance modes series*

In this section, it is easier to use a single index for the resonance modes and the resonance wavenumbers. Assume that the sets of resonance wavenumbers λ_n and resonance modes $w_n(M)$ (assumed to have a unit norm) are known. Theorem 2.4 states that, when it exists, $w(M)$ can be expanded into a series of resonance modes

$$w(M) = \sum_{n=1}^{\infty} a_n w_n(M)$$

Let us introduce this expansion into equation (2.47), multiply both members by $w_m(M)$ and integrate them over the plate domain. We obtain:

$$\sum_{n=1}^{\infty} a_n \int_{\Sigma} \big[\Delta^2 - \lambda^4\big] w_n(M) w_m(M)\,\mathrm{d}\Sigma = \frac{1}{D}\int_{\Sigma} f(M) w_m(M)\,\mathrm{d}\Sigma \tag{2.48}$$

Using the equation satisfied by each resonance mode, and accounting for the orthogonality of the modes, equation (2.48) becomes:

$$a_m(\lambda_m^4 - \lambda^4) = \frac{1}{D}\int_{\Sigma} f(M) w_m(M)\,\mathrm{d}\Sigma = \frac{f_m}{D} \tag{2.48$'$}$$

As far as λ is different from all the resonance wavenumbers, the coefficients a_m have a uniquely determined value and the response of the plate is given by:

$$w(M) = \frac{1}{D} \sum_{m=1}^{\infty} \frac{f_m}{\lambda_m^4 - \lambda^4} w_m(M) \tag{2.49}$$

If λ is equal to a resonance wavenumber, say λ_j, the coefficient a_j is in general equal to infinity – and the solution of (2.47) does not exist – unless $f_j = 0$ ($f(M)$ orthogonal to $w_j(M)$). In that case, there exists a non-unique solution given by:

$$w(M) = \frac{1}{D} \sum_{m \neq j,\, m=1}^{\infty} \frac{f_m}{\lambda_m^4 - \lambda^4} w_m(M) + a_j w_j(M) \tag{2.49$'$}$$

where a_j is arbitrary. This series is still valid if the excitation is described by a distribution (for example, a Dirac measure δ_S located at a point S): the integrals which give the coefficients f_m are replaced by the duality product (for a point excitation, $f_m = \langle \delta_S, w_m \rangle = w_m(S)$).

The convergence rate of series (2.49) depends on point M. Furthermore, this series does not point out the influence of the boundaries on the plate displacement. This influence is more easily seen if the plate displacement is decomposed into an "incident displacement" – that is, the displacement of an infinite plate – and a "diffracted displacement" – that is, the perturbation introduced by the plate boundaries.

2. *Series of resonance modes for the "diffracted" displacement*

The plate displacement is decomposed into the sum of the infinite plate displacement and a "diffracted" displacement, representing the boundaries effect, which is expanded into a resonance modes series:

$$w(M) = u_0(M) + u_1(M)$$

$$\text{with} \quad u_0(M) = \int_\Sigma f(M')\Gamma(|MM'|)\,\mathrm{d}\Sigma(M')\,, \; u_1(M) = \sum_{n=1}^{\infty} b_n w_n(M) \tag{2.50}$$

where $\Gamma(|MM'|)$ is Green's function of the plate equation as given by equation (2.26), $|MM'|$ being the distance between the points M and M'. The function $u_1(M)$ satisfies the homogenous plate equation and non-homogenous boundary conditions. The coefficients of its expansion cannot be obtained by the simple procedure used in the previous section.

The equations which give the b_n depend on the boundary conditions. Let us assume that the plate is clamped along $\partial\Sigma$ (for other boundary conditions, the expressions of the coefficients are established in an absolutely similar way). Thus, the function $u_1(M)$ satisfies the following system of boundary conditions:

$$u_1(M) = -u_0(M) \quad \text{and} \quad \partial_n u_1(M) = -\partial_n u_0(M) \quad \text{for} \quad M \in \partial\Sigma \qquad (2.51)$$

Let us consider the integral (variational form (1.33′) of the plate equation):

$$I = \int_\Sigma \Big[\Delta u_1 \Delta w_m + (1-\nu)\big(2u_{1,12}\, w_{m,12} - u_{1,11}\, w_{m,22} - u_{1,22}\, w_{m,11}\big)\Big] \mathrm{d}\Sigma$$

A first double integration by parts, which accounts for the equation satisfied by u_1 and for the boundary conditions satisfied by the resonance modes, gives:

$$I = \lambda^4 \int_\Sigma u_1 w_m \, \mathrm{d}\Sigma = \lambda^4 b_m$$

A second double integration by parts, which accounts for the equation satisfied by w_m and for the boundary conditions satisfied by u_1, gives:

$$I = \lambda_m^4 b_m - g_m$$

$$\text{with} \quad g_m = \int_{\partial\Sigma} \Big\{ \big[\operatorname{Tr} \Delta w_m - (1-\nu) \operatorname{Tr} \partial_{s^2} w_m \big] \operatorname{Tr} \partial_n u_0 - \big[(1-\nu)\partial_{\bar{s}} \operatorname{Tr} \partial_n \partial_s w_m + \operatorname{Tr} \partial_n \Delta w_m \big] \operatorname{Tr} u_0 \Big\} \mathrm{d}\bar{s}$$

These two expressions of I give the coefficients b_m and the expansion of u_1 is thus:

$$u_1(M) = \sum_{n=1}^{\infty} \frac{g_m}{\lambda_m^4 - \lambda^4} w_m(M) \qquad (2.52)$$

This series is defined under the same conditions as (2.49).

2.3.7. *Boundary Integral Equations and the Boundary Element Method*

The methods that we have described up to now are valid for simple shapes of plates and uniform boundary conditions. There are less simple cases for which such methods can be used, as, for example, rectangular plates with different boundary conditions on the different sides of the plate. Similar methods have been developed for elliptic plates, or triangular plates. For plates with complex geometries and boundary conditions, only numerical methods can be used. Finite Element Methods are commonly used in mechanics. Though Boundary Element Methods are very efficient, they remain less popular. We will give an overview of this method. The reader who is interested in a more complete study can easily find specialized textbooks on Boundary Integral Equations and Boundary Elements.

The basic idea is the following. For a bounded plate, the solution of the non-homogenous equation is sought as the sum of the displacement of an infinite plate ("incident displacement") and a solution of the homogenous equation ("reflected displacement") which describes the influence of the boundaries. This last term is described as the displacement generated by sources supported by the plate boundary. The source densities are the solution of a system of integral equations along the plate boundary, which are solved numerically. The simplest way to establish such equations is based on the use of Green's formula.

1. *Green's formula for the plate operator*

Let us again consider the following integral:

$$I = \int_\Sigma \Big[\Delta w \Delta v + (1-\nu)\big(2w_{,12}\, v_{,12} - w_{,11}\, v_{,22} - w_{,22}\, v_{,11}\big)\Big] \mathrm{d}\Sigma \qquad (2.53)$$

where w and v are functions which are twice continuously differentiable on Σ. This integral is integrated by parts in two different ways and two expressions of the integral I are obtained:

$$\begin{aligned} I = \int_\Sigma \Delta^2 w\; v \,\mathrm{d}\Sigma + \int_{\partial\Sigma} \Big[&\big(\mathrm{Tr}\,\Delta w - (1-\nu)\,\mathrm{Tr}\,\partial_{s^2} w\big)\,\mathrm{Tr}\,\partial_n v \\ &-\big(\mathrm{Tr}\,\partial_n \Delta w + (1-\nu)\partial_{\bar{s}}\,\mathrm{Tr}\,\partial_n\partial_s w\big)\,\mathrm{Tr}\,v\Big]\,\mathrm{d}\bar{s} \end{aligned}$$

$$\begin{aligned} I = \int_\Sigma w\; \Delta^2 v \,\mathrm{d}\Sigma + \int_{\partial\Sigma} \Big[&\mathrm{Tr}\,\partial_n w\big(\mathrm{Tr}\,\Delta v - (1-\nu)\,\mathrm{Tr}\,\partial_{s^2} v\big) \\ &-\mathrm{Tr}\,w\big(\mathrm{Tr}\,\partial_n \Delta v + (1-\nu)\partial_{\bar{s}}\,\mathrm{Tr}\,\partial_n\partial_s v\big)\Big]\,\mathrm{d}\bar{s} \end{aligned}$$

These expressions are similar to (1.6) and are valid if $\partial\Sigma$ has no angular points. If angular points are present, additional terms appear (see section 1.2.3). By subtracting these two expressions, we obtain Green's formula for the plate operator:

$$\begin{aligned} &\int_\Sigma \Big[\Delta^2 w\; v - w\; \Delta^2 v\Big]\,\mathrm{d}\Sigma + \int_{\partial\Sigma} \Big[\big(\mathrm{Tr}\,\Delta w - (1-\nu)\,\mathrm{Tr}\,\partial_{s^2} w\big)\,\mathrm{Tr}\,\partial_n v \\ &-\big(\mathrm{Tr}\,\partial_n \Delta w + (1-\nu)\partial_{\bar{s}}\,\mathrm{Tr}\,\partial_n\partial_s w\big)\,\mathrm{Tr}\,v - \mathrm{Tr}\,\partial_n w\big(\mathrm{Tr}\,\Delta v - (1-\nu)\,\mathrm{Tr}\,\partial_{s^2} v\big) \\ &\qquad + \mathrm{Tr}\,w\big(\mathrm{Tr}\,\partial_n \Delta v + (1-\nu)\partial_{\bar{s}}\,\mathrm{Tr}\,\partial_n\partial_s v\big)\Big]\,\mathrm{d}\bar{s} = 0 \qquad (2.54) \end{aligned}$$

2. *Green's representation of the plate displacement and the direct Boundary Element Method*

Let now $w(M)$ be the solution of the non-homogenous equation (2.47), satisfying a set of boundary conditions, which will be explicitly given when necessary. Green's formula applied to $w(M')$ and to the Green's kernel $\Gamma(M, M')$ leads to:

$$\begin{aligned}
&\int_{\Sigma}\Big[\Delta^2 w(M')\;\Gamma(M,M') - w(M')\;\Delta^2\Gamma(M,M')\Big]\,\mathrm{d}\Sigma(M')\\
&+\int_{\partial\Sigma}\Big[\big(\operatorname{Tr}\Delta w(P) - (1-\nu)\operatorname{Tr}\partial_{s^2} w(P)\big)\operatorname{Tr}\partial_n\Gamma(M,P)\\
&\quad-\big(\operatorname{Tr}\partial_n\Delta w(P) + (1-\nu)\partial_{\bar{s}}\operatorname{Tr}\partial_n\partial_s w(P)\big)\operatorname{Tr}\Gamma(M,P)\\
&\qquad-\operatorname{Tr}\partial_n w(P)\big(\operatorname{Tr}\Delta\Gamma(M,P) - (1-\nu)\operatorname{Tr}\partial_{s^2}\Gamma(M,P)\big)\\
&+\operatorname{Tr} w(P)\big(\operatorname{Tr}\partial_n\Delta\Gamma(M,P) + (1-\nu)\partial_{\bar{s}}\operatorname{Tr}\partial_n\partial_s\Gamma(M,P)\big)\Big]\,\mathrm{d}\bar{s}(P) = 0 \qquad (2.55)
\end{aligned}$$

In the boundary integral, all derivatives are taken with respect to the point P. Accounting for the equations satisfied by w and Γ, this equation leads to the following expression of plate displacement:

$$\begin{aligned}
w(M) = w_0(M) + D\int_{\partial\Sigma}\Big[&\operatorname{Tr} w(P)\big(\operatorname{Tr}\partial_n\Delta\Gamma(M,P) + (1-\nu)\partial_{\bar{s}}\operatorname{Tr}\partial_n\partial_s\Gamma(M,P)\big)\\
&-\operatorname{Tr}\partial_n w(P)\big(\operatorname{Tr}\Delta\Gamma(M,P) - (1-\nu)\operatorname{Tr}\partial_{s^2}\Gamma(M,P)\big)\\
&+\big(\operatorname{Tr}\Delta w(P) - (1-\nu)\operatorname{Tr}\partial_{s^2} w(P)\big)\operatorname{Tr}\partial_n\Gamma(M,P)\\
&-\big(\operatorname{Tr}\partial_n\Delta w(P) + (1-\nu)\partial_{\bar{s}}\operatorname{Tr}\partial_n\partial_s w(P)\big)\operatorname{Tr}\Gamma(M,P)\Big]\,\mathrm{d}\bar{s}(P)
\end{aligned}$$

$$\text{with}\quad w_0(M) = \int_{\Sigma} f(M')\Gamma(M,M')\,\mathrm{d}\Sigma(M') \qquad (2.56)$$

where $w_0(M)$ is the displacement of an infinite plate excited by the force $f(M)$. This last expression involves four unknown functions:

$$\begin{aligned}
\varpi_0(P) &= \operatorname{Tr} w(P) \quad,\quad \varpi_1(P) = \operatorname{Tr}\partial_n w(P)\\
\varpi_2(P) &= \operatorname{Tr}\Delta w(P) - (1-\nu)\operatorname{Tr}\partial_{s^2} w(P)\\
\varpi_3(P) &= \operatorname{Tr}\partial_n\Delta w(P) + (1-\nu)\partial_{\bar{s}}\operatorname{Tr}\partial_n\partial_s w(P)
\end{aligned}$$

Assume that the plate is clamped along its boundary, thus, $\varpi_0(P)$ and $\varpi_1(P)$ are equal to zero. Then the Green's representation of the plate displacement reduces to:

$$w(M) = w_0(M) + D \int_{\partial\Sigma} \left[\varpi_2(P) \operatorname{Tr} \partial_n \Gamma(M,P) - \varpi_3(P) \Gamma(M,P) \right] \mathrm{d}\bar{s}(P) \quad (2.57)$$

Using this relationship to express $w(M)$ and $\partial_n w(M)$ on $\partial\Sigma$, a system of two integral equations is obtained, which determines the unknown functions. Other boundary integral equations can be obtained by expressing the values, on $\partial\Sigma$, of higher order normal derivatives of $w(M)$.

Some care must be taken. The Green's kernel $\Gamma(M, M')$ and its first order derivative are regular functions. But its second order derivative has a logarithmic singularity and, thus, the function

$$\vec{n} \cdot \vec{\nabla}_M \int_{\partial\Sigma} \varpi_2(P) \operatorname{Tr} \partial_n \Gamma(M,P) \, \mathrm{d}\bar{s}$$

presents a discontinuity across $\partial\Sigma$. A limit procedure must be used to obtain the correct value of this function on the boundary. Expressions involving higher order derivatives imply introducing the notion of "finite part" of an integral (the definition of this notion is beyond the scope of this course).

For other boundary conditions, the procedure is exactly the same. Moreover, it is possible to consider discontinuous boundary conditions as, for example, a plate clamped along a part of its boundary, simply supported along a second part and free along the remaining part.

The resonance wavenumbers are the values of λ for which the homogenous system of integral equations has a non-zero solution.

There exist many numerical algorithms to solve integral equations. The simplest one is to approximate the unknown functions by piecewise constant functions and apply a collocation method. More precisely, the boundary is divided into N arc elements on which the two unknown functions are approximated by constants. Then the integral equations are written at the center of each arc element. Thus, a system of $2N$ linear equations with $2N$ unknown constants is obtained. Approximations of the resonance wavenumbers of the plate are given by the values of the wavenumber λ for which the determinant of the linear system is zero. Theoretically, an infinite sequence of such values can be found. But the number of resonance wavenumbers which are correctly approximated is much less than the rank of the linear system which approximates the system of boundary integral equations.

3. *The indirect Boundary Element Method*

It can be shown that the following functions

$$\begin{aligned} u_0(M) &= \Gamma * \chi_0 \otimes \delta_\Sigma = \int_{\partial\Sigma} \chi_0(P)\Gamma(M,P)\,\mathrm{d}\bar{s}(P) \\ u_1(M) &= \Gamma * \chi_1 \otimes \delta'_\Sigma = -\int_{\partial\Sigma} \chi_1(P)\partial_{n(P)}\Gamma(M,P)\,\mathrm{d}\bar{s}(P) \\ u_2(M) &= \Gamma * \chi_2 \otimes \delta''_\Sigma = \int_{\partial\Sigma} \chi_2(P)\partial_{n^2(P)}\Gamma(M,P)\,\mathrm{d}\bar{s}(P) \\ u_3(M) &= \Gamma * \chi_3 \otimes \delta'''_\Sigma = -\int_{\partial\Sigma} \chi_3(P)\partial_{n^3(P)}\Gamma(M,P)\,\mathrm{d}\bar{s}(P) \end{aligned} \tag{2.58}$$

are four independent solutions of the plate equation in Σ (the symbol "$*$" stands for the space convolution product). The source terms – $\chi_0 \otimes \delta_\Sigma$, $\chi_1 \otimes \delta'_\Sigma$, $\chi_2 \otimes \delta''_\Sigma$, $\chi_3 \otimes \delta'''_\Sigma$ – represent sources supported by the plate boundary. By analogy with electrostatics, they are respectively called simple layer, double layer, layer of order 2 and layer of order 3. The solution of equation (2.47) can be expressed with two of these functions, that is:

$$w(M) = w_0(M) + (-1)^i \int_{\partial\Sigma} \chi_i(P)\partial_{n^i(P)}\Gamma(M,P)\,\mathrm{d}\bar{s}(P) + (-1)^j \int_{\partial\Sigma} \chi_j(P)\partial_{n^j(P)}\Gamma(M,P)\,\mathrm{d}\bar{s}(P)$$

with $i \neq j$ and $w_0(M)$ being as in (2.56). By writing the boundary conditions, we obtain a system of two Boundary Integral Equations which determines the two layer densities χ_i and χ_j. As in the direct method, the resonance wavenumbers are the values of λ for which the homogenous system of integral equations has a non-zero solution. The same numerical methods apply.

The use of layer potentials to model the plate displacement (including the Green's representation) requires their value or the value of various of their derivatives along the plate boundary to determine the unknown functions. As we will show, it is necessary to take some care to express the value of such integrals along the layer support. More precisely, a layer potential is an infinitely derivable function out of the layer support. But the function itself, or a more or less high order normal derivative, can present a discontinuity across the layer support. The layer potentials discontinuities are easily pointed out when the plate operator is written in the distribution sense.

4. *The plate operator in the distribution sense*

The distributions theory enables us to define the derivatives of a discontinuous function. Let $w(M)$ be a function which is indefinitely differentiable inside a bounded domain Σ and inside its complementary domain $\overline{\Sigma}$. We assume that $w(M)$ and its successive derivatives can have a discontinuity across $\partial\Sigma$ the boundary of Σ. The plate operator in the distribution sense is defined by:

$$\begin{aligned}\langle \mathcal{A}w, \phi\rangle &= \langle w, {}^t\!\mathcal{A}\phi\rangle \\ &= D \iint_{\mathbf{R}^2} w\Big[\big(\phi_{,1111} + \phi_{,1122} + \phi_{,2211} + \phi_{,2222}\big) \\ &\quad + (1-\nu)\big(\phi_{,1221} + \phi_{,2112} - \phi_{,2211} - \phi_{,1122}\big)\Big]\end{aligned} \tag{2.59}$$

In this expression, ${}^t\!\mathcal{A}$ is the operator transposed of $\mathcal{A}$ and the symbol $\langle w, \phi\rangle$ is the duality product (see, for example, [CRI 92] or [SCH 61]). The function ϕ is an indefinitely differentiable function, with compact support.

As we did to establish equality (1.6), the integral is split into integrals over $\overline{\Sigma}$ and over Σ:

$$\begin{aligned}\langle \mathcal{A}w, \phi\rangle = D \int_{-\infty}^{+\infty} \mathrm{d}x_2 \Bigg[&\int_{-\infty}^{x_1^{\min}(x_2)} \mathrm{d}x_1 w\Big[\big(\phi_{,1111} + \phi_{,1122} + \phi_{,2211} + \phi_{,2222}\big) \\ &\quad + (1-\nu)\big(\phi_{,1221} + \phi_{,2112} - \phi_{,2211} - \phi_{,1122}\big)\Big] \\ + &\int_{x_1^{\min}(x_2)}^{x_1^{\max}(x_2)} \mathrm{d}x_1 w\Big[\big(\phi_{,1111} + \phi_{,1122} + \phi_{,2211} + \phi_{,2222}\big) \\ &\quad + (1-\nu)\big(\phi_{,1221} + \phi_{,2112} - \phi_{,2211} - \phi_{,1122}\big)\Big] \\ + &\int_{x_1^{\max}(x_2)}^{+\infty} \mathrm{d}x_1 w\Big[\big(\phi_{,1111} + \phi_{,1122} + \phi_{,2211} + \phi_{,2222}\big) \\ &\quad + (1-\nu)\big(\phi_{,1221} + \phi_{,2112} - \phi_{,2211} - \phi_{,1122}\big)\Big]\Bigg]\end{aligned} \tag{2.59'}$$

The abscissae $x_1^-(x_2)$ and $x_1^+(x_2)$ are the points where the line x_2 = constant cuts the curve $\partial\Sigma$.

Now, integrations by parts can be performed. For example, we have:

$$\int_{-\infty}^{+\infty} \mathrm{d}x_2 \int_{-\infty}^{x_1^{\min}(x_2)} \mathrm{d}x_1 w\phi_{,1111} = -\int_{-\infty}^{+\infty} \mathrm{d}x_2 \int_{-\infty}^{x_1^{\min}(x_2)} \mathrm{d}x_1 w_{,1}\,\phi_{,111}$$

$$+ \int_{-\infty}^{+\infty} \mathrm{d}x_2 w^-(x_1^{\min}(x_2), x_2)\phi_{,111}\,(x_1^{\min}(x_2), x_2)$$

with $w^-(x_1^{\min}(x_2), x_2) = \lim\limits_{x_1 \to x_1^{\min}(x_2)} w(x_1, x_2)$

By proceeding in the same way on the other integration domains, we obtain:

$$\iint_{\mathbf{R}^2} w\phi_{,1111} = -\iint_{\mathbf{R}^2} \{w_{,1}\}\phi_{,111} - \int_{\Sigma} (\,\mathrm{Tr}^+ w - \mathrm{Tr}^- w)\phi_{,111}$$

Here $\{w_{,1}\}$ is the classical derivative of w, which is defined everywhere but on $\partial\Sigma$. The functions $\mathrm{Tr}^+ w$ and $\mathrm{Tr}^- w$ are respectively the exterior and interior limits of w. The difference $\mathrm{Tr}^+ w - \mathrm{Tr}^- w = \overline{\mathrm{Tr}\, w}$ is the step of w across the curve $\partial\Sigma$. When all the integrations by parts are performed, the following result is obtained:

$$\langle \mathcal{A}w, \phi\rangle = \iint_{\mathbf{R}^2} \{\mathcal{A}w\}\phi + D\int_{\partial\Sigma} \Big[-\overline{\mathrm{Tr}\, w}\Big(\partial_n \Delta\phi + (1-\nu)\partial_{\bar{s}^2}\partial_n\phi\Big)$$
$$+ \overline{\mathrm{Tr}\,\partial_n w}\Big(\Delta\phi - (1-\nu)\partial_{\bar{s}^2}\phi\Big) - \Big(\overline{\mathrm{Tr}\,\Delta w} - (1-\nu)\,\overline{\mathrm{Tr}\,\partial_{s^2} w}\Big)\partial_n\phi$$
$$+ \Big(\overline{\mathrm{Tr}\,\partial_n \Delta w} + (1-\nu)\partial_{\bar{s}}\,\overline{\mathrm{Tr}\,\partial_n\partial_s w}\Big)\phi\Big] \quad (2.60)$$

where $\{\mathcal{A}w\}$ is the plate operator applied to w in the classical sense (defined everywhere but along $\partial\Sigma$). The functions $\overline{\mathrm{Tr}\,\partial_n w}$, $\overline{\mathrm{Tr}\,\Delta w}$, $\overline{\mathrm{Tr}\,\partial_{s^2} w}$ and $\overline{\mathrm{Tr}\,\partial_n\partial_s w}$ are the steps of, respectively, $\partial_n w$, Δw, $\partial_{s^2} w$ and $\partial_n\partial_s w$. Expression (2.60) is, of course, quite similar to Green's formula for the plate operator.

Now, integrations by parts with respect to the curvilinear variable $\bar{s}$ are performed and the various terms are reordered:

$$\langle \mathcal{A}w, \phi\rangle = \iint_{\mathbf{R}^2} \{\mathcal{A}w\}\phi + D\int_{\partial\Sigma} \Big[-\overline{\mathrm{Tr}\, w}\,\partial_{n^3}\phi + \overline{\mathrm{Tr}\,\partial_n w}\,\partial_{n^2}\phi$$
$$-\Big(\overline{\mathrm{Tr}\,\partial_{n^2} w} - 2(1-\nu)\partial_{\bar{s}^2}\,\overline{\mathrm{Tr}\, w}\Big)\partial_n\phi + \Big(\overline{\mathrm{Tr}\,\partial_{n^3} w} + 2\partial_{\bar{s}^2}\,\overline{\mathrm{Tr}\,\partial_n w}\Big)\phi\Big] \quad (2.61)$$

This implies that the plate operator in the distribution sense is defined by:

$$\mathcal{A}w = \{\mathcal{A}w\} + D\Big[\overline{\operatorname{Tr} w} \otimes \delta'''_{\partial\Sigma} + \overline{\operatorname{Tr} \partial_n w} \otimes \delta''_{\partial\Sigma} + \Big(\overline{\operatorname{Tr} \partial_{n^2} w} - 2(1-\nu)\partial_{\bar{s}^2} \overline{\operatorname{Tr} w}\Big) \otimes \delta'_{\partial\Sigma} + \Big(\overline{\operatorname{Tr} \partial_{n^3} w} + 2\partial_{\bar{s}^2} \overline{\operatorname{Tr} \partial_n w}\Big) \otimes \delta_{\partial\Sigma}\Big] \quad (2.62)$$

In this expression, $\delta_{\partial\Sigma}$ is the Dirac measure supported by the curve $\partial\Sigma$ and $\delta^{(j)}_{\partial\Sigma}$ is the j-th transverse derivative defined by the following integral expression:

$$\langle \delta^{(j)}_{\partial\Sigma}, \phi \rangle = (-1)^j \int_{\partial\Sigma} \partial^j_n \phi$$

For layer potentials, it can be shown that the discontinuities occur on the normal derivatives only. We can thus assert that the layer potentials as defined in (2.58) present the following discontinuities:

- the simple layer potential u_0 has a discontinuity on its third order normal derivative $\partial_{n^3} u_0$, with the following limits

$$\lim_{M\in\Sigma\to P\in\partial\Sigma} \partial_{n^3} u_0(M) = -\frac{\chi_0(P)}{2D} + \int_{\partial\Sigma} \chi_0(P)\, \partial_{n^3} \Gamma(P, P')\, \mathrm{d}\bar{s}(P')$$

$$\lim_{M\in\overline{\Sigma}\to P\in\partial\Sigma} \partial_{n^3} u_0(M) = +\frac{\chi_0(P)}{2D} + \int_{\partial\Sigma} \chi_0(P)\, \partial_{n^3} \Gamma(P, P')\, \mathrm{d}\bar{s}(P')$$

- the double layer potential u_1 has a discontinuity on its second order normal derivative $\partial_{n^2} u_1$, with the following limits

$$\lim_{M\in\Sigma\to P\in\partial\Sigma} \partial_{n^2} u_1(M) = -\frac{\chi_1(P)}{2D} - \int_{\partial\Sigma} \chi_1(P)\, \partial_{n^2}\partial_{n'} \Gamma(P, P')\, \mathrm{d}\bar{s}(P')$$

$$\lim_{M\in\overline{\Sigma}\to P\in\partial\Sigma} \partial_{n^2} u_1(M) = +\frac{\chi_1(P)}{2D} - \int_{\partial\Sigma} \chi_1(P)\, \partial_{n^2}\partial_{n'} \Gamma(P, P')\, \mathrm{d}\bar{s}(P')$$

- the layer potential u_2 of order 2 has a discontinuity on its first order normal derivative $\partial_n u_2$, with the following limits

$$\lim_{M\in\Sigma\to P\in\partial\Sigma} \partial_n u_2(M) = -\frac{\chi_2(P)}{2D} + \int_{\partial\Sigma} \chi_2(P)\, \partial_n\partial_{n'^2} \Gamma(P, P')\, \mathrm{d}\bar{s}(P')$$

$$\lim_{M\in\overline{\Sigma}\to P\in\partial\Sigma} \partial_n u_2(M) = +\frac{\chi_2(P)}{2D} + \int_{\partial\Sigma} \chi_2(P)\, \partial_n\partial_{n'^2} \Gamma(P, P')\, \mathrm{d}\bar{s}(P')$$

◇ the layer potential u_3 of order 3 is a discontinuous function, with the following limits

$$\lim_{M\in\Sigma\to P\in\partial\Sigma} u_3(M) = -\frac{\chi_3(P)}{2D} - \int_{\partial\Sigma} \chi_3(P)\,\partial_{n'^3}\Gamma(P,P')\,\mathrm{d}\bar{s}(P')$$

$$\lim_{M\in\overline{\Sigma}\to P\in\partial\Sigma} u_3(M) = +\frac{\chi_3(P)}{2D} - \int_{\partial\Sigma} \chi_3(P)\,\partial_{n'^3}\Gamma(P,P')\,\mathrm{d}\bar{s}(P')$$

All the integrals in the preceding formulae are Cauchy principal values. The derivatives of a layer potential $u_j(M)$ which involve derivatives of the Green's kernel of the plate larger than 3 are expressed by integrals which are not convergent when the point M is taken on the layer support. Nevertheless, the limit for $M \in \Sigma \to P \in \partial\Sigma$ (or $M \in \overline{\Sigma} \to P \in \partial\Sigma$ as well) exists but is a Hadamard finite part of integral (denoted by the symbol "P.f.").

5. *Discontinuities of the layer potentials and of their derivatives: analytic proof*

A more direct, but rather tedious, analytic calculation of the above limits was given in [VIV 72]. Such detailed analytic calculations are beyond the scope of this monograph. We will only consider the simple layer case and give the main lines of the proof.

Let us consider a simple layer potential

$$u_0(M) = \int_{\partial\Sigma} \chi(P')\Gamma(M,P')\,\mathrm{d}\bar{s}(P') \quad , \quad M \text{ not on } \partial\Sigma \tag{2.63}$$

Asymptotic expansions (2.27) show that the plate Green's kernel and its first order derivative are continuous functions. The second order derivative has a logarithmic singularity and the third order one is singular like ρ^{-1}.

Thus, no care is required to obtain the value, along $\partial\Sigma$, of the simple layer potential and of its first and second order normal derivatives. For third order derivatives, the singularity of the Green's kernel derivative involved requires some caution. Furthermore, numerical approximations can require to evaluate analytically integrals of the kernel and their limits on the layer support. Thus, the following calculations have two interests: the proof of the discontinuity formulae and applications to numerical approximations.

Our aim is to look at the behavior of the simple layer potential $u_0(M)$ and of its successive derivatives – both normal and tangent – up to order 3, when the point M tends to a point P on $\partial\Sigma$.

In a first step, a small vicinity $\delta(P)$, with length 2ε, around the point P is isolated and the integral involved is split into two terms. The first one is the integral along $\partial\Sigma - \delta(P)$ which can be evaluated everywhere, and, in particular, at P. The second term is the integral $u_0^\delta(M)$ along $\delta(P)$.

At the point P, let us define a Cartesian coordinate system (x, y), where the x-axis is tangent to $\partial\Sigma$ in P and the y-axis is normal. The boundary arc $\delta(P)$ is replaced by the line segment $(-\varepsilon < x < \varepsilon, y = 0)$. The source density is replaced by its value at P (this is equivalent to approximate it by the first term of its Taylor series). The contribution of $\delta(P)$ to $u_0(M)$ is approximated by:

$$u_0^\delta(M) \simeq \chi(P) \int_{-\varepsilon}^{\varepsilon} \Gamma(\rho(M, P'))\, \mathrm{d}x' \tag{2.64}$$

where the coordinates of M are (x, y), those of P' being $(x', 0)$, and where $\rho(M, P')$ is the distance between M and P'. We now recall the Taylor-like expansion of $\Gamma\big(\rho(M, P')\big)$ in terms of ρ:

$$\begin{aligned}\Gamma(\rho(M, P')) = \frac{1}{16\lambda^2 \pi D}\Big[&\big(\lambda^2\rho^2 + 4\big)\ln(\imath\lambda) + \big(\lambda^2\rho^2 - 4\big)\ln\lambda \\ &+ \lambda^2\rho^2\big(2\gamma - 2 - \imath\pi - \ln 4 + 2\ln\rho\big)\Big] + \mathcal{O}\big(\rho^4\big)\end{aligned} \tag{2.65}$$

Expansion (2.65) is introduced into (2.64) with $\rho = \sqrt{(x - x')^2 + y^2}$ and the different integrals are calculated analytically. The contribution of the arc element $\delta(P)$ is approximated by:

$$\begin{aligned}u_0^\delta(M) \simeq \chi(P)\frac{2}{144\, D\pi\lambda^2}\Big\{&\varepsilon\lambda^2\Big(\varepsilon^2\left(-8 + 6\gamma - 3\imath\pi - 3\ln 4\right) \\ &- 3x^2\big(8 - 6\gamma + 3\imath\pi + \ln 64\big) - 3y^2\big(10 - 6\gamma + 3\imath\pi + \ln 64\big)\Big) \\ &- 3\big(3y^2 + (x-\varepsilon)^2\big)\big(x-\varepsilon\big)\lambda^2\ln\big(y^2 + (x-\varepsilon)^2\big) + 6\imath\, y^3\lambda^2\ln\Big(\frac{\imath x + y - \imath\varepsilon}{y}\Big) \\ &+ 3\Big[-2\imath\, y^3\lambda^2\ln\Big(\frac{-\imath x + y + \imath\varepsilon}{y}\Big) - 2\imath\, y^3\lambda^2\ln\Big(\frac{\imath x + y + \imath\varepsilon}{y}\Big) \\ &+ 2\imath\, y^3\lambda^2\ln\Big(-\frac{x + \imath y + \varepsilon}{\imath y}\Big) + x^3\lambda^2\ln\big(y^2 + (x+\varepsilon)^2\big) + 3xy^2\lambda^2\ln\big(y^2 + (x+\varepsilon)^2\big) \\ &+ 3\varepsilon\lambda^2\big(x^2 + y^2\big)\ln\big(y^2 + (x+\varepsilon)^2\big) + \varepsilon^3\lambda^2\ln\big(y^2 + (x+\varepsilon)^2\big) \\ &+ \varepsilon\big(24 + 6x^2\lambda^2 + 6y^2\lambda^2 + 2\varepsilon^2\lambda^2\big)\ln(\imath\lambda) \\ &+ 2\varepsilon\Big(-12 + \big(3(x^2 + y^2) + \varepsilon^2\big)\lambda^2\Big)\ln\lambda\Big]\Big\}\end{aligned} \tag{2.66}$$

This expression can be derivated in both the normal and the tangent directions, and the limit of the resulting expressions for $y \to 0$ are easy to obtain.

The value of this expression at the center of the segment $(-\varepsilon < x < \varepsilon, y = 0)$ is:

$$u_0^\delta(P) \simeq \chi(P)\frac{\varepsilon}{72\pi\lambda^2 D}\Big[\varepsilon^2\lambda^2(6\gamma - 8 - 3\imath\pi - 3\ln 4) + 36\big(\ln(\imath\lambda) - \ln\lambda\big) + 3\varepsilon^2\lambda^2\big(\ln\varepsilon^2 + \ln(\imath\lambda) + \ln\lambda\big)\Big]$$

It is independent of the direction in which the limit is taken.

The normal derivative, at $x = 0$, has a simple expression:

$$\partial_y u_0^\delta\big(M(0, y)\big) \simeq \chi(P)\frac{y}{4\pi D}\Big[\imath y\Big(\ln\big(1 - \frac{\imath\varepsilon}{y}\big) - \ln\big(1 + \frac{\imath\varepsilon}{y}\big)\Big) + \varepsilon\big(2\gamma - 3 - \imath\pi + \ln\big(y^2 + \varepsilon^2\big) + \ln\frac{\imath\lambda}{4} + \ln\lambda\big)\Big]$$

It is equal to 0 for $y = 0$. This implies that the third order derivative $\partial_{x^2} \operatorname{Tr} \partial_y u_0^\delta(M)$ is zero.

The second order normal derivative, at $x = 0$, takes the form

$$\partial_{y^2} u_0^\delta\big(M(0, y)\big) \simeq \chi(P)\frac{1}{4\pi D}\Big[2\imath y\Big(\ln\big(1 - \frac{\imath\varepsilon}{y}\big) - \ln\big(1 + \frac{\imath\varepsilon}{y}\big)\Big) + \varepsilon\Big(2\gamma - 3 - \imath\pi + \ln\big(y^2 + \varepsilon^2\big) + \ln(\imath\lambda/4) + \ln\lambda\Big)\Big]$$

and has a finite value for $y = 0$.

The third order normal derivative is given by:

$$\partial_{y^3} u_0^\delta\big(M(0, y)\big) \simeq \chi(P)\frac{\imath}{2\pi D\big(y^2 + \varepsilon^2\big)}\Big[\imath y\varepsilon + \big(y^2 + \varepsilon^2\big)\Big(\ln\big(1 - \frac{\imath\varepsilon}{y}\big) - \ln\big(1 + \frac{\imath\varepsilon}{y}\big)\Big)\Big]$$

This leads to two different limits which depend on the direction in which y tends to 0:

$$\lim_{y\to 0^+} \partial_{y^3} u_0^\delta\big(M(0, y)\big) \simeq \chi(P)\frac{\imath}{2\pi D}\Big[\ln\big(1 - \imath\infty\big) - \ln\big(1 + \imath\infty\big)\Big] = +\frac{\chi(P)}{2D}$$

$$\lim_{y\to 0^-} \partial_{y^3} u_0^\delta\big(M(0, y)\big) \simeq \chi(P)\frac{\imath}{2\pi D}\Big[\ln\big(1 + \imath\infty\big) - \ln\big(1 - \imath\infty\big)\Big] = -\frac{\chi(P)}{2D}$$

This proves that the third order normal derivative of the simple layer potential is a discontinuous function. The same method applies for the study of discontinuities of higher order layer potentials.

6. Resonance frequencies of plates with "complex" shape or "complex" boundary conditions: examples of results obtained by the indirect Boundary Element Method

These examples are detailed in [VIV 72] and [VIV 74]. The first case is a plate which has a non-classical shape and is clamped all along its boundary. The second case deals with a circular plate with discontinuous boundary conditions: it is clamped along an arc α and free along the other part of its boundary (see Figure 2.9).

The plate displacement is described by the vibration field due to two layer potentials of different orders i and j, say:

$$w(M) = \int_{\partial\Sigma} \left[(-1)^i \chi^{(i)}(P') \partial_{n'^i} \Gamma(M, P') + (-1)^j \chi^{(j)}(P') \partial_{n'^j} \Gamma(M, P') \right] \mathrm{d}\bar{s}(P')$$

It must satisfy two boundary conditions along the plate contour:

$$\begin{aligned} L_1 w(P) = \lim_{M\in\Sigma\to P\in\partial\Sigma} L_1(P) \int_{\partial\Sigma} & \left[(-1)^i \chi^{(i)}(P') \partial_{n'^i} \Gamma(M, P') \right. \\ & \left. + (-1)^j \chi^{(j)}(P') \partial_{n'^j} \Gamma(M, P') \right] \mathrm{d}\bar{s}(P') = 0 \\ L_2 w(P) = \lim_{M\in\Sigma\to P\in\partial\Sigma} L_2(P) \int_{\partial\Sigma} & \left[(-1)^i \chi^{(i)}(P') \partial_{n'^i} \Gamma(M, P') \right. \\ & \left. + (-1)^j \chi^{(j)}(P') \partial_{n'^j} \Gamma(M, P') \right] \mathrm{d}\bar{s}(P') = 0 \end{aligned}$$

The boundary conditions can be either constant all along the plate boundary, or discontinuous (in the example outlined, the plate is clamped along a first part, and free along the remaining part). The system of homogenous equations thus obtained has a non-zero solution if the angular frequency ω, which the Green's kernel depends on, is a resonance frequency of the plate. We will now briefly describe the numerical method used to approximate such frequencies.

First, the plate contour $\partial\Sigma$ is divided into N arc elements $\partial\Sigma_n$. Each arc element is replaced by the segment σ_n joining its two bounds, with center P_n. Thus, the plate is approximated by a polygon the area of which is, in general, smaller than the plate area. So, the computed resonance frequencies are a little higher than the exact frequencies.

In a second step, the layer densities $\chi^{(i)}$ and $\chi^{(j)}$ are approximated by constants $\chi_n^{(i)}$ and $\chi_n^{(j)}$ on each segment σ_n. The plate displacement takes the following approximate expression:

$$w(M) \simeq \sum_{n=1}^{N} \int_{\sigma_n} \Big[(-1)^i \chi_n^{(i)} \partial_{n'^i} \Gamma(M, P') + (-1)^j \chi_n^{(j)} \partial_{n'^j} \Gamma(M, P')\Big] \mathrm{d}\sigma(P')$$

Then, the two boundary conditions are written at each point P_m, centers of the segments σ_m to obtain a system of $2N$ equations with $2N$ unknowns.

For $M \to P_m$, the integrals

$$(-1)^i \chi_n^{(i)} \int_{\sigma_n} \partial_{n'^i} \Gamma(M, P') \quad \text{and} \quad (-1)^j \chi_n^{(j)} \int_{\sigma_n} \partial_{n'^j} \Gamma(M, P') \, \mathrm{d}\sigma(P')$$

do not present any particular analytical difficulty for $m \neq n$. It is possible to invert the application of the boundary conditions operators and the integration and take the value of the result for $M = P_m$, that is, to evaluate integrals of the form

$$(-1)^i \chi_n^{(i)} \int_{\sigma_n} L_{(1,2)} \partial_{n'^i} \Gamma(P_m, P') \quad \text{and} \quad (-1)^j \chi_n^{(j)} \int_{\sigma_n} L_{(1,2)} \partial_{n'^j} \Gamma(P_m, P')$$

Different methods of quadrature can be considered. The simplest one is to use a Gauss-Legendre algorithm with a very small number of integration points.

The major difficulty concerns the diagonal terms, that is, the terms for which the point M will tend to P_n, which, of course have the main importance. Analytical integrations and limit procedures are necessary to obtain an accurate result. In [VIV 72], the author used a high order Taylor-like series expansion of the plate Green's kernel up to terms in ρ^{10}:

$$\begin{aligned}
\Gamma\big(\rho(M, Q')\big) = {} & \frac{1}{1\,769\,472\,000 \pi \lambda^2 D} \\
& \Big\{30\Big[14\,745\,600 + \lambda^2\rho^2\big(3\,686\,400 + 230\,400\lambda^2\rho^2 \\
& \qquad + 6\,400\lambda^4\rho^4 + 100\lambda^6\rho^6 + \lambda^8\rho^8\big)\Big] \ln \imath\lambda \\
& + 30\Big[-14\,745\,600 + \lambda^2\rho^2\big(3\,686\,400 - 230\,400\lambda^2\rho^2 \\
& \qquad + 6\,400\lambda^4\rho^4 - 100\lambda^6\rho^6 + \lambda^8\rho^8\big)\Big] \ln \lambda \\
& + \lambda^2\rho^2\Big[30\big(2\gamma - \imath\pi\big)\big(640 + \lambda^4\rho^4\big)\big(5\,670 + \lambda^4\rho^4\big) \\
& - 221\,184\,000(1 + \ln 2) - \lambda^4\rho^4\big(\lambda^4\rho^4(137 + 60\ln 2) + 64\,000(11 + \ln 64)\big) \\
& \qquad + 60\big(640 + \lambda^4\rho^4\big)\big(5\,760 + \lambda^4\rho^4\big)\ln\rho\Big]\Big\} + \mathcal{O}(\rho^{12})
\end{aligned}$$

Such an expansion can, now, be established with the help of a mathematics software, like Mathematica, the only work left to the user being to reorder the various terms.

The normal derivatives of order i and j (with respect to the coordinates of P') of this expansion are taken and analytically integrated. Then, the boundary operators L_1 and L_2 are applied to the resulting expressions. And, finally, the limits, for $M \to P_n$ are taken. This is, of course, a very tedious calculation when it is done by hand. But that is the best way to get an excellent approximation of the diagonal terms.

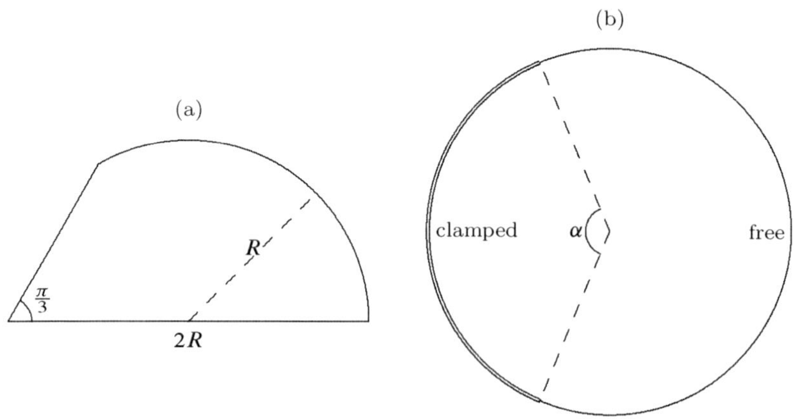

Figure 2.9. *Boundary Element Method – geometry of the plates: clamped plate with complex shape (a); circular plate clamped along an arc α and free along the remaining part (b)*

In both examples, the computed first resonance frequencies were compared with experiments. Comparisons between numerical predictions and experimental results are shown in Tables 2.4, 2.5 and 2.6. The relative error on the resonance frequencies, which increases with the order of the mode, does not exceed 6%: it is, of course, due to numerical approximation but also to errors on the data (mainly the Young's modulus, the plate thickness which is not perfectly constant and a small non-isotropy of the material used). When the ratio of the successive resonance frequencies to one of them (here f_1 or f_2) is considered, the error between numerical predictions and experimental results is much smaller (less than 3%): indeed, this comparison partly eliminates the influence of the errors on the physical data.

	Numerical results			Experimental results					
m	f_m (Hz)	f_m/f_1	f_m/f_2	ν_m	$\varepsilon\%$	ν_m/ν_1	$\varepsilon\%$	ν_m/ν_2	$\varepsilon\%$
0	408								
1	631	1.00	0.70	601	4.75	1.00		0.69	1.45
2	898	1.42	1.00	867	3.57	1.41	1.42	1.00	
3	1 038	1.64	1.16	981	5.81	1.63	0.61	1.13	2.65
4	1 200	1.90	1.34	1 160	3.45	1.93	2.12	1.34	0.00
5	1 430	2.25	1.59	1 394	2.58	2.32	1.58	1.61	1.26
6	1 532	2.43	1.71	1 492	2.68	2.48	2.06	1.72	0.58
7	1 910	3.03	2.13	1 798	6.23	2.99	1.34	2.07	2.90

Table 2.4. *Clamped plate: comparison between experimental resonance frequencies ν_m and BEM predictions f_m ($\varepsilon\%$ is the relative error)*

	Numerical results		Experimental results			
m	f_m (Hz)	f_m/f_0	ν_m	$\varepsilon\%$	ν_m/ν_0	$\varepsilon\%$
0	62	1.00	60	3.33	1.00	
1	195	3.15	190	2.63	3.17	0.63
2	358	5.77	342	4.68	5.70	1.23
3	418	6.74	394	6.09	6.57	2.59

Table 2.5. *Clamped–free plate with $\alpha = 3\pi/4$: comparison between experimental resonance frequencies ν_m and BEM predictions f_m ($\varepsilon\%$ is the relative error)*

The experiments were conducted with plates made of a commercial laminated material (stainless steel). To conduct the calculations, it was necessary to introduce values of the physical constants as accurate as possible. The surface mass of the plate was determined by weighting a piece of material with a known area. A mean thickness, measured over about 20 randomly distributed points, was used. The Young's modulus and the Poisson's ratio were more difficult to estimate. To do this, the first ten resonance frequencies of a circular plate with free boundary was measured. Then the Young's modulus and the Poisson's ratio were chosen to obtain the best fit between the first ten theoretical resonance frequencies and the experimental frequencies.

Experimentally, the resonance frequencies are measured as follows. The plate is covered with a layer of light sand and excited by a harmonic force with a slowly sliding frequency. When a resonance frequency is reached, the sand accumulates along the nodal lines and the nodal patterns are thus visualized.

Figure 2.10 shows the experimental assembly used for both plates. The plate is supported by a heavy concrete block which is isolated from the floor of the experiment room. The clamping of the plates is realized by thick pieces

	Numerical results		Experimental results			
m	f_m (Hz)	f_m/f_0	ν_m	$\varepsilon\%$	ν_m/ν_0	$\varepsilon\%$
0	98	1.00	96	2.08	1.00	
1	292	2.98	286	2.10	2.98	0.00
2	433	4.42	416	4.08	4.33	2.08
3	603	6.15	584	3.25	6.08	1.15

Table 2.6. *Clamped–free plate with $\alpha = \pi$: comparison between experimental resonance frequencies ν_m and BEM predictions f_m ($\varepsilon\%$ is the relative error)*

Figure 2.10. *General experimental assembly: (1-2) frequency generator and amplifier; (3) driver; (4) heavy concrete support and clamping device; (5) frequency-meter*

Figure 2.11. *Experimental assembly for the clamped plate: (1) bolts on the concrete support used to clamp the plate; (2-4) clamping device; (3) stainless steel plate*

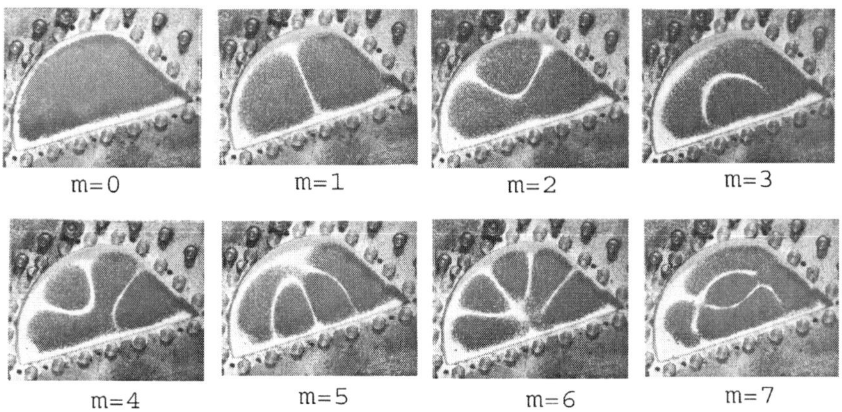

Figure 2.12. *Clamped plate of "complex shape": experimental Chladni's patterns of the first eight resonance modes*

Figure 2.13. *Experimental assembly for the clamped-free circular plate: (1) plate on its support; (2) the stainless steel plate; (3) clamping devices*

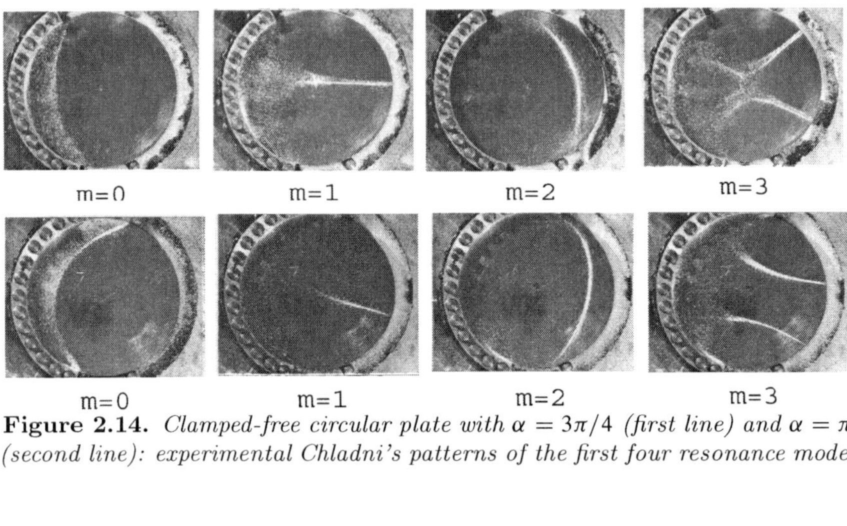

Figure 2.14. *Clamped-free circular plate with $\alpha = 3\pi/4$ (first line) and $\alpha = \pi$ (second line): experimental Chladni's patterns of the first four resonance modes*

of iron, regularly bolted with a given moment. To avoid constraints on the plate which could change its natural resonance frequencies, the driver excites the plate support and not the plate directly.

Figures 2.11 and 2.13 show the details of the assembly used for each kind of plate (clamped and cantilever). Figures 2.12 and 2.14 show the experimental nodal patterns – or Chladni's patterns – of the resonance modes corresponding to Tables 2.4, 2.5 and 2.6 respectively.

2.3.8. *Resonance Frequencies of Plates with Variable Thickness*

From both theoretical and numerical points of view, the determination of the resonance frequencies and the resonance modes of a plate with variable thickness does not present any particular difficulty. For plates with a simple shape (circular, rectangular, etc.), and simple boundary conditions (in particular, boundary conditions which are the same along the whole boundary), the

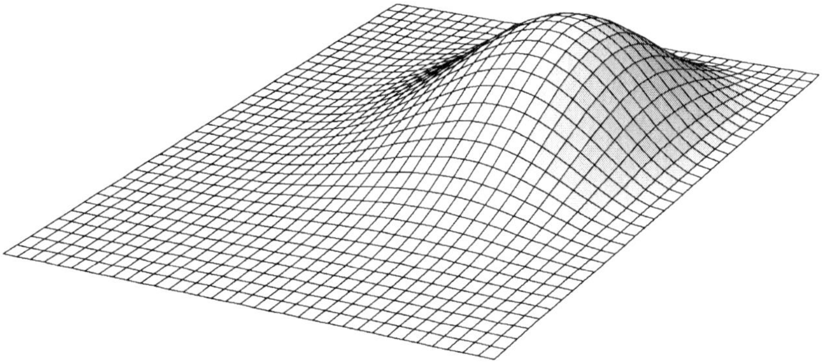

Figure 2.15. *Variation of the plate thickness*

Ritz-Galerkin method can be used in the same way as before. For simplicity, let us consider the case of a rectangular plate, clamped along its boundary and with a thickness defined by:

$$h = h_0\big(1 + \varepsilon(x, y)\big)$$

The rigidity of the plate and its surface density are written respectively as $D = D_0\big(1 + \varepsilon(x, y)\big)^3$ and $\bar{\mu} = \bar{\mu}_0\big(1 + \varepsilon(x, y)\big)$, with $D_0 = Eh_0^3/12(1 - \nu^2)$ and $\bar{\mu}_0 = \mu_s h_0$.

The procedure to establish the resonance modes equation is identical to that developed in section 2.3.5, paragraph 2. Let U_{mn} be the same basis of functions used in the case of a constant thickness plate to expand the resonance modes. Equation (2.42) must be written in a slightly different form, which accounts for the variation of the thickness:

$$\sum_{mn} w^{mn} \int_{-1/2}^{+1/2} \int_{-L/2}^{+L/2} \Big\{ (1+\varepsilon)^3 \Big[\Delta U_{mn} \Delta U_{pq} + (1-\nu)\big(2U_{mn,12}\, U_{pq,12} - U_{mn,11}\, U_{pq,22} - U_{mn,22}\, U_{pq,11}\big)\Big] - \Lambda^4 (1+\varepsilon) U_{mn} U_{pq} \Big\} \, \mathrm{d}x_1 \, \mathrm{d}x_2 = 0 \quad \text{for all} \quad (p,q) \qquad (2.67)$$

with $\Lambda^4 = \bar{\mu}_0 \omega^2 / D_0$. To obtain approximations of the resonance wavenumbers and modes, system (2.67) is truncated.

Warburton's approximation – system (2.67) replaced by its diagonal only – can also be used: the accuracy for the resonance reduced wavenumbers is almost the same as that obtained for the constant thickness plate, but the resonance modes can be rather badly described.

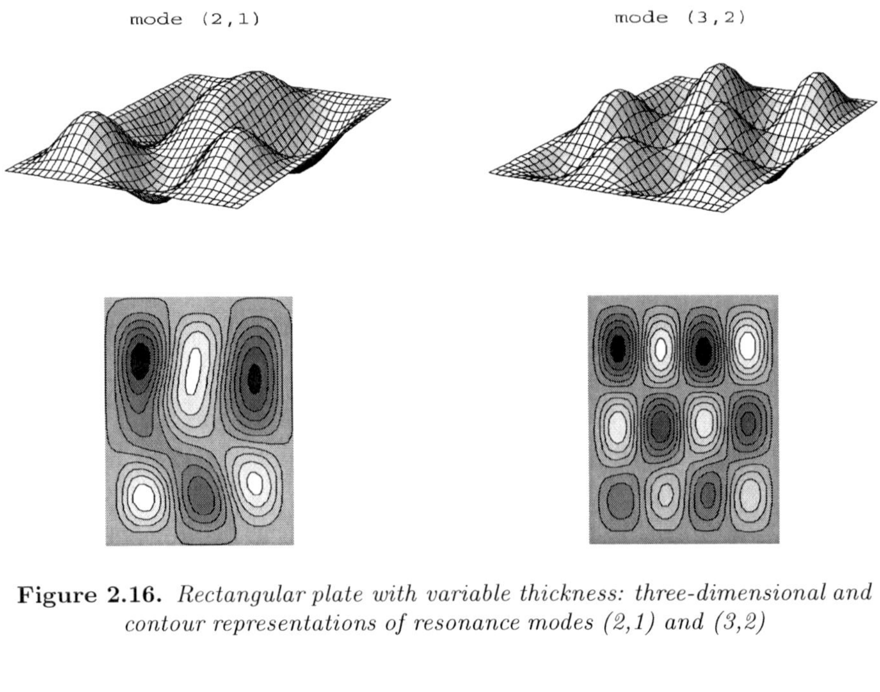

Figure 2.16. *Rectangular plate with variable thickness: three-dimensional and contour representations of resonance modes (2,1) and (3,2)*

To illustrate this point, let us consider a plate with reduced length $L = \pi/2$ and a thickness variation given by:

$$\varepsilon(x, y) = \chi\big[2x + 1\big]^2\big[1 + \cos(2\pi x)\big]\big[2y/L + 1\big]^2\big[1 + \cos(2\pi y/L)\big]$$

This function is zero along the plate contour. It has a maximum at ($x = 0.29$, $y = 0.46$), which, for $\chi = 1/14$, is about 0.51. It is represented in Figure 2.15.

Modes (2,1) and (3,2) are represented in Figure 2.16. It can be seen that the mode shapes notably differ from the mode shapes of the constant thickness plate. As a consequence, the corresponding Warburton's approximation – in which a mode shape is approximated by the product of two beam modes – is rather far from the exact function. Finally, we must mention that the value of the resonance wavenumbers of the variable thickness plate described here increases with χ: the reason is that the mean value of the ratio "mass per unit area / rigidity" decreases as the mean thickness of the plate is increased.

2.3.9. *Transient Response of an Infinite Plate with Constant Thickness*

When we established the equation governing the motion of thin structures, we mentioned that the final approximate equation is of the parabolic type: this implies that the wave velocity is infinite, though the initial exact equations are of the hyperbolic type (finite wave velocities). To show this property explicitly, let us look at the response of an infinite plate excited by a transient point force.

Let us consider a plate with constant thickness. The ratio $\bar{\mu}/D$ of the mass per unit area to the rigidity is denoted by ϖ. It is excited by a point impulse force $\delta_O \otimes \delta_t$ acting at the coordinate origin O. The plate displacement $\tilde{w}(M, t)$ satisfies the following equation:

$$\Delta^2 \tilde{w}(M, t) + \varpi \ddot{\tilde{w}}(M, t) = \frac{\delta_O \otimes \delta_t}{D} \tag{2.68}$$

It is assumed that the plate is at rest for $t < 0$. Because the excitation is axisymmetric, the plate displacement depends on the radial distance ρ only, and not on the two space variables. The plate displacement will thus be denoted by $\tilde{w}(\rho, t)$. To solve this equation, use is made of the Laplace transform. Let $w(\rho, p) = \mathcal{L}\tilde{w}(\rho, t)$ be the Laplace transform of the plate displacement which satisfies

$$\Delta^2 w(\rho, p) + \varpi p^2 w(\rho, p) = \frac{\delta_O}{D} \tag{2.69}$$

We then apply the space Fourier transform to equation (2.69) and the result depends on the radial variable $\xi = \sqrt{\xi_1^2 + \xi_2^2}$ only (see section 2.3.2). This leads to:

$$\left[16\pi^4\xi^4 + \varpi p^2\right]\hat{w}(\xi, p) = \frac{1}{D} \quad \Longrightarrow \quad \hat{w}(\xi, p) = \frac{1}{D}\frac{1}{16\pi^4\xi^4 + \varpi p^2} \tag{2.70}$$

where $w(\rho, p)$ and $\hat{w}(\xi, p)$ are related by:

$$\hat{w}(\xi, p) = 2\pi \int_0^\infty w(\rho, p) J_0(2\pi\xi\rho)\rho \, \mathrm{d}\rho$$

$$w(\rho, p) = 2\pi \int_0^\infty \hat{w}(\xi, p) J_0(2\pi\xi\rho)\xi \, \mathrm{d}\xi = \pi \int_{\infty e^{\imath\pi}}^\infty \hat{w}(\xi, p) H_0(2\pi\xi\rho)\xi \, \mathrm{d}\xi$$

The residue method is well suited to calculate the function $w(\rho, p)$. For that purpose, the location of the four roots $\xi^{(i)}$ of the denominator of $\hat{w}(\xi, p)$ must be determined. Recalling that $\Re(p) > 0$, we have:

$$\begin{aligned}
2\pi\xi^{(1)} &= \imath^{1/2}\varpi^{1/4}p^{1/2} && \Longrightarrow && \arg(\xi^{(1)}) \in [0, \pi/2] \\
2\pi\xi^{(2)} &= -\imath^{1/2}\varpi^{1/4}p^{1/2} && \Longrightarrow && \arg(\xi^{(2)}) \in [\pi, 3\pi/2] \\
2\pi\xi^{(3)} &= \imath^{3/2}\varpi^{1/4}p^{1/2} && \Longrightarrow && \arg(\xi^{(3)}) \in [\pi/2, \pi] \\
2\pi\xi^{(4)} &= -\imath^{3/2}\varpi^{1/4}p^{1/2} && \Longrightarrow && \arg(\xi^{(4)}) \in [3\pi/2, 2\pi]
\end{aligned}$$

The integration contour is composed of the real axis $\Im(\xi) = 0$, and of the half-circle $\xi = Re^{\imath\theta}$ with $0 < \theta < \pi$ and $R \to \infty$. Thus, the integral along this half-circle decreases to zero as $R \to \infty$. The only roots to be accounted for are $\xi^{(1)}$ and $\xi^{(2)}$. The result is:

$$\begin{aligned}
w(\rho, p) &= \frac{1}{8D\varpi^{1/2}p}\left[H_0(\imath^{1/2}\varpi^{1/4}p^{1/2}) - H_0(\imath^{3/2}\varpi^{1/4}p^{1/2})\right] \\
&= \frac{1}{4D\varpi^{1/2}\imath\pi p}\left[K_0(\imath^{1/2}\varpi^{1/4}p^{1/2}) - K_0(\imath^{3/2}\varpi^{1/4}p^{1/2})\right]
\end{aligned} \tag{2.71}$$

where $K_0(z) = 2/\imath\pi H_0(z)$ is the modified Hankel function. From tables of integral transforms (see, for example, [ERD 54-2]) we have that the inverse Laplace transform $\mathcal{L}^{-1}K_0(a^{1/2}p^{1/2})$ of $K_0(a^{1/2}p^{1/2})$ is $e^{-a/4t}/2t$ for $\Re(a) \geq 0$ and $a \neq 0$. Thus

$$\mathcal{L}^{-1}\frac{K_0(a^{1/2}p^{1/2})}{p} = \int \frac{e^{-a/4t}}{2t}\,\mathrm{d}t$$

which is determined up to an additive constant. Finally, the plate displacement

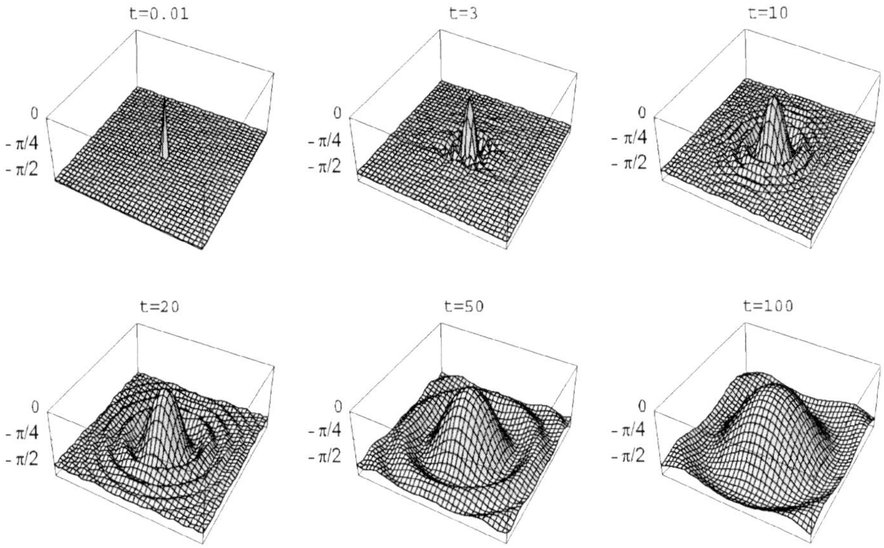

Figure 2.17. *Deflection of an infinite plate, excited by an impulse point force, for increasing time steps*

$\tilde{w}(\rho, t)$ is obtained as:

$$\begin{aligned}\tilde{w}(\rho, t) &= Y(t)\frac{1}{4\imath\pi\sqrt{\bar{\mu}D}}\int\left\{\frac{e^{\imath\sqrt{\mu/D}\rho^2/4t}}{2t} - \frac{e^{-\imath\sqrt{\mu/D}\rho^2/4t}}{2t}\right\}\mathrm{d}t \\ &= Y(t)\frac{1}{4\pi\sqrt{\bar{\mu}D}}\int\frac{\sin\left(\sqrt{\mu/D}\rho^2/4t\right)}{t}\mathrm{d}t \\ &= -Y(t)\frac{1}{4\pi\sqrt{\bar{\mu}D}}\int\frac{\sin\left(\sqrt{\mu/D}\rho^2\tau/4\right)}{\tau}\mathrm{d}\tau \\ &= -Y(t)\frac{1}{4\pi\sqrt{\bar{\mu}D}}\mathrm{Si}\left(\frac{\rho^2}{4\sqrt{D/\mu}t}\right)\end{aligned} \tag{2.72}$$

where $\mathrm{Si}(z)$ is the sine integral function (see [ABR 65] or [ERD 54-1] for definition and properties). The third equality is obtained by the change of variable $\tau = 1/t$. By choosing an integration constant equal to zero, the plate displacement tends to zero when t tends to infinity: that is, the plate goes back to its initial state when time is indefinitely increased.

The important point to remark is that expression (2.17) proves that the wave propagates with a velocity which is infinity: indeed, it is not of the form $f(ct - \rho)$. All the points of the plate reach the maximum deflection — $\tilde{w}(\rho, t) = -1/8\sqrt{\bar{\mu}D}$ — at the same time $t = 0$. Figure 2.17 shows the displacement of a plate (multiplied by $4\pi\sqrt{\bar{\mu}D}$) impinged by an impulse point force, at various time steps. The plate domain which is represented is $\{-15 < x < 15, -15 < y < 15\}$. For $t = 0.01$, the immediate vicinity of the impact point has already recovered a position close to the rest position ($\tilde{w} \simeq 0$), while the remainder of the plate keeps almost the maximum displacement. As time increases, the plate goes back to its initial position, with a time delay which increases with the distance to the impact point, and the plate deflection looks as if there were waves propagating at a finite speed.

2.4. Vibrations of Cylindrical Shells

In the first section we look for the free vibrations of a shell of infinite length. Then, the Green's tensor of the cylindrical shell operator is calculated. The third section deals with the resonance frequencies and modes of shells of finite length. Finally, we show how the response of a shell to a time harmonic excitation can be calculated.

2.4.1. *Free Oscillations of Cylindrical Shells of Infinite Length*

Let us consider a cylindrical shell with radius R, with infinite length – that is $-\infty < z < +\infty$ – and with thickness h. The material is characterized by a density μ_s, a Young's modulus E and a Poisson's ratio ν. A free oscillation displacement vector (u, v, w) of the shell is a solution of the homogenous vector equation (see section 1.4):

$$\frac{Eh}{1-\nu^2}\mathcal{M}^c\begin{pmatrix} u \\ v \\ w \end{pmatrix} - \omega^2\mu_s h\begin{pmatrix} u \\ v \\ w \end{pmatrix} = \begin{pmatrix} 0 \\ 0 \\ 0 \end{pmatrix} \tag{2.73}$$

where the matrix operator $\mathcal{M}^c$ is:

$$\begin{pmatrix} -\frac{\partial^2}{\partial z^2} - \frac{1-\nu}{2R^2}\frac{\partial^2}{\partial\varphi^2} & -\frac{1+\nu}{2R}\frac{\partial^2}{\partial z\partial\varphi} & -\frac{\nu}{R}\frac{\partial}{\partial z} \\ -\frac{1+\nu}{2R}\frac{\partial^2}{\partial z\partial\varphi} & -\frac{1-\nu}{2}\frac{\partial^2}{\partial z^2} - \frac{1}{R^2}\frac{\partial^2}{\partial\varphi^2} & -\frac{1}{R^2}\frac{\partial}{\partial\varphi} \\ \frac{\nu}{R}\frac{\partial}{\partial z} & \frac{1}{R^2}\frac{\partial}{\partial\varphi} & \frac{1}{R^2} + \frac{h^2}{12}\left[\frac{\partial^4}{\partial z^4} + \frac{2}{R^2}\frac{\partial^4}{\partial z^2\partial\varphi^2} + \frac{1}{R^4}\frac{\partial^4}{\partial\varphi^4}\right] \end{pmatrix} \tag{2.73'}$$

As we did for the plate equation, it is better to use a reduced shell equation. To this end, the following change of variable is adopted:

$$\zeta = \frac{z}{R}$$

Introducing the variable $\Omega^2 = \omega^2 R^2 \mu_s (1-\nu^2)/E$, the shell equation becomes:

$$\widehat{\mathcal{M}}^c \begin{pmatrix} u \\ v \\ w \end{pmatrix} - \Omega^2 \begin{pmatrix} u \\ v \\ w \end{pmatrix} = \begin{pmatrix} 0 \\ 0 \\ 0 \end{pmatrix} \tag{2.74}$$

where the matrix operator $\widehat{\mathcal{M}}^c$ is:

$$\begin{pmatrix} -\dfrac{\partial^2}{\partial \zeta^2} - \dfrac{1-\nu}{2}\dfrac{\partial^2}{\partial \varphi^2} & -\dfrac{1+\nu}{2}\dfrac{\partial^2}{\partial \zeta \partial \varphi} & -\nu \dfrac{\partial}{\partial \zeta} \\ -\dfrac{1+\nu}{2}\dfrac{\partial^2}{\partial \zeta \partial \varphi} & -\dfrac{1-\nu}{2}\dfrac{\partial^2}{\partial \zeta^2} - \dfrac{\partial^2}{\partial \varphi^2} & -\dfrac{\partial}{\partial \varphi} \\ \nu \dfrac{\partial}{\partial \zeta} & \dfrac{\partial}{\partial \varphi} & 1 + \dfrac{h^2}{12R^2}\left[\dfrac{\partial^4}{\partial \zeta^4} + 2\dfrac{\partial^4}{\partial \zeta^2 \partial \varphi^2} + \dfrac{\partial^4}{\partial \varphi^4}\right] \end{pmatrix} \tag{2.74'}$$

The solutions of equation (2.74) are 2π-periodic in φ and we look for functions which depend exponentially on ζ. Thus, the displacement of the shell has the form

$$\begin{pmatrix} \alpha \\ \beta \\ \gamma \end{pmatrix} e^{\imath \Lambda \zeta} e^{\imath n \varphi}$$

The system of partial differential equations (2.74) is replaced by a sequence of algebraic equations:

$$\begin{pmatrix} \Lambda^2 + \dfrac{1-\nu}{2} n^2 - \Omega^2 & \dfrac{1+\nu}{2}\Lambda n & -\imath \nu \Lambda \\ \dfrac{1+\nu}{2}\Lambda n & \dfrac{1-\nu}{2}\Lambda^2 + n^2 - \Omega^2 & -\imath n \\ \imath \nu \Lambda & \imath n & 1 + \dfrac{h^2}{12R^2}\left[\Lambda^2 + n^2\right]^2 - \Omega^2 \end{pmatrix} \begin{pmatrix} \alpha \\ \beta \\ \gamma \end{pmatrix} = \begin{pmatrix} 0 \\ 0 \\ 0 \end{pmatrix} \tag{2.75}$$

There exists a solution if the determinant $D(n, \Omega, \Lambda)$ of the left hand side matrix is zero. An elementary calculation shows that this determinant is a polynomial of degree 4 in Λ^2. In general, it has 4 roots $\Lambda_j (j = 1, \cdots, 4)$ with positive real parts, the 4 other roots being $-\Lambda_j$. It can also be easily seen that this determinant is even in n: it is thus sufficient to compute the roots Λ_j for $n > 0$ only.

Let $(\bar{\alpha}_j, \bar{\beta}_j, \bar{\gamma}_j)$ be the vector which corresponds to (Λ_j, n) with $\Re(\Lambda_j) > 0$ and $n > 0$. Four different solutions of equation (2.74) are obtained:

$$\begin{pmatrix} \bar{\alpha}_j \\ \bar{\beta}_j \\ \bar{\gamma}_j \end{pmatrix} e^{\imath \Lambda_j \zeta} e^{\imath n\varphi}, \quad \begin{pmatrix} \bar{\alpha}_j \\ -\bar{\beta}_j \\ \bar{\gamma}_j \end{pmatrix} e^{\imath \Lambda_j \zeta} e^{-\imath n\varphi}$$
$$\begin{pmatrix} \bar{\alpha}_j \\ -\bar{\beta}_j \\ -\bar{\gamma}_j \end{pmatrix} e^{-\imath \Lambda_j \zeta} e^{\imath n\varphi}, \quad \begin{pmatrix} \bar{\alpha}_j \\ \bar{\beta}_j \\ -\bar{\gamma}_j \end{pmatrix} e^{-\imath \Lambda_j \zeta} e^{-\imath n\varphi} \tag{2.76}$$

The first two solutions behave like waves propagating in the $\zeta > 0$ direction, while the other two behave like waves which propagate in the $\zeta < 0$ direction. For that reason, the Λ_j are called "wavenumbers" though the shell equation is not, properly speaking, a wave equation.

These solutions can also be combined together to obtain solutions which are real when Λ_j is real:

$$\begin{pmatrix} \alpha_j \cos \Lambda_j\zeta \cos n\varphi \\ \beta_j \sin \Lambda_j\zeta \sin n\varphi \\ \gamma_j \sin \Lambda_j\zeta \cos n\varphi \end{pmatrix}, \quad \begin{pmatrix} \alpha_j \cos \Lambda_j\zeta \sin n\varphi \\ -\beta_j \sin \Lambda_j\zeta \cos n\varphi \\ \gamma_j \sin \Lambda_j\zeta \sin n\varphi \end{pmatrix}$$
$$\begin{pmatrix} \alpha_j \sin \Lambda_j\zeta \cos n\varphi \\ -\beta_j \cos \Lambda_j\zeta \sin n\varphi \\ -\gamma_j \cos \Lambda_j\zeta \cos n\varphi \end{pmatrix}, \quad \begin{pmatrix} \alpha_j \sin \Lambda_j\zeta \sin n\varphi \\ \beta_j \cos \Lambda_j\zeta \cos n\varphi \\ -\gamma_j \cos \Lambda_j\zeta \sin n\varphi \end{pmatrix} \tag{2.76'}$$

The free oscillation wavenumbers of an infinite shell depend on n and on Ω. Remember that, for the plate equation, we found two wavenumbers, one real and one imaginary, both proportional to the square root of the angular frequency. In the case of a shell, the dependence on Ω is much more complicated, as shown in Figure 2.18.

There exist values of the angular frequency Ω for which $\Lambda = 0$ is one of the roots of the determinant $D(n, \Omega, \Lambda)$: in that case, the shell displacement does not depend on the variable ζ. The corresponding equation in Ω is:

$$D(n, \Omega, 0) = \Big(\frac{1-\nu^2}{2}n^2 - \Omega^2\Big)\Big[\big(n^2 - \Omega^2\big)\big(1 + n^4\frac{h^2}{12R^2} - \Omega^2\big) - n^2\Big] = 0$$

This equation, together with (2.75), shows that:

1. for $n = 0$, a root $\Lambda = 0$ occurs for $\Omega = 0$ (which corresponds to no deformation of the shell), and for $\Omega = 1$ which corresponds to a purely radial displacement (the two tangential components being zero);

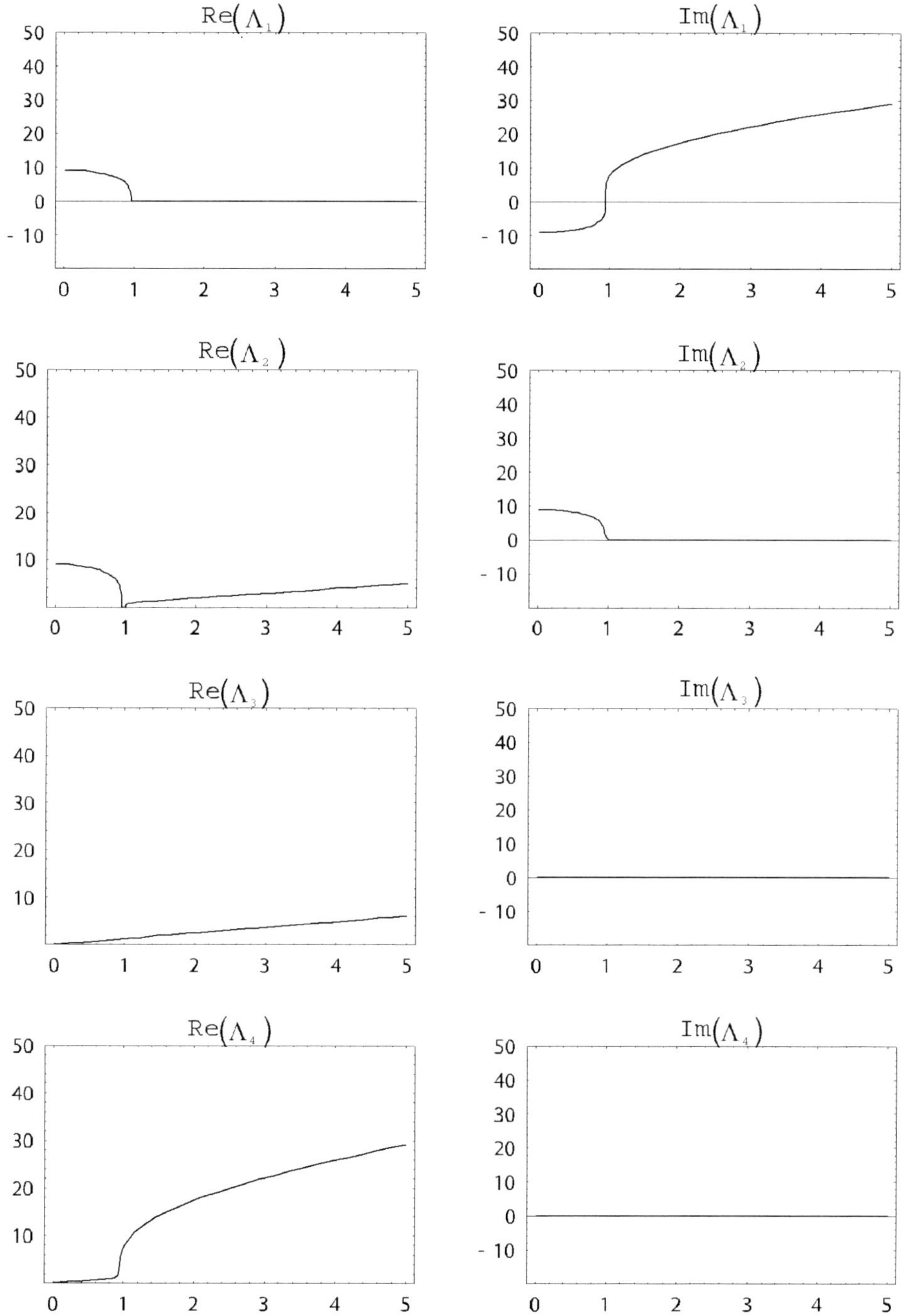

Figure 2.18. *Variation of the 4 roots $\Lambda_j (j = 1, \cdots, 4)$ as functions of Ω, for $n = 0$*

2. for $n \neq 0$, the value $\Omega = n\sqrt{(1-\nu^2)/2}$ corresponds to a pure translation of the shell in the direction of its axis ($\alpha \neq 0, \beta = \gamma = 0$), with no deformation;

3. for $n \neq 0$, there are two other values of Ω for which the shell displacement has two non-zero components ($\alpha = 0, \beta \neq 0, \gamma \neq 0$).

2.4.2. *Green's Tensor for the Cylindrical Shell Equation*

For a matrix operator, such as the one involved in the vector shell equation, the equivalent of the Green's function for scalar equations is a tensor called the *Green's tensor* of the matrix operator.

Let $\mathcal{M}^c_M$ be the matrix operator defined by equation (2.73′), where the subscript M means that the derivatives are taken with respect to the coordinates of the point M. A Green's tensor $\Gamma^c(M, M')$ of the cylindrical shell operator is a matrix defined by:

$$\left\{\frac{Eh}{1-\nu^2}\mathcal{M}^c_M - \omega^2\mu_s h\mathbf{I}\right\}\Gamma^c(M, M') = \begin{pmatrix} \delta_{M'}(M) & 0 & 0 \\ 0 & \delta_{M'}(M) & 0 \\ 0 & 0 & \delta_{M'}(M) \end{pmatrix} \tag{2.77}$$

where $\mathbf{I}$ is the identity matrix, and $\delta_{M'}(M)$ is the Dirac measure located at the point M' of the shell.

The components of the first column of $\Gamma^c(M, M')$ represents the components of a shell displacement due to a tangential point unit force in the z-direction located at the point M': to determine this uniquely and obtain a physically meaningful result, we impose that the principle of energy conservation is respected (how to do this will be shown later). The second and third columns of the Green's tensor have a similar interpretation.

It does not seem possible to give a close form of $\Gamma^c(M, M')$ in terms, for example, of special functions.

This tensor is 2π-periodic with respect to φ; it is thus natural to look for a Fourier series of the form:

$$\Gamma^c(M, M') = \sum_{n=-\infty}^{+\infty} \Gamma^{cn}(z, z')e^{\imath n(\varphi-\varphi')}$$

where (φ, z) and (φ', z') are the coordinates of M and M' respectively. The Fourier components of the Green's tensor satisfy the following set of equations:

$$\left\{\frac{Eh}{1-\nu^2}\mathcal{M}^{cn}_z - \omega^2\mu_s h\mathbf{I}\right\}\Gamma^{cn}(z, z') = \frac{1}{2\pi R}\begin{pmatrix} \delta_{z'}(z) & 0 & 0 \\ 0 & \delta_{z'}(z) & 0 \\ 0 & 0 & \delta_{z'}(z) \end{pmatrix} \tag{2.78}$$

where the operator $\mathcal{M}_z^{cn}$ is defined by:

$$\mathcal{M}_z^{cn} = \begin{pmatrix} -\frac{\partial^2}{\partial z^2}+n^2\frac{1-\nu}{2R^2} & -\imath n\frac{1+\nu}{2R}\frac{\partial}{\partial z} & -\frac{\nu}{R}\frac{\partial}{\partial z} \\ -\imath n\frac{1+\nu}{2R}\frac{\partial}{\partial z} & -\frac{1-\nu}{2}\frac{\partial^2}{\partial z^2}+\frac{n^2}{R^2} & -\frac{\imath n}{R^2} \\ \frac{\nu}{R}\frac{\partial}{\partial z} & \frac{\imath n}{R^2} & \frac{1}{R^2}+\frac{h^2}{12}\left[\frac{\partial^4}{\partial z^4} - \frac{2n^2}{R^2}\frac{\partial^2}{\partial z^2}+\frac{n^4}{R^4}\right] \end{pmatrix} \quad (2.78')$$

(the subscript z recalls that the derivatives are taken with respect to this variable).

The components $\Gamma_{ij}^{cn}(z,z'), (i,j=z,\varphi,r)$ of the tensor $\Gamma^{cn}(z,z')$ are determined by Fourier transform with respect to z as follows. Let $\hat{f}(\zeta)$ be the Fourier transform of any function $f(z)$ defined by:

$$\hat{f}(\zeta) = \int_{-\infty}^{+\infty} f(z)\mathrm{e}^{-2\imath\pi z\zeta}\,dz \quad , \quad f(z) = \int_{-\infty}^{+\infty} \hat{f}(\zeta)\mathrm{e}^{+2\imath\pi z\zeta}\,d\zeta$$

The matrix operator $\mathcal{M}_z^{cn}$ becomes $\widehat{\mathcal{M}}^{cn}$ by replacing $\partial/\partial z$ by $2\imath\pi\zeta$. The Fourier transform of $\Gamma^{cn}(z,z')$ is of the form $\widehat{\Gamma}^{cn}(\zeta)\mathrm{e}^{-2\imath\pi\zeta z'}$, and $\widehat{\Gamma}^{cn}(\zeta)$ is the solution of:

$$\left\{\frac{Eh}{1-\nu^2}\widehat{\mathcal{M}}^{cn} - \omega^2\mu_s h\mathbf{I}\right\}\widehat{\Gamma}^{cn}(\zeta) = \frac{1}{2\pi R}\begin{pmatrix} 1 & 0 & 0 \\ 0 & 1 & 0 \\ 0 & 0 & 1 \end{pmatrix} \quad (2.79)$$

The inverse Fourier transform of the solution of this equation is not uniquely defined: a condition must be added to ensure that the result is physically meaningful. This is achieved by applying the limit absorption principle: the angular frequency ω is first replaced by $\omega_\varepsilon = \omega(1+\imath\varepsilon)$ with $\varepsilon > 0$. The inverse Fourier transform $\Gamma_\varepsilon^{cn}(z,z')$ of the corresponding solution $\widehat{\Gamma}_\varepsilon^{cn}(\zeta)\mathrm{e}^{-2\imath\pi\zeta z'}$ is uniquely determined by integration in the complex plane (residue method). Then, the limit for $\varepsilon \to 0$ of $\Gamma_\varepsilon^{cn}(z,z')$ is taken. It can be shown that this limit is unique.

The components of the corresponding solution to system (2.79) have the following form:

$$\widehat{\Gamma}_\varepsilon{}^{cn}{}_{ij}(\zeta) = \frac{1}{2\pi R}\frac{\mathbf{D}_{ij}^n(\zeta,\varepsilon)}{\mathbf{D}^n(\zeta,\varepsilon)}$$

where $\mathbf{D}^n(\zeta,\varepsilon)$ is the determinant of the matrix $[Eh/(1-\nu^2)\widehat{\mathcal{M}}^{cn} - \mu_s\omega_\varepsilon^2\mathbf{I}]$ and $\mathbf{D}_{ij}^n(\zeta,\varepsilon)$ is the minor of the term (ij). The determinant $\mathbf{D}^n(\zeta,\varepsilon)$ is a polynomial of degree 4 in ζ^2: thus, it has 4 zeros $\zeta_q^n(\varepsilon)$ with positive imaginary parts and 4 zeros $-\zeta_q^n(\varepsilon)$ with negative imaginary parts, which, all, depend on ε. The inverse Fourier transform of $\widehat{\Gamma}_{\varepsilon}^{cn}{}_{ij}(\zeta)e^{-2\imath\pi\zeta z'}$ is uniquely determined by applying the residue theorem, which gives:

$$\begin{aligned}\Gamma_{\varepsilon}^{c\,n}{}_{ij}(z,z') &= \int_{-\infty}^{+\infty} \widehat{\Gamma}_{\varepsilon}^{c\,n}{}_{ij}(\zeta)e^{2\imath\pi(z-z')\zeta}\,d\zeta \\ &= 2\imath\pi\sum_{q=1}^{4}\frac{\mathbf{D}_{ij}^{n}(\zeta_q^n(\varepsilon),\varepsilon)}{\mathbf{D}^{n\prime}(\zeta_q^n(\varepsilon),\varepsilon)}e^{2\imath\pi\zeta_q^n(\varepsilon)|z-z'|}\end{aligned} \tag{2.80}$$

where $\mathbf{D}^{n\prime}(\zeta,\varepsilon)$ is the derivative of $\mathbf{D}^n(\zeta,\varepsilon)$ with respect to ζ. The components of the Green's tensor are obtained by taking the value, for $\varepsilon = 0$, of the functions defined by equation (2.80), that is:

$$\Gamma_{ij}^{c\,n}(M,M') = 2\imath\pi\sum_{q=1}^{4}\frac{\mathbf{D}_{ij}^{n}(\zeta_q^n,0)}{\mathbf{D}^{n\prime}(\zeta_q^n,0)}e^{2\imath\pi\zeta_q^n|z-z'|} \quad \text{with } i,j=1,2,3 \tag{2.81}$$

where ζ_q^n stands for $\zeta_q^n(0)$.

The parameters $\lambda_q^n = 2\pi\zeta_q^n$ have the dimension of wavenumbers which, as in the case of a thin plate, depend on the excitation frequency. Expression (2.81) points out that, away from the excitation point, the components of the Green's tensor are identical to the free oscillations which have been studied in the previous section. The components of the Green's tensor of the shell operator have the following expression:

$$\Gamma_{ij}^{c}(M,M') = 2\imath\pi\sum_{n=-\infty}^{\infty}\sum_{q=1}^{4}\frac{\mathbf{D}_{ij}^{n}(\zeta_q^n,0)}{\mathbf{D}^{n\prime}(\zeta_q^n,0)}e^{2\imath\pi\zeta_q^n|z-z'|}e^{\imath n(\varphi-\varphi')} \tag{2.81'}$$

REMARK 1.– As we mentioned previously, the first column of the Green's tensor $\Gamma^c(M,M')$ is the vector solution of the following equation:

$$\left\{\frac{Eh}{1-\nu^2}\mathcal{M}_M^c - \mu_s\omega^2\mathbf{I}\right\}\Gamma_z^c(M,M') = \begin{pmatrix}\delta_{M'}(M)\\0\\0\end{pmatrix} \tag{2.77'}$$

Its components are denoted by $(\Gamma_{zz}^c, \Gamma_{z\varphi}^c, \Gamma_{zr}^c)$, which is preferred to $(\Gamma_{11}^c, \Gamma_{21}^c, \Gamma_{31}^c)$. Similarly, the second column is the vector $\Gamma_{\varphi}^c(M,M')$, with components $(\Gamma_{\varphi z}^c, \Gamma_{\varphi\varphi}^c, \Gamma_{\varphi r}^c)$, corresponding to the excitation force, with components

$(0, \delta_{M'}(M), 0)$; and the third column is the response $\Gamma_r^c(M, M')$, with components $(\Gamma_{rz}^c, \Gamma_{r\varphi}^c, \Gamma_{rr}^c)$, to the excitation $(0, 0, \delta_{M'}(M))$.

REMARK 2.– Like the matrix $\mathcal{M}^c$, the Green's tensor has only six independent components, the non-diagonal ones satisfying the following relationships:

$$\begin{aligned} \Gamma_{z\varphi}^c &= \Gamma_{\varphi z}^c \\ \Gamma_{zr}^c = -\Gamma_{rz}^c \quad &, \quad \Gamma_{r\varphi}^c = -\Gamma_{\varphi r}^c \end{aligned} \tag{2.82}$$

2.4.3. *Harmonic Vibrations of a Cylindrical Shell of Finite Dimensions: General Definition and Theorems*

The definition of the resonance modes of a shell is identical to that already given for beams and plates. The theorem for the existence and uniqueness of the response of a shell to a harmonic excitation is also the same.

Let us consider a cylindrical shell occupying a domain Σ, limited by two curves $\partial\Sigma_1$ and $\partial\Sigma_2$ of any shape.

DEFINITION 2.3.– *A* resonance mode *is a vector function which satisfies the homogenous shell equation*

$$\frac{Eh}{1-\nu^2}\mathcal{M}^c\begin{pmatrix} u \\ v \\ w \end{pmatrix} - \omega^2\mu_s h\begin{pmatrix} u \\ v \\ w \end{pmatrix} = \begin{pmatrix} 0 \\ 0 \\ 0 \end{pmatrix} \tag{2.83}$$

and a given set of boundary conditions along $\partial\Sigma_1$ and $\partial\Sigma_2$. The corresponding value of ω is called a resonance angular frequency.

THEOREM 2.5 (EXISTENCE OF RESONANCE MODES).– *A purely elastic cylindrical shell, submitted to conservative boundary conditions has a countable set of resonance modes and real resonance frequencies. The set of resonance modes is an orthogonal basis on which any solution of the non-homogenous harmonic shell equation (when it exists) can be expanded.*

If there is any energy loss (by material damping or through the boundaries), the resonance frequencies are complex.

The existence and uniqueness of the response of a shell to a harmonic excitation is governed by the following theorem.

THEOREM 2.6 (EXISTENCE AND UNIQUENESS OF THE SOLUTION).– *Let $\vec{f}(M)$ be the harmonic (vector) excitation of a shell and ω be the corresponding angular frequency.*

◇ *If ω is different from all the resonance angular frequencies, then the solution $\vec{U}$, with components (u, v, w), of the non-homogenous shell equation exists and is unique for any excitation $\vec{f}(M)$.*

◇ *If ω is equal to one resonance angular frequency, say ω_n, the solution of the non-homogenous shell equation does not exist, except if $\vec{f}(M)$ is orthogonal to the corresponding resonance mode $\vec{U}_n$.*

◇ *If ω is equal to ω_n and $\vec{f}(M)$ is orthogonal to $\vec{U}_n$, a non-unique solution to the non-homogenous shell equation exists, which can be decomposed into a uniquely determined component $\vec{U}^{(1)}$ orthogonal to $\vec{U}_n$, and a component $\vec{U}^{(2)} = \alpha\vec{U}_n$ with arbitrary amplitude α.*

Let us recall that two vector functions $\vec{U}_1$ (with components (u_1, v_1, w_1)) and $\vec{U}_2$ (with components (u_2, v_2, w_2)) are said to be orthogonal if the following integral

$$\int_\Sigma \big(u_1(M)u_2(M) + v_1(M)v_2(M) + w_1(M)w_2(M)\big)\,\mathrm{d}\Sigma(M)$$

is zero.

The proof of these theorems is classical and can be found in any textbook on mathematical analysis.

2.4.4. *Resonance Modes of a Cylindrical Shell Closed by Shear Diaphragms at Both Ends*

It seems that this is the only very simple case which can be solved analytically. Let us consider, as in section 1.4, a cylindrical shell of radius R, extending from $z = -L/2$ to $z = +L/2$ and with thickness h. The shear diaphragm boundary conditions at $z = \pm L/2$

$$\mathrm{Tr}\, v = 0\,,\ \ \mathrm{Tr}\, w = 0\,,\ \ \mathrm{Tr}\, u_{,z} + \frac{\nu}{R}\,\mathrm{Tr}\,\big[v_{,\varphi} + w\big] = 0\,,\ \ \mathrm{Tr}\, w_{,zz} + \frac{\nu}{R^2}\,\mathrm{Tr}\, w_{,\varphi\varphi} = 0$$

take the following simple form:

$$\mathrm{Tr}\, u_{,z} = \mathrm{Tr}\, v = \mathrm{Tr}\, w = \mathrm{Tr}\, w_{,zz} = 0 \quad \text{for} \quad z = -L/2 \text{ and } z = +L/2 \tag{2.84}$$

We adopt the reduced shell equation given by equations (2.74) and (2.74′) and set $\zeta = z/R$.

1. *Longitudinal displacement even in ζ*

Let us consider the vector

$$u = \alpha \cos \Lambda\zeta \cos n\varphi \ , \ v = \beta \sin \Lambda\zeta \sin n\varphi \ , \ w = \gamma \sin \Lambda\zeta \cos n\varphi \qquad (2.85)$$

It will satisfy the boundary conditions if

$$\sin \Lambda L/2R = 0 \quad \Longrightarrow \quad \Lambda_m^e = 2m\pi R/L \quad , \quad m = 0, 1, \cdots \infty \qquad (2.85')$$

The resonance angular frequencies Ω are obtained by introducing this solution into equation (2.74). We obtain:

$$\begin{pmatrix} \Lambda_m^{e\,2} + \frac{1-\nu}{2} n^2 - \Omega^2 & -\frac{1+\nu}{2} \Lambda_m^e n & -\nu \Lambda_m^e \\ -\frac{1+\nu}{2} \Lambda_m^e n & \frac{1-\nu}{2} \Lambda_m^{e\,2} + n^2 - \Omega^2 & n \\ -\nu \Lambda_m^e & n & 1 + \frac{h^2}{12R^2} \left(\Lambda_m^{e\,2} + n^2\right)^2 - \Omega^2 \end{pmatrix} \begin{pmatrix} \alpha \\ \beta \\ \gamma \end{pmatrix} = \begin{pmatrix} 0 \\ 0 \\ 0 \end{pmatrix} \qquad (2.86)$$

This system has a non-zero solution if its determinant is zero. This determinant is a polynomial of degree 3 in Ω^2, which has three roots, which can be obtained analytically. Theorem 2.5 states that these roots are all positive (the proof for this simple case is easy and left as an exercise). Thus, three resonance angular frequencies Ω_{mni}^e, $(i = 1, 2, 3)$ are associated with the couple (n, Λ_m^e). The coefficients $(\alpha_{mni}, \beta_{mni}, \gamma_{mni})$ which define each mode shape

$$\begin{gathered} u = \alpha_{mni} \cos \Lambda_m^e \zeta \ \cos n\varphi \ , \ v = \beta_{mni} \sin \Lambda_m^e \zeta \ \sin n\varphi \\ w = \gamma_{mni} \sin \Lambda_m^e \zeta \ \cos n\varphi \end{gathered} \qquad (2.87)$$

depend on n, Λ_m and the angular frequency Ω_{mni}^e. As a solution of a homogenous system, they are not uniquely determined: it is of common use to choose them so that the mode has a unit $\mathbf{L}^2$-norm, that is

$$\int_\Sigma \Big[\left(\alpha_{mni} \cos \Lambda_m^e \zeta \cos n\varphi\right)^2 + \left(\beta_{mni} \sin \Lambda_m^e \zeta \sin n\varphi\right)^2 + \left(\gamma_{mni} \sin \Lambda_m^e \zeta \cos n\varphi\right)^2 \Big] \mathrm{d}\Sigma = 1$$

The solution

$$u = \alpha \cos \Lambda\zeta \sin n\varphi \quad , \quad v = -\beta \sin \Lambda\zeta \cos n\varphi \quad , \quad w = \gamma \sin \Lambda\zeta \sin n\varphi$$

leads to the same resonance angular frequencies, but the mode shapes are, of course, different: they are deduced from expression (2.87) by replacing $n\varphi$ with

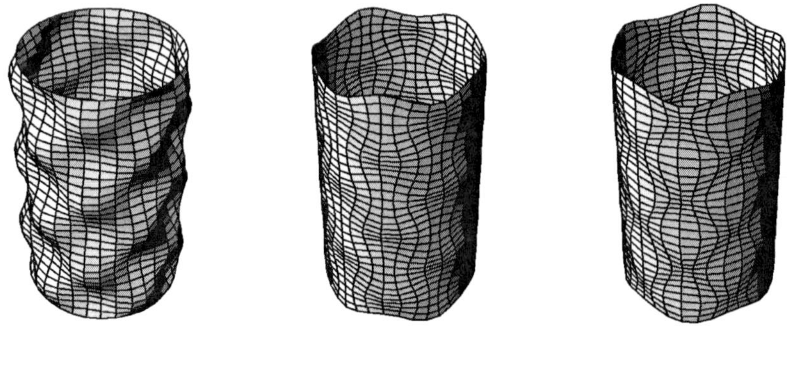

Figure 2.19. *Representation of the three resonance modes corresponding to solution* (2.87) *with* $m = 3$ *and* $n = 5$ *(increasing frequency from left to right) for ratios* $L/R = 3$ *and* $h/R = 1/50$

$n\varphi - \pi/2$. So, for a given couple (n, Λ_m^e) we have three resonance frequencies of multiplicity order 2 and, as a consequence, six even resonance modes.

Figure 2.19 shows the three modes corresponding to solution (2.87) with $m = 3$ and $n = 5$ for ratios $L/R = 3$ and $h/R = 1/50$; the Poisson's ratio is equal to 1/3. The values of the reduced angular frequencies are 0.736, 5.893 and 8.056.

2. *Longitudinal displacement odd in* ζ

The solution

$$u = \alpha \,\sin \Lambda\zeta\, \cos n\varphi \;,\; v = \beta\, \cos \Lambda\zeta\, \sin n\varphi \;,\; w = \gamma\, \cos \Lambda\zeta\, \cos n\varphi \qquad (2.88)$$

satisfies the boundary conditions if

$$\cos \Lambda L/2R = 0 \quad \Longrightarrow \quad \Lambda_m^o = (2m+1)\pi R/L \quad , \quad m = 0, 1, \cdots \infty \qquad (2.88')$$

The resonance angular frequencies are the roots of the determinant of the following matrix:

$$\begin{pmatrix} {\Lambda_m^o}^2 + \dfrac{1-\nu}{2} n^2 - \Omega^2 & \dfrac{1+\nu}{2}\Lambda_m^o n & \nu \Lambda_m^o \\ \dfrac{1+\nu}{2}\Lambda_m^o n & \dfrac{1-\nu}{2}{\Lambda_m^o}^2 + n^2 - \Omega^2 & n \\ \nu\Lambda_m^o & n & 1 + \dfrac{h^2}{12R^2}\left({\Lambda_m^o}^2 + n^2\right)^2 - \Omega^2 \end{pmatrix} \qquad (2.89)$$

These roots are different from the former ones, because the values of Λ are different. The solution

$$u = \alpha \sin \Lambda\zeta \sin n\varphi \quad , \quad v = -\beta \cos \Lambda\zeta \cos n\varphi \quad , \quad w = \gamma \cos \Lambda\zeta \sin n\varphi$$

leads to the same resonance frequencies, but generates different modes.

The final conclusion is that, for the same couple (n, Λ_m^o), there are three resonance frequencies and six odd resonance modes.

2.4.5. *Resonance Modes of a Cylindrical Shell Clamped at Both Ends*

We consider again the same geometry and we adopt the reduced shell equation (2.74-2.74′) or its variational form. The resonance modes can be expressed in terms of the linearly independent solutions of the homogenous shell equation which we have obtained earlier. For example, using the reduced variable $\zeta = z/R$, we can look for a mode (u, v, w) as a linear combination of the eight functions

$$U_j(\zeta,\varphi) = \begin{pmatrix} \alpha_j \cos \Lambda_j\zeta \cos n\varphi \\ \beta_j \sin \Lambda_j\zeta \sin n\varphi \\ \gamma_j \sin \Lambda_j\zeta \cos n\varphi \end{pmatrix} , \quad V_j(\zeta,\varphi) = \begin{pmatrix} \alpha_j \sin \Lambda_j\zeta \cos n\varphi \\ -\beta_j \cos \Lambda_j\zeta \sin n\varphi \\ -\gamma_j \cos \Lambda_j \zeta \cos n\varphi \end{pmatrix}$$

$$j = 1, 2, 3, 4$$

$$\begin{pmatrix} u \\ v \\ w \end{pmatrix} = \sum_{j=1}^{4} \left[A_j U_j + B_j V_j \right]$$

which first and third components are even functions of φ, the second one being an odd function of φ. The eight boundary conditions lead to a linear system of homogenous equations to determine the unknown coefficients A_j and B_j. Because of the symmetry of the problem with respect to the plane $z = 0$ (same boundary conditions at $z = -L/2$ and $z = +L/2$), this system can be split into two systems of order 4: one of them to determine the coefficients A_j, the other one to determine the coefficients B_j. This shows that the modes can be classified into two groups: in the first one, $u(\zeta)$ is an even function of ζ, while $v(\zeta)$ and $w(\zeta)$ are odd functions; in the second one, $u(\zeta)$ is an odd function of ζ, while $v(\zeta)$ and $w(\zeta)$ are even functions.

Any of these systems has a non-zero solution if its determinant is zero. The cancellation of the determinant is obtained for a sequence of values for the reduced angular frequency Ω. Because the relationships between the parameters $(\Lambda_j, \alpha_j, \beta_j, \gamma_j)$ and Ω are extremely complex (remember that the Λ_j are the roots of a polynomial of order 4, the coefficients of which depend on Ω), the numerical calculation of the resonance angular frequencies by this method is somewhat difficult. It seems easier to use the Ritz-Galerkin method which has the advantage of applying to more general cases (as, for example, shells with non-constant thickness).

We can first notice that for a shell limited by circles, the boundary value problem, defined by the shell equation and the boundary conditions, is separable: this implies that any solution of the homogenous problem is the product of a function of ζ by a function of φ. More precisely, using the results of section 2.4.1, any resonance mode can be sought in one of the following two forms:

$$\begin{array}{l} u = \alpha(\zeta)\cos n\varphi \quad , \quad v = \beta(\zeta)\sin n\varphi \quad , \quad w = \gamma(\zeta)\cos n\varphi \quad \text{(a)} \\ \text{or} \\ u = \hat{\alpha}(\zeta)\ \sin n\varphi \quad , \quad v = \hat{\beta}(\zeta)\ \cos n\varphi \quad , \quad w = \hat{\gamma}(\zeta)\ \sin n\varphi \quad \text{(b)} \end{array} \tag{2.90}$$

Let Ω be a resonance frequency associated with a mode of form (2.90-a). Corresponding to the same resonance frequency, we obtain a second resonance mode of form (2.90-b) with:

$$\hat{\alpha}(\zeta) = \alpha(\zeta) \quad , \quad \hat{\beta}(\zeta) = -\beta(\zeta) \quad , \quad \hat{\gamma}(\zeta) = \gamma(\zeta)$$

This result is very simply obtained by writing the system of partial differential equations which each mode must satisfy.

Let us now give a few details on the Ritz-Galerkin method. We consider solutions of form (2.90-a). The boundary conditions lead to:

$$\alpha(\pm L/2R) = 0 \quad , \quad \beta(\pm L/2R) = 0 \quad , \quad \gamma(\pm L/2R) = 0 \quad , \quad \gamma'(\pm L/2R) = 0$$

The functions $\alpha(\zeta)$ and $\beta(\zeta)$ must be square integrable together with their first order derivatives and be zero at $\zeta = -L/2R$ and $\zeta = +L/2R$. These properties define a Hilbert space and we denote by $U_p(\zeta)$ a basis of this space. The function $\gamma(\zeta)$ and its first and second order derivatives are square integrable, and $\gamma(\zeta)$ and $\gamma'(\zeta)$ cancel at $\zeta = -L/2R$ and $\zeta = +L/2R$. These properties define another Hilbert space and we denote by $W_p(\zeta)$ a basis of this space. The resonance modes are sought as series expansions of the following form:

$$\begin{pmatrix} u(\zeta,\varphi) = \cos n\varphi \sum\limits_{p=1}^{\infty} a_p U_p(\zeta) \\ v(\zeta,\varphi) = \sin n\varphi \sum\limits_{p=1}^{\infty} b_p U_p(\zeta) \\ w(\zeta,\varphi) = \cos n\varphi \sum\limits_{p=1}^{\infty} c_p W_p(\zeta) \end{pmatrix} \tag{2.91}$$

The reduced variational equation governing a resonance mode is deduced from equations (1.34-1.34′):

$$\mathcal{A}^C(u, v, w; \hat{u}, \hat{v}, \hat{w}) - \Omega^2 \int_0^{2\pi} \mathrm{d}\varphi \int_{-L/2R}^{L/2R} \mathrm{d}\zeta (u\hat{u}^* + v\hat{v}^* + w\hat{w}^*) = 0 \qquad (2.92)$$

with:

$$\mathcal{A}^C(u, v, w; \hat{u}, \hat{v}, \hat{w}) = \int_0^{2\pi} \mathrm{d}\varphi \int_{-L/2R}^{L/2R} \mathrm{d}\zeta \Big[\nu\Big(u,_\zeta + v,_\varphi + w\Big)\Big(\hat{u},_\zeta^* + \hat{v},_\varphi^* + \hat{w}^*\Big)$$

$$+(1-\nu)\Big(u,_\zeta \hat{u},_\zeta^* + (v,_\varphi + w)(\hat{v},_\varphi^* + \hat{w}^*)\Big) + \frac{1-\nu}{2}\Big(u,_\varphi + v,_\zeta\Big)\Big(\hat{u},_\varphi^* + \hat{v},_\zeta^*\Big)$$

$$+\frac{h^2}{12R^2}\Big[\nu\Big(w,_{\zeta\zeta} + w,_{\varphi\varphi}\Big)\Big(\hat{w},_{\zeta\zeta}^* + \hat{w},_{\varphi\varphi}^*\Big)$$

$$+(1-\nu)\Big(w,_{\zeta\zeta}\hat{w},_{\zeta\zeta}^* + w,_{\zeta\varphi}\hat{w},_{\zeta\varphi}^* + w,_{\varphi\zeta}\hat{w},_{\varphi\zeta}^* + w,_{\varphi\varphi}\hat{w},_{\varphi\varphi}^*\Big)\Big]\Big] \qquad (2.92')$$

Let us introduce expansion (2.91) into equation (2.92) which must be satisfied for any vector $(\hat{u}, \hat{v}, \hat{w})$ having the same properties as (u, v, w) (differentiability, integrability and boundary conditions). It is sufficient that the equation is satisfied for all the basis vectors, which are:

$$\begin{pmatrix} \cos n\varphi \; U_q(\zeta) \\ \sin n\varphi \; U_q(\zeta) \\ \cos n\varphi \; W_q(\zeta) \end{pmatrix} \qquad q = 0, 1, 2, \cdots, \infty \qquad (2.93)$$

We obtain an infinite system of algebraic linear equations to determine the resonance frequencies Ω and coefficients (a_p, b_p, c_p):

$$2\pi \sum_{p=1}^{\infty} \int_{-L/2R}^{+L/2R} \mathrm{d}\zeta \Big\{ \nu\big(a_p U_p'(\zeta) + n b_p U_p(\zeta) + c_p W_p(\zeta)\big)\big(U_q'(\zeta) + nU_q(\zeta) + W_q(\zeta)\big)$$

$$+ (1-\nu)\Big[a_p U_p'(\zeta)U_q'(\zeta) + \big(n b_p U_p(\zeta) + c_p W_p(\zeta)\big)\big(nU_q(\zeta) + W_q(\zeta)\big)\Big]$$

$$+ \frac{1-\nu}{2}\big(-n a_p U_p(\zeta) + b_p U_p'(\zeta)\big)\big(-nU_q(\zeta) + U_q'(\zeta)\big)$$

$$+ \frac{h^2}{12R^2}\Big[\nu\big(c_p W_p''(\zeta) - n^2 W_p(\zeta)\big)\big(W_q''(\zeta) - n^2 W_q(\zeta)\big)$$

$$+ (1-\nu)c_p\big(W_p''(\zeta)W_q''(\zeta) + 2n^2 W_p'(\zeta)W_q'(\zeta) + n^4 W_p(\zeta)W_q(\zeta)\big)\Big]\Big\}$$

$$= 2\pi \Omega^2 \sum_{p=1}^{\infty} \int_{-L/2R}^{+L/2R} d\zeta \Big[a_p U_p(\zeta) U_q(\zeta) + b_p U_p(\zeta) U_q(\zeta) + c_p W_p(\zeta) W_q(\zeta) \Big]$$

$$\text{for} \qquad q = 1, 2, \cdots \infty \qquad (2.94)$$

This system has, of course, an infinite sequence of eigenvalues Ω_{mn}^2 and eigenvectors $(a_p^{mn}, b_p^{mn}, c_p^{mn})$ which define two sets of resonance modes, one of form (2.90-a) and the other of form (2.90-b). Approximations of the resonance angular frequencies and modes are obtained by solving a truncated form of this system.

A simple example of basis functions is obtained with the following functions:

◇ the functions $U_n(\zeta)$ are the resonance modes of a string with length L/R and clamped at both ends, that is:

$$U_{2n}(\zeta) = \sqrt{\frac{2R}{L}} \sin \frac{2n\pi R\zeta}{L} \quad , \qquad n = 1, 2, \ldots, \infty$$

$$U_{2n+1}(\zeta) = \sqrt{\frac{2R}{L}} \cos \frac{(2n+1)\pi R\zeta}{L} \quad , \qquad n = 0, 1, \ldots, \infty$$

◇ the functions $W_n(\zeta)$ are the resonance modes of a beam with length L/R and clamped at both ends as defined in section 2.2.3 (paragraph 4).

Some simplifications can be introduced into the numerical computations. We have said previously that when $u(\zeta)$ is an even function, $v(\zeta)$ and $w(\zeta)$ are odd functions, and *vice versa*. It is important to account for these properties because the size of the linear system which approximates the exact equations is thus reduced by a factor 2. Furthermore, the accuracy of the numerical results is increased: indeed, if a general expansion is used to determine a mode which first component is an even (resp. odd) function, the coefficients of the odd (resp. even) basis functions will never be found exactly equal to zero by a computer. Let $U_n^e(\zeta)$ and $W_n^e(\zeta)$ (resp. $U_n^o(\zeta)$ and $W_n^o(\zeta)$) be the even (resp. odd) functions of the bases $U_n(\zeta)$ and $W_n(\zeta)$. The following expansions must be used for the resonance modes:

$$\begin{pmatrix} u^e(\zeta,\varphi) = \cos n\varphi \sum_{p=1}^{\infty} a_p U_p^e(\zeta) \\ v^o(\zeta,\varphi) = \sin n\varphi \sum_{p=1}^{\infty} b_p U_p^o(\zeta) \\ w^o(\zeta,\varphi) = \cos n\varphi \sum_{p=1}^{\infty} c_p W_p^o(\zeta) \end{pmatrix} \quad , \quad \begin{pmatrix} u^o(\zeta,\varphi) = \cos n\varphi \sum_{p=1}^{\infty} a_p U_p^o(\zeta) \\ v^e(\zeta,\varphi) = \sin n\varphi \sum_{p=1}^{\infty} b_p U_p^e(\zeta) \\ w^e(\zeta,\varphi) = \cos n\varphi \sum_{p=1}^{\infty} c_p W_p^e(\zeta) \end{pmatrix}$$

The resonance modes of form (2.90-b) have the same resonance frequencies and are thus directly deduced from the resonance modes of form (2.90-a) by changing $\cos n\varphi$ for $\sin n\varphi$ and $\sin n\varphi$ for $\cos n\varphi$.

2.4.6. *Response of a Cylindrical Shell to a Harmonic Excitation: Resonance Modes Representation*

Let Σ be a bounded cylindrical surface, with boundary $\partial\Sigma$ (composed of two curves $\partial\Sigma_1$ and $\partial\Sigma_2$) which is occupied by a thin shell of constant thickness and composed of a homogenous material.

As done before, the reduced shell equation is used. It is assumed that the shell displacement (u, v, w) satisfies a set of natural boundary conditions. The shell is excited by a harmonic force with components (f_z, f_φ, f_ρ), the factor $e^{-\imath\Omega t}$ being omitted (Ω is the reduced angular frequency). Let (u_n, v_n, w_n) be the resonance modes of the shell and Ω_n the corresponding resonance frequencies. We will show that the shell displacement can be expressed as a series of resonance modes for any $\omega \neq \Omega_n$: for that purpose, we must first recall that the Ω_n^2 are positive (see exercise 16 at the end of the chapter), and show that the resonance modes are orthogonal.

Let $\mathbf{H}$ be the functional space which the shell displacement belongs to, and let $(\hat{u}, \hat{v}, \hat{w})$ be any function in $\mathbf{H}$. The resonance modes satisfy the following variational equation:

$$\mathcal{A}^C(u_n, v_n, w_n; \hat{u}, \hat{v}, \hat{w}) = \Omega_n^2 \int_\Sigma (u_n\hat{u}^* + v_n\hat{v}^* + w_n\hat{w}^*)\,\mathrm{d}\Sigma \qquad (2.95)$$

Equation (2.95) is satisfied for $(\hat{u}, \hat{v}, \hat{w})$ replaced by (u_m, v_m, w_m):

$$\mathcal{A}^C(u_n, v_n, w_n; u_m, v_m, w_m) = \Omega_n^2 \int_\Sigma (u_n u_m^* + v_n v_m^* + w_n w_m^*)\,\mathrm{d}\Sigma \qquad (2.95\text{-a})$$

The term $\mathcal{A}^C(u_n, v_n, w_n; u_m, v_m, w_m)$ is equal to $\mathcal{A}^C(u_m, v_m, w_m; u_n, v_n, w_n)^*$, and we have:

$$\mathcal{A}^C(u_m, v_m, w_m; u_n, v_n, w_n)^* = \left\{\Omega_m^2 \int_\Sigma (u_m u_n^* + v_m v_n^* + w_m w_n^*)\,\mathrm{d}\Sigma\right\}^* \qquad (2.95\text{-b})$$

We recall that Ω_m^2 is real and we remark that

$$\left\{\int_\Sigma (u_m u_n^* + v_m v_n^* + w_m w_n^*)\,\mathrm{d}\Sigma\right\}^* = \int_\Sigma (u_n u_m^* + v_n v_m^* + w_n w_m^*)\,\mathrm{d}\Sigma$$

By subtracting equality (2.95-b) from equality (2.95-a), we obtain:

$$\left(\Omega_n^2 - \Omega_m^2\right) \int_\Sigma \left(u_n u_m^* + v_n v_m^* + w_n w_m^*\right) \mathrm{d}\Sigma = 0$$

This shows that the resonance modes are orthogonal.

The non-homogenous variational shell equation is:

$$\mathcal{A}^C (u, v, w; \hat{u}, \hat{v}, \hat{w}) - \Omega^2 \int_\Sigma \left(u\hat{u}^* + v\hat{v}^* + w\hat{w}^*\right) \mathrm{d}\Sigma = \int_\Sigma \left(f_z \hat{u}^* + f_\varphi \hat{v}^* + f_\rho \hat{w}^*\right) \mathrm{d}\Sigma \qquad (2.96)$$

The shell displacement is sought as a series of resonance modes:

$$\begin{pmatrix} u \\ v \\ w \end{pmatrix} = \sum_n \alpha_n \begin{pmatrix} u_n \\ v_n \\ w_n \end{pmatrix}$$

An elementary calculation leads to the result:

$$\alpha_n = \frac{1}{\Omega_n^2 - \Omega^2} \int_\Sigma \left(f_z u_n^* + f_\varphi v_n^* + f_\rho w_n^*\right) \mathrm{d}\Sigma$$

$$\begin{pmatrix} u \\ v \\ w \end{pmatrix} = \sum_n \frac{1}{\Omega_n^2 - \Omega^2} \left[\int_\Sigma \left(f_z u_n^* + f_\varphi v_n^* + f_\rho w_n^*\right) \mathrm{d}\Sigma \right] \begin{pmatrix} u_n \\ v_n \\ w_n \end{pmatrix} \qquad (2.97)$$

The coefficients are uniquely determined if $\Omega \neq \Omega_n$ for all n. If the excitation angular frequency is equal to one resonance frequency, say Ω_m, the corresponding term of the expansion is infinite except if the excitation force is orthogonal to the resonance mode number m: in that case, the solution of equation (2.96) is uniquely determined up to a vector $\alpha_m(u_m, v_m, w_m)$ with arbitrary amplitude α_m. This result is quite similar to the result obtained for a plate.

2.4.7. *Boundary Integral Equations and Boundary Element Method*

The determination of the resonance modes for a cylindrical shell or the response of the shell to a harmonic excitation can be obtained using Boundary Integral Equations because the Green's tensor of the shell operator is known. The shell displacement is expressed as the sum of the displacement of an infinite shell (incident displacement) and the displacement due to sources supported by the shell boundaries. The boundary conditions provide a system of Boundary Integral Equations which determines the sources densities. The resonance frequencies are those for which the homogenous system (no incident displacement) has a non-zero solution. Though the method is quite general, we will present it on the simple case of a cylindrical shell limited by two circles and clamped at both ends.

In this section, the physical variables are used. There is no difficulty in converting the forthcoming formulae into reduced variables, if desired.

1. *Green's formula*

We must start by establishing the Green's formula for the shell operator. Let us consider a shell as in the previous section. Let $\mathcal{A}^c(u,v,w;\hat{u},\hat{v},\hat{w})$ be the bilinear form defined by equation (1.34′) associated with two vectors with components (u,v,w) and $(\hat{u},\hat{v},\hat{w})$. An integration by parts leads to:

$$\mathcal{A}^c(u,v,w;\hat{u},\hat{v},\hat{w}) = \int_0^{2\pi} R\,\mathrm{d}\varphi \int_{-L/2}^{+L/2} \mathcal{M}^c\begin{pmatrix} u \\ v \\ w \end{pmatrix} \cdot \begin{pmatrix} \hat{u}^* \\ \hat{v}^* \\ \hat{w}^* \end{pmatrix}$$
$$+ \int_0^{2\pi} R\,\mathrm{d}\varphi\big(F_{zz}\hat{u}^* + F_{z\varphi}\hat{v}^* + F_{zr}\hat{w}^* + M_{z\varphi}\hat{w}^*_{,\varphi} + M_{zz}\hat{w}^*_{,z}\big)(+L/2,\varphi)$$
$$- \int_0^{2\pi} R\,\mathrm{d}\varphi\big(F_{zz}\hat{u}^* + F_{z\varphi}\hat{v}^* + F_{zr}\hat{w}^* + M_{z\varphi}\hat{w}^*_{,\varphi} + M_{zz}\hat{w}^*_{,z}\big)(-L/2,\varphi)$$

In the first integral of this expression, the dot stands for the scalar product of the two vectors $\mathcal{M}^c\begin{pmatrix} u \\ v \\ w \end{pmatrix}$ and $\begin{pmatrix} \hat{u}^* \\ \hat{v}^* \\ \hat{w}^* \end{pmatrix}$. The quantities F_{zz}, $F_{z\varphi}$ and F_{zr} are the boundary forces defined in equations (1.22-a); the quantities $M_{z\varphi}$ and M_{zz} are the moments given by (1.22-b) and (1.22-c). Similarly, we have:

$$\mathcal{A}^c(u,v,w;\hat{u},\hat{v},\hat{w}) = \int_0^{2\pi} R\,\mathrm{d}\varphi \int_{-L/2}^{+L/2} \begin{pmatrix} u \\ v \\ w \end{pmatrix} \cdot \mathcal{M}^c\begin{pmatrix} \hat{u}^* \\ \hat{v}^* \\ \hat{w}^* \end{pmatrix}$$
$$+ \int_0^{2\pi} R\,\mathrm{d}\varphi\big(u\hat{F}^*_{zz} + v\hat{F}^*_{z\varphi} + w\hat{F}^*_{zr} + w_{,\varphi}\hat{M}^*_{z\varphi} + w_{,z}\hat{M}^*_{zz}\big)(+L/2,\varphi)$$
$$- \int_0^{2\pi} R\,\mathrm{d}\varphi\big(u\hat{F}^*_{zz} + v\hat{F}^*_{z\varphi} + w\hat{F}^*_{zr} + w_{,\varphi}\hat{M}^*_{z\varphi} + w_{,z}\hat{M}^*_{zz}\big)(-L/2,\varphi)$$

By subtracting these two equalities, we obtain the Green's formula for the shell operator:

$$\begin{aligned}
0 = &\int_0^{2\pi} R\,\mathrm{d}\varphi \int_{-L/2}^{+L/2} \left[\mathcal{M}^c \begin{pmatrix} u \\ v \\ w \end{pmatrix} \cdot \begin{pmatrix} \hat{u}^* \\ \hat{v}^* \\ \hat{w}^* \end{pmatrix} - \begin{pmatrix} u \\ v \\ w \end{pmatrix} \cdot \mathcal{M}^c \begin{pmatrix} \hat{u}^* \\ \hat{v}^* \\ \hat{w}^* \end{pmatrix} \right] \\
&+ \int_0^{2\pi} R\,\mathrm{d}\varphi \Big[\big(F_{zz}\hat{u}^* + F_{z\varphi}\hat{v}^* + F_{zr}\hat{w}^* + M_{z\varphi}\hat{w}^*{,_\varphi} + M_{zz}\hat{w}^*{,_z} \big)(+L/2, \varphi) \\
&\qquad - \big(u\hat{F}^*_{zz} + v\hat{F}^*_{z\varphi} + w\hat{F}^*_{zr} + w{,_\varphi}\,\hat{M}^*_{z\varphi} + w{,_z}\,\hat{M}^*_{zz} \big)(+L/2, \varphi) \Big] \\
&- \int_0^{2\pi} R\,\mathrm{d}\varphi \Big[\big(F_{zz}\hat{u}^* + F_{z\varphi}\hat{v}^* + F_{zr}\hat{w}^* + M_{z\varphi}\hat{w}^*{,_\varphi} + M_{zz}\hat{w}^*{,_z} \big)(-L/2, \varphi) \\
&- \int_0^{2\pi} R\,\mathrm{d}\varphi \big(u\hat{F}^*_{zz} + v\hat{F}^*_{z\varphi} + w\hat{F}^*_{zr} + w{,_\varphi}\,\hat{M}^*_{z\varphi} + w{,_z}\,\hat{M}^*_{zz} \big)(-L/2, \varphi) \Big]
\end{aligned} \tag{2.98}$$

2. *Boundary Integral Equations for a shell clamped at both ends*

The shell is assumed to be excited by a harmonic force with components (f_z, f_φ, f_ρ). Let us first remark that the boundary condition $w = 0$ implies that $w,_\varphi$ is zero along the boundaries. In the same way, due to the boundary condition $w,_z = 0$, $w,_{z\varphi}$ is zero along the boundaries, which implies that $M_{z\varphi}$ is zero too.

Let us apply the Green's formula to the shell displacement and to the first component $\Gamma_z^{c*}(M, M')$ of the complex conjugate of the Green's tensor as it has been defined in equations (2.77) and (2.77′). The following result is easy to obtain:

$$\begin{aligned}
u(M) = &\int_\Sigma \begin{pmatrix} f_z(M') \\ f_\varphi(M') \\ f_\rho(M') \end{pmatrix} \cdot \begin{pmatrix} \Gamma^c_{zz}(M, M') \\ \Gamma^c_{z\varphi}(M, M') \\ \Gamma^c_{zr}(M, M') \end{pmatrix} \mathrm{d}\Sigma(M') \\
&+ \int_0^{2\pi} \Big[F_{zz}(M'^+)\Gamma^c_{zz}(M, M'^+) + F_{z\varphi}(M'^+)\Gamma^c_{z\varphi}(M, M'^+) \\
&+ F_{zr}(M'^+)\Gamma^c_{zr}(M, M'^+) + M_{zz}\Gamma^c_{zr},_{z'}(M, M'^+) \Big] R\,\mathrm{d}\varphi' \\
&- \int_0^{2\pi} \big[F_{zz}(M'^-)\Gamma^c_{zz}(M, M'^-) + F_{z\varphi}(M'^-)\Gamma^c_{z\varphi}(M, M'^-) \\
&\qquad + F_{zr}(M'^-)\Gamma^c_{zr}(M, M'^-) + M_{zz}\Gamma^c_{zr},_{z'}(M, M'^-) \Big] R\,\mathrm{d}\varphi'
\end{aligned} \tag{2.99}$$

where M'^+ (resp. M'^-) is the point with coordinates $(+L/2, \varphi')$ (resp. $(-L/2,$

φ')). This is the Green's representation of the first component of the shell displacement. The representations of the second and third components are obtained by introducing in the Green's formula $\Gamma_{\varphi}^{c*}(M, M')$ and $\Gamma_{r}^{c*}(M, M')$. This is left to the reader as an exercise. The final result can be put into a compact form:

$$\begin{aligned}
\begin{pmatrix} u(M) \\ v(M) \\ w(M) \end{pmatrix} &= \int_{\Sigma} {}^t\Gamma^c(M, M') \begin{pmatrix} f_z(M') \\ f_\varphi(M') \\ f_\rho(M') \end{pmatrix} \mathrm{d}\Sigma(M') \\
&+ \int_0^{2\pi} \left[{}^t\Gamma^c(M, M'^+) \begin{pmatrix} F_{zz}(M'^+) \\ F_{z\varphi}(M'^+) \\ F_{zr}(M'^+) \end{pmatrix} \right. \\
&\qquad \left. + M_{zz}(M'^+) \begin{pmatrix} \Gamma_{zr,z}(M, M'^+) \\ \Gamma_{\varphi r,z}(M, M'^+) \\ \Gamma_{rr,z}(M, M'^+) \end{pmatrix} \right] R\,\mathrm{d}\varphi' \\
&- \int_0^{2\pi} \left[{}^t\Gamma^c(M, M'^-) \begin{pmatrix} F_{zz}(M'^-) \\ F_{z\varphi}(M'^-) \\ F_{zr}(M'^-) \end{pmatrix} \right. \\
&\qquad \left. + M_{zz}(M'^-) \begin{pmatrix} \Gamma_{zr,z}(M, M'^-) \\ \Gamma_{\varphi r,z}(M, M'^-) \\ \Gamma_{rr,z}(M, M'^-) \end{pmatrix} \right] R\,\mathrm{d}\varphi' \qquad (2.100)
\end{aligned}$$

Here, ${}^t\Gamma^c(M, M')$ is the matrix operator transposed of the matrix operator $\Gamma^c(M, M')$.

The shell displacement depends on eight unknown functions. By writing the values of u, v, w and $w_{,z}$ given by (2.100) for $z = +L/2$ and $z = -L/2$, one obtains a system of eight Boundary Integral Equations which determines these functions uniquely, unless the angular frequency is equal to a resonance frequency (the establishment of this system is left as an exercise). The system of Boundary Integral Equations is easily solved in the present case. Indeed, the unknown functions can be expanded into Fourier series of the angular variable φ'. Then, using the Fourier series of the Green's tensor, we finally obtain a sequence of systems of eight linear algebraic equations, which determines the Fourier components of the unknown functions.

2.5. Vibrations of Spherical Shells

2.5.1. *General Definition and Theorems*

We can first state the following general definition and theorems. Let us consider a shell occupying any domain Σ of a sphere. As for the plate and the cylindrical shell, we have:

DEFINITION 2.4.– *A resonance mode is a vector function which satisfies the homogenous shell equation*

$$\frac{Eh}{1-\nu^2}\mathcal{M}^s\begin{pmatrix}u\\v\\w\end{pmatrix}-\omega^2\mu_s h\begin{pmatrix}u\\v\\w\end{pmatrix}=\begin{pmatrix}0\\0\\0\end{pmatrix} \qquad (2.101)$$

and a given set of boundary conditions along the boundaries (and at the apexes if they are included in Σ). The corresponding value of ω is called resonance angular frequency.

THEOREM 2.7 (EXISTENCE OF RESONANCE MODES).– *A purely elastic spherical shell, submitted to conservative boundary conditions has a countable set of resonance modes and real resonance frequencies. The set of resonance modes is an orthogonal basis on which any solution of the non-homogenous harmonic shell equation (when it exists) can be expanded.*

If there is any energy loss (by material damping or through the boundaries), the resonance frequencies are complex.

The following theorem gives the conditions of existence and uniqueness of the response of a spherical shell to a time harmonic excitation.

THEOREM 2.8 (EXISTENCE AND UNIQUENESS OF THE SOLUTION).– *Let $\vec{f}(M)$ be the harmonic (vector) excitation of the shell and ω be the corresponding angular frequency.*

- ◇ *If ω is different from all the resonance angular frequencies, then the solution $\vec{U}$, with components (u, v, w), of the non-homogenous shell equation exists and is unique for any excitation $\vec{f}(M)$.*

- ◇ *If ω is equal to any resonance angular frequency, say ω_n, the solution of the non-homogenous shell equation does not exist unless $\vec{f}(M)$ is orthogonal the corresponding resonance mode $\vec{U}_n$.*

- ◇ *If $\omega = \omega_m$ and $\vec{f}(M)$ is orthogonal to $\vec{U}_n$, a non-unique solution of the non-homogenous shell equation exists, which can be decomposed into a uniquely determined component $\vec{U}^{(1)}$ orthogonal to $\vec{U}_n$, and a component $\vec{U}^{(2)} = \alpha\vec{U}_n$ with arbitrary amplitude α.*

2.5.2. *Solution of the Time Harmonic Spherical Shell Equation*

The spherical shell equation is much more difficult to solve. It is possible to separate the variables by looking for a solution as a series of terms of the form $\begin{pmatrix} u_n(\theta) \\ v_n(\theta) \\ w_n(\theta) \end{pmatrix} e^{\imath n\varphi}$. But the equations obtained for the components $\big(u_n(\theta), v_n(\theta), w_n(\theta)\big)$ have non-constant coefficients. Thus, it appears that, in general, only numerical methods can be used.

In this section we look at the simplest case but the method developed can be applied to more general cases with the appropriate changes.

1. *The variational equation*

Let us consider a spherical shell with radius R, occupying the domain $\Sigma \equiv (0 < \varphi \leq 2\pi, \theta_1 < \theta < \theta_2)$ and clamped at both ends. Assume that it is excited by a harmonic ($e^{-\imath\omega t}$) force with components $(f_\theta, f_\varphi, f_\rho)$. Let $(\hat{u}, \hat{v}, \hat{w})$ be any vector which satisfies the same boundary conditions as the shell displacement and has the same regularity properties, that is: $\hat{u}$ and $\hat{v}$ are square integrable together with their first order derivatives with respect to θ and φ; $\hat{w}$ is square integrable together with its first and second derivatives with respect to θ and φ. The variational form of the spherical shell equations is (see (1.34′)):

$$\mathcal{A}^s(u, v, w; \hat{u}, \hat{v}, \hat{w}) - \omega^2 \int_\Sigma (u\hat{u}^* + v\hat{v}^* + w\hat{w}^*)\, d\Sigma = \int_\Sigma (f_\theta\hat{u}^* + f_\varphi\hat{v}^* + f_\rho\hat{w}^*)\, d\Sigma \quad (2.102)$$

Because of the geometry considered here, the shell displacement is 2π-periodic in φ and can be expanded into a Fourier series

$$\begin{pmatrix} u \\ v \\ w \end{pmatrix} = \sum_{n=-\infty}^{+\infty} \begin{pmatrix} u_n \\ v_n \\ w_n \end{pmatrix} e^{\imath n\varphi}$$

Let $(\hat{u}_n^*, \hat{v}_n^*, \hat{w}_n^*)$ be the Fourier components of $(\hat{u}^*, \hat{v}^*, \hat{w}^*)$, and $(f_{\theta n}, f_{\varphi n}, f_{rn})$ those of $(f_\theta, f_\varphi, f_\rho)$. Equation (2.102) is replaced by the sequence of equations:

$$\mathcal{A}^{sn}(u_n, v_n, w_n; \hat{u}_n, \hat{v}_n, \hat{w}_n) - \omega^2 \int_\Sigma (u_n\hat{u}_n^* + v_n\hat{v}_n^* + w_n\hat{w}_n^*)\, d\Sigma =$$
$$\int_\Sigma (f_{\theta n}\hat{u}_n^* + f_{\varphi n}\hat{v}_n^* + f_{rn}\hat{w}_n^*)\, d\Sigma \qquad n = -\infty, \ldots, -1, 0, 1, \ldots, +\infty \quad (2.103)$$

where the bilinear form $\mathcal{A}^{sn}(u_n, v_n, w_n; \hat{u}_n, \hat{v}_n, \hat{w}_n)$ is defined by:

$$\mathcal{A}^{sn}(u_n, v_n, w_n; \hat{u}_n, \hat{v}_n, \hat{w}_n) = \frac{Eh}{(1-\nu^2)} \int_{\theta_1}^{\theta_2} \sin\theta \, d\theta$$

$$\Big\{ \nu \big(u_n \operatorname{cotg}\theta + u_{n,\theta} + \frac{\imath n v_n}{\sin\theta} + 2w_n\big)\big(\hat{u}_n^* \operatorname{cotg}\theta + \hat{u}^*_{n,\theta} - \frac{\imath n \hat{v}_n^*}{\sin\theta} + 2\hat{w}_n^*\big)$$

$$+(1-\nu)\Big[\big(u_n \operatorname{cotg}\theta + \frac{\imath n v_n}{\sin\theta} + w_n\big)\big(\hat{u}_n^* \operatorname{cotg}\theta - \frac{\imath n \hat{v}_n^*}{\sin\theta} + \hat{w}_n^*\big)$$

$$+\big(u_{n,\theta} + w_n\big)\big(\hat{u}^*_{n,\theta} + \hat{w}_n^*\big)$$

$$+\frac{1}{2}\big(\frac{\imath n u_n}{\sin\theta} - v_n \operatorname{cotg}\theta + v_{n,\theta}\big)\big(-\frac{\imath n \hat{u}_n^*}{\sin\theta} - \hat{v}_n^* \operatorname{cotg}\theta + \hat{v}^*_{n,\theta}\big)\Big]$$

$$+\frac{h^2}{12R^2}\Big[\nu\big(w_{n,\theta} \operatorname{cotg}\theta - \frac{n^2 w_n}{\sin^2\theta} + w_{n,\theta\theta}\big)\big(\hat{w}^*_{n,\theta} \operatorname{cotg}\theta - \frac{n^2 \hat{w}_n^*}{\sin^2\theta} + \hat{w}^*_{n,\theta\theta}\big)$$

$$+(1-\nu)\Big(\big(w_{n,\theta} \operatorname{cotg}\theta - \frac{n^2 w_n}{\sin^2\theta}\big)\big(\hat{w}^*_{n,\theta} \operatorname{cotg}\theta - \frac{n^2 \hat{w}_n^*}{\sin^2\theta}\big) + w_{n,\theta\theta}\, \hat{w}^*_{n,\theta\theta}\Big)$$

$$+2\frac{(1-\nu)}{\sin^2\theta}\big(\imath n w_n \operatorname{cotg}\theta - \imath n w_{n,\theta}\big)\big(-\imath n \hat{w}_n^* \operatorname{cotg}\theta + \imath n \hat{w}^*_{n,\theta}\big)\Big]\Big\} \qquad (2.103')$$

Equation (2.103) must be satisfied for any vector $(\hat{u}_n, \hat{v}_n, \hat{w}_n)$ which satisfies the boundary conditions and the regularity properties given previously.

2. *General method for solving equation (2.103)*

Let $U_p(\theta)$ be a basis of the Hilbert space of functions, defined on $(\theta_1 < \theta < \theta_2)$, which are square integrable together with their first order derivatives on this domain and which are zero for $\theta = \theta_1$ and $\theta = \theta_2$. Let $W_p(\theta)$ be a basis of the Hilbert space of functions, defined on $(\theta_1 < \theta < \theta_2)$, which are square integrable together with their first and second order derivatives on this domain and which satisfy $W_p = W_{p,\theta} = 0$ for $\theta = \theta_1$ and $\theta = \theta_2$. The components of the vector $\begin{pmatrix} u_n \\ v_n \\ w_n \end{pmatrix}$ can be expanded into series of these functions:

$$u_n = \sum_p \alpha_n^p U_p \quad , \quad v_n = \sum_p \beta_n^p U_p \quad , \quad w_n = \sum_p \gamma_n^p W_p \qquad (2.104)$$

By introducing these expansions into equation (2.103) and replacing $(\hat{u}_n, \hat{v}_n, \hat{w}_n)$ successively by the three sets of vectors $(U_q, 0, 0)$, $(0, U_q, 0)$ and $(0, 0, W_q)$, we obtain an infinite system of linear algebraic equations which determines the coefficients $(\alpha_n^q, \beta_n^q, \gamma_n^q)$ unless ω is a resonance frequency of the shell (see theorems 2.7 and 2.8). The resonance angular frequencies are the values of ω for which the homogenous system has a non-zero solution and the corresponding solutions of the homogenous system are the resonance modes. To get a numerical approximation of the solution (forced regime or resonance modes and frequencies), the system is truncated.

3. *Example of bases*

It is useful to show now an example of basis functions U_p and W_p. Let $P_s(x)$, $(s = 0, 1, \ldots)$ be a complete set of orthogonal polynomials defined on the interval $-1 < x < +1$ (for example, the Legendre polynomials).

First, consider the sequence of functions

$$U_p(\theta) = a_p P_s\Big(\frac{2\theta - (\theta_1 + \theta_2)}{\theta_2 - \theta_1}\Big) + b_p P_{s+2}\Big(\frac{2\theta - (\theta_1 + \theta_2)}{\theta_2 - \theta_1}\Big) \quad s = 0, 1, \ldots$$

where the coefficients a_p and b_p are chosen so that the function $U_p(\theta)$ is zero at θ_1 and θ_2 and has a $\mathbf{L}^2$-norm equal to one. It is easy to show that these coefficients are uniquely determined. These functions are linearly independent and constitute a basis on which the u_n and v_n can be expanded.

Then, consider the sequence of functions

$$W_p(\theta) = a'_p P_s\Big(\frac{2\theta - (\theta_1 + \theta_2)}{\theta_2 - \theta_1}\Big)$$

$$+ b'_p P_{s+2}\Big(\frac{2\theta - (\theta_1 + \theta_2)}{\theta_2 - \theta_1}\Big) + c'_p P_{s+4}\Big(\frac{2\theta - (\theta_1 + \theta_2)}{\theta_2 - \theta_1}\Big) \quad , \quad s = 0, 1, \ldots$$

where the coefficients a'_p, b'_p and c'_p are chosen so that $W_p(\theta)$ and its first derivative cancel at θ_1 and θ_2; they are uniquely determined by adding that the $\mathbf{L}^2$-norm of $W_p(\theta)$ is unity. This set of functions is a basis on which the w_n can be expanded. For any other boundary conditions, very similar bases can be built up.

In the next chapter, an example of such bases will be defined explicitly: it concerns the more complicated case of a *Line 2'* shell which is composed of a cylindrical shell closed by two hemi-spherical end caps.

2.6. Exercises

1.Even resonance modes of the clamped beam: establish (2.13) and (2.13″).

2.Odd resonance modes of the clamped beam: establish (2.15) and (2.15″).

3.Find the resonance wavenumber equation for a beam of unit length, clamped at $X = -1/2$ and simply supported at $X = +1/2$: instead of using the basis functions $\cos \Lambda X$, etc., it is much more convenient to use the translated functions $\cos \Lambda(X - 1/2)$, etc. Establish the expression of the resonance modes and calculate the normalization factors.

4.Show that the resonance wavenumber equation for a beam of unit length, clamped at $X = -1/2$ and free at $X = +1/2$ is $\cos \Lambda \cosh \Lambda = -1$: instead of using the basis functions $\cos \Lambda X$, etc., it is much more convenient to use the translated functions $\cos \Lambda(X + 1/2)$, etc. By drawing the curves $z_1 = \cos \Lambda \cosh \Lambda$ and $z_2 = -1$, find a first order approximation of the roots. Using the same procedure as in section 2.2.3 (paragraph 4), find an approximation of the solutions. Establish the expression of the resonance modes and calculate the normalization factors.

5.Let us consider the three beams represented in the figure below, which have the same cross-section area. Compare the flexural rigidity of the H-shape beam

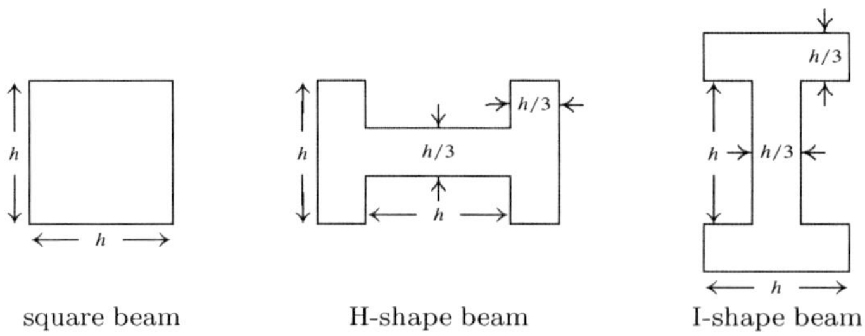

and of the I-shape beam to that of the square one. Compare their respective resonance frequencies.

6.Resonance frequencies of a plate submitted to natural boundary conditions: using the variational form (1.33) of the plate equation, prove that the resonance frequencies are real ($\lambda^4 > 0$).

7.Establish the resonance wavenumber equations of circular plates (2.32′), (2.34) and (2.35′).

8.Write the equation governing the nodal radii of a circular plate with a clamped, simply supported and completely free boundary.

9.Numerical programming .– Compute the resonance wavenumbers and frequencies ($n = 0$ to 5, $s = 0$ to 5) for a circular steel plate having the following characteristics:

$$E = 2 \times 10^{11}\text{Pa};\ \mu = 7\ 800\text{kg/m}^3;\ \nu = 0.33;\ h = 10^{-3}\text{m};\ R = 3 \times 10^{-1}\text{m}$$

Compute the numerical values of corresponding nodal radii. Compare the results obtained for the 3 fundamental boundary conditions.

10. Resonance modes of rectangular plates: establish formulae (2.43) and (2.46).

11. Forced regime: prove that the existence and uniqueness conditions of the solution given by equations (2.49) and (2.52) are identical.

12. Free oscillations of an infinite shell: prove that the determinant $D(n, \Omega, \Lambda)$ of the matrix given in equation (2.75) is even in n and in Λ.

13. Establish formulae (2.76) and (2.76′).

14. Prove the symmetry properties of the cylindrical shell Green's tensor given in equation (2.82).

15. Resonance frequencies of a cylindrical shell closed by shear diaphragms: by introducing the solution (2.85-2.85′) into the variational form of the shell equation (1.34-1.34′), prove that the resonance angular frequencies are real ($\Omega^2 > 0$).

16. Resonance frequencies of a cylindrical shell submitted to any kind of natural boundary conditions: use the variational form of the shell equation (1.34-1.34′) to prove that the resonance angular frequencies are real ($\Omega^2 > 0$).

17. Establish result (2.97).

18. Similarly to equation (2.99), establish the Green's representations for the components $v(M)$ and $w(M)$ of the displacement of a cylindrical shell clamped at both ends.

19. Establish the system of Boundary Integral Equations for a cylindrical shell clamped at both ends. Write the sequence of systems of algebraic equations which determine the Fourier components of the unknown functions.

Chapter 3

Acoustic Radiation and Transmission by Thin Structures

3.1. Introduction

This chapter is devoted to an important problem of acoustical engineering: sound generation by vibrating structures and noise transmission through elastic structures. Though the study is restricted to thin structures, the different aspects of the physical phenomenon appear.

Fluid loading is strongly related to the ratio of fluid density to the structure surface density (the volume density, in the case of three-dimensional solids). When the fluid is air, the fluid loading, which is experimentally known to have a small effect on the vibrations of the structure, appears as a perturbation term in the equations, which, as a consequence, can often be solved by classical perturbation methods. Conversely, for a liquid, the fluid loading is strong and, if accurate results are required, the exact equations have to be solved directly by specific methods. Nevertheless, even in this case, the classical perturbation methods are a good approach to the phenomenon and provide rough approximations, which correctly describe the qualitative behavior of the physical systems.

In a first section, a very simple example is examined in detail. We consider two semi-infinite waveguides, separated by a spring suspended mass and filled either by the same fluid or by two different fluids. Three problems are studied: a/ the existence of free oscillations of the system; b/ the generation of sound when the mass is excited by an external force; c/ the transmission of acoustic energy across the mass when an incident acoustic wave is emitted in one of the two waveguides. As a first result, for identical fluids, at the resonance of the mass-spring system, the mass has no effect on the acoustic wave, which is transmitted across it as if there were no obstacle. This phenomenon occurs when the system is an infinite plate separating two semi-infinite spaces filled with the same fluid. Another interest of this example is that it enables us to define rigorously the so-called *insertion loss* which characterizes the acoustic attenuation due to the presence of the mass. This notion plays a basic role in acoustical engineering in qualifying the insulation properties of wall and floor materials.

The second example is a little more complex. The system which is looked at is a one-dimensional cavity connected to a semi-infinite waveguide by a spring-suspended mass. Attention is paid to the transmission of sound from the exterior of the cavity to the interior, or from the cavity to the semi-infinite waveguide. This is, of course, an elementary example of a room coupled to an external infinite domain. The analysis shows that the resonance frequency of the mass-spring system plays a fundamental role, but the phenomenon is also governed by the resonance frequencies of the cavity itself. These fundamental aspects exist in a realistic three-dimensional system composed of a room limited by vibrating boundaries (walls, ceilings and floors, windows, etc.).

The chapter continues with a short section on theoretical acoustics in which fundamental elementary basic concepts and formulae are recalled.

The next section is devoted to the vibro-acoustic response of infinite structures (plates and cylindrical shells) immersed in fluids. The existence of possible free oscillations and the response to a harmonic excitation (incident acoustic wave in the fluid or external force applied to the structure) are developed. It is shown that the basic aspects pointed out in the one-dimensional examples appear in these more complex situations.

More realistic situations are then examined. The first example is a baffled plate excited by a harmonic, random or transient source. The second structure is a plate closing a cavity: the system is excited by an incident harmonic acoustic wave and we describe the transmission of acoustic energy through the plate.

Structures involving cylindrical and spherical shells are then considered. First, the example of a cylindrical shell excited by an internal turbulent flow is developed. Then follows the example of a finite length cylindrical cavity excited by an internal harmonic acoustic wave and radiating into a heavy fluid extending up to infinity. The last structure is a finite length cylindrical shell, closed by two hemi-spherical shells and immersed in an infinite heavy fluid: the system is excited by an incident transient plane wave and the time-dependent diffracted field is calculated.

For all these examples, analytical (whenever possible) and numerical methods are proposed and numerical results are given. Most examples are illustrated by a comparison between numerical predictions and experimental results.

3.2. Sound Transmission across a Piston in a One-Dimensional Waveguide

We start this chapter with a simple one-dimensional example, which points out the basic phenomena that occur when a vibrating structure is coupled to one or two fluids. We consider two semi-infinite waveguides, with identical cross-sections, occupying respectively the domains $x < 0$ and $x > 0$, and containing fluids characterized by densities μ_1 and μ_2, and sound velocities c_1 and c_2. They are separated by a piston, with mass (per unit area) μ_s, elastically supported by a spring with stiffness (per unit area) r.

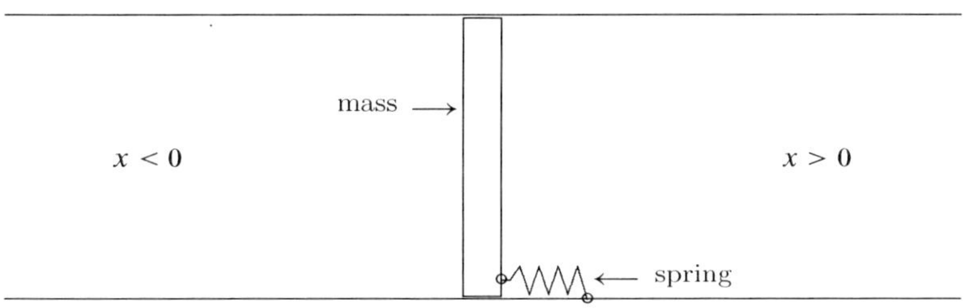

Figure 3.1. *Coupling of two semi-infinite waveguides by an elastically supported mass*

3.2.1. *Governing Equations*

The mass displacement $\tilde{u}(t)$ satisfies the following equation

$$\mu_s \frac{\mathrm{d}^2\tilde{u}(t)}{\mathrm{d}t^2} + r\tilde{u}(t) = \tilde{F}(t) \tag{3.1}$$

where $\tilde{F}(t)$ is the pressure acting on it. In general, this pressure is the sum of an external pressure $\tilde{F}_e(t)$ (external force per unit area) and the difference

$\tilde{P}(t)$ between the acoustic pressure exerted by the fluid contained in the domain $x < 0$ and the acoustic pressure exerted by the fluid contained in the domain $x > 0$.

Let $\tilde{p}_1^i(x,t)$ be an incident wave propagating in the waveguide $x < 0$, in the direction $x > 0$. The total pressure in this waveguide is the sum of $\tilde{p}_1^i(x,t)$ and of a term $\tilde{p}_1^r(x,t)$, which corresponds to both reflection and radiation by the piston. Similarly, in the waveguide $x > 0$, we define an incident pressure $\tilde{p}_2^i(x,t)$ and a reflected/radiated pressure $\tilde{p}_2^r(x,t)$. The incident acoustic pressures $\tilde{p}_1^i(x,t)$ and $\tilde{p}_2^i(x,t)$, and the reflected/radiated pressures $\tilde{p}_1^r(x,t)$ and $\tilde{p}_2^r(x,t)$ satisfy the following homogenous wave equations:

$$\begin{aligned} &\frac{\partial^2 \tilde{p}_1^i(x,t)}{\partial x^2} - \frac{1}{c_1^2}\frac{\partial^2 \tilde{p}_1^i(x,t)}{\partial t^2} = 0 \quad , \quad \frac{\partial^2 \tilde{p}_1^r(x,t)}{\partial x^2} - \frac{1}{c_1^2}\frac{\partial^2 \tilde{p}_1^r(x,t)}{\partial t^2} = 0 \\ &\frac{\partial^2 \tilde{p}_2^i(x,t)}{\partial x^2} - \frac{1}{c_2^2}\frac{\partial^2 \tilde{p}_2^i(x,t)}{\partial t^2} = 0 \quad , \quad \frac{\partial^2 \tilde{p}_2^r(x,t)}{\partial x^2} - \frac{1}{c_2^2}\frac{\partial^2 \tilde{p}_2^r(x,t)}{\partial t^2} = 0 \end{aligned} \tag{3.2}$$

Thus, we have

$$\tilde{P}(t) = \tilde{p}_1^i(0,t) - \tilde{p}_2^i(0,t) + \tilde{p}_1^r(0,t) - \tilde{p}_2^r(0,t) \tag{3.3}$$

We must express that the fluid particles remain in contact with the piston. This is achieved by writing that the particle accelerations $\tilde{\gamma}^1(x,t)$ and $\tilde{\gamma}^2(x,t)$ of both fluids are equal, at $x = 0$, to the mass acceleration $\partial^2\tilde{u}(t)/\partial t^2$. Using the momentum equation which relates the pressure gradient $\partial_x \tilde{p}(x,t)$ in a fluid to the particle acceleration $\tilde{\gamma}(x,t)$

$$\mu\tilde{\gamma}(x,t) + \partial_x \tilde{p}(x,t) = 0$$

we obtain

$$\frac{\partial^2 \tilde{u}(t)}{\partial t^2} = -\frac{1}{\mu_1}\frac{\partial \tilde{p}_1^i(0,t)}{\partial x} - \frac{1}{\mu_1}\frac{\partial \tilde{p}_1^r(0,t)}{\partial x} = -\frac{1}{\mu_2}\frac{\partial \tilde{p}_2^i(0,t)}{\partial x} - \frac{1}{\mu_2}\frac{\partial \tilde{p}_2^r(0,t)}{\partial x} \tag{3.4}$$

The energy conservation principle will be satisfied by requiring that the acoustic waves reflected and radiated by the piston are purely outgoing waves. The statement of the problem is completed by initial conditions, the simplest ones being that the system is at rest for $t < 0$. An easy method to solve equations (3.1) to (3.4) is to first solve their time Fourier transforms, that is, to look at the response of the system to a harmonic excitation.

The physical phenomenon is more easily understood by considering separately the two simpler cases:

1. the radiation of sound by the piston, which occurs for $\tilde{F}_e(t) \neq 0$ and $\tilde{p}_1^i(x,t) = \tilde{p}_2^i(x,t) = 0$;

2. the transmission of sound through the piston, which corresponds to $\tilde{p}_1^i(t) \neq 0$ and $\tilde{F}_e(t) = \tilde{p}_2^i(x,t) = 0$.

3.2.2. *Time Fourier Transform of the Equations – Response of the System to a Harmonic Excitation*

We introduce the Fourier transform $f(\omega)$ of a function $\tilde{f}(t)$ and its inverse defined by:

$$f(\omega) = \int_{-\infty}^{+\infty} \tilde{f}(t)\mathrm{e}^{\imath\omega t}\,\mathrm{d}t \quad , \quad \tilde{f}(t) = \frac{1}{2\pi}\int_{-\infty}^{+\infty} f(\omega)\mathrm{e}^{-\imath\omega t}\,\mathrm{d}\omega$$

From equations (3.1) to (3.4), the functions $u(\omega)$, $p_1^r(x,\omega)$ and $p_2^r(x,\omega)$ satisfy the following system of equations:

$$\begin{aligned}
\left(\omega_0^2 - \omega^2\right)u(\omega) &= \frac{1}{\mu_s}\Big[F_e(\omega) + p_1^i(\omega,0) + p_1^r(\omega,0) \\
&\qquad\qquad -p_2^i(\omega,0) - p_2^r(\omega,0)\Big] \\
&\frac{\mathrm{d}^2 p_1^r(x,\omega)}{\mathrm{d}x^2} + k_1^2 p_1^r(x,\omega) = 0 \\
&\frac{\mathrm{d}^2 p_2^r(x,\omega)}{\mathrm{d}x^2} + k_2^2 p_2^r(x,\omega) = 0 \\
\omega^2 u(\omega) &= \frac{\partial_x p_1^i(\omega,0) + \partial_x p_1^r(\omega,0)}{\mu_1} = \frac{\partial_x p_2^i(\omega,0) + \partial_x p_2^i(\omega,0)}{\mu_2} \\
&\text{with } \omega_0^2 = r/\mu_s, \quad k_1^2 = \frac{\omega^2}{c_1^2}, \quad k_2^2 = \frac{\omega^2}{c_2^2}
\end{aligned} \tag{3.5}$$

From the second and third equations of (3.5), we can conclude that the pressures $p_1^r(x,\omega)$ and $p_2^r(x,\omega)$, which must be outgoing waves, are of the form:

$$p_1^r(x,\omega) = A\mathrm{e}^{-\imath k_1 x} \quad , \quad p_2^r(x,\omega) = B\mathrm{e}^{\imath k_2 x} \tag{3.6}$$

where the amplitudes A and B, which are independent of x, must be determined.

1. *Free oscillations of the system*

Let us first assume that the system does not receive any external energy, that is, $F_e(\omega) = p_1^i(x,\omega) = p_2^i(x,\omega) = 0$. Do particular values of the angular frequency ω exist for which oscillations can take place? By introducing (3.6) into the last two equations of (3.5), A and B are obtained as functions of $u(\omega)$. Then, the first equation of (3.5) provides an equation for $u(\omega)$ only:

$$\left[\omega_0^2 - \imath\omega\frac{\mu_1 c_1 + \mu_2 c_2}{\mu_s} - \omega^2\right]u(\omega) = 0 \tag{3.7}$$

It has a non-zero solution if the angular frequency is one of the roots of the polynomial $[\omega_0^2 - \imath\omega(\rho_1 c_1 + \rho_2 c_2)/\mu_s - \omega^2]$ which appears in equation (3.7), that is, if ω takes one of the following two values:

$$\begin{aligned}\Omega^+ &= -\imath\frac{\mu_1 c_1 + \mu_2 c_2}{2\mu_s} + \sqrt{\omega_0^2 - \left(\frac{\mu_1 c_1 + \mu_2 c_2}{2\mu_s}\right)^2} \\ \Omega^- &= -\imath\frac{\mu_1 c_1 + \mu_2 c_2}{2\mu_s} - \sqrt{\omega_0^2 - \left(\frac{\mu_1 c_1 + \mu_2 c_2}{2\mu_s}\right)^2}\end{aligned} \tag{3.8}$$

These values are called the *resonance angular frequencies of the fluid loaded piston*. If the quantity $\delta = \omega_0^2 - ((\rho_1 c_1 + \rho_2 c_2)/2\mu)^2$ is positive (which is the more realistic situation), the resonance angular frequencies Ω^+ and Ω^- have the same negative imaginary part and opposite real parts; for $\delta \leq 0$, Ω^+ and Ω^- have negative imaginary values. Whatever the sign of δ is, the free oscillations of the system have an amplitude which decreases exponentially with time: the time damping of the free oscillations is a consequence of the acoustic energy loss at infinity in the two semi-infinite waveguides.

2. *Acoustic radiation of the piston*

In this section, we assume that the incident pressures are zero and that the external pressure exerted on the piston is non-zero. Equations (3.5) lead to the following result:

$$\begin{aligned}u(\omega) &= \frac{F_e(\omega)}{\mu_s\big(\omega_0^2 - \imath\omega(\mu_1 c_1 + \mu_2 c_2)/\mu_s - \omega^2\big)} \\ p_1^r(x,\omega) &= \imath\omega\mu_1 c_1\frac{F_e(\omega)e^{-\imath k_1 x}}{\mu_s\big(\omega_0^2 - \imath\omega(\mu_1 c_1 + \mu_2 c_2)/\mu_s - \omega^2\big)} \\ p_2^r(x,\omega) &= -\imath\omega\mu_2 c_2\frac{F_e(\omega)e^{\imath k_2 x}}{\mu_s\big(\omega_0^2 - \imath\omega(\mu_1 c_1 + \mu_2 c_2)/\mu_s - \omega^2\big)}\end{aligned} \tag{3.9}$$

The amplitudes of these three quantities have a maximum for $\omega = \omega_0$, the resonance frequency of the *in vacuo* mass-spring system.

When the fluids are gases, their influence on the piston motion is "small" and results (3.9) can be replaced by approximations. Let us assume that the two quantities $\varepsilon_1 = \mu_1 c_1/\mu_s$ and $\varepsilon_2 = \mu_2 c_2/\mu_s$ are both small compared to 1. It is tempting to approximate the piston displacement by

$$u(\omega) \simeq \frac{F_e(\omega)}{\mu_s\left(\omega_0^2 - \omega^2\right)}\left[1 + \frac{\imath\omega}{\omega_0^2 - \omega^2}\left(\varepsilon_1 + \varepsilon_2\right)\right] \tag{3.10}$$

This is the classical *light-fluid approximation* . Expression (3.10) is the Taylor expansion, limited to order 1, of $u(\omega)$ in terms of the parameter $\varepsilon = \omega\left(\varepsilon_1 + \varepsilon_2\right)/\left(\omega_0^2 - \omega^2\right)$: this series is convergent for $\varepsilon \ll 1$. Under this condition – which is never fulfilled for $\omega = \omega_0$ – approximation (3.10) is valid. Similar approximations can be obtained for the radiated pressures.

Let us evaluate the mean power (per unit area) $\mathcal{P}_0$ over one period $T = \omega/2\pi$ admitted by the piston and the mean power flows (per unit area) $\mathcal{P}_1$ and $\mathcal{P}_2$ radiated in the fluids. They are defined by:

$$\mathcal{P}_0 = \frac{1}{T}\int_0^T \Re\left(F_e(\omega)\mathrm{e}^{-\imath\omega t}\right)\Re\left(-\imath\omega u(\omega)\mathrm{e}^{-\imath\omega t}\right)$$

$$\mathcal{P}_1 = \frac{1}{T}\int_0^T \Re\left(p_1^r(x,\omega)\mathrm{e}^{-\imath\omega t}\right)\Re\left(-\frac{1}{\imath\omega\mu_1}\partial_x p_1^r(x,\omega)\mathrm{e}^{-\imath\omega t}\right)$$

$$\mathcal{P}_2 = \frac{1}{T}\int_0^T \Re\left(p_2^r(x,\omega)\mathrm{e}^{-\imath\omega t}\right)\Re\left(\frac{1}{\imath\omega\mu_2}\partial_x p_2^r(x,\omega)\mathrm{e}^{-\imath\omega t}\right)$$

The following result is easy to obtain:

$$\begin{aligned}
\mathcal{P}_0 &= \frac{|F_e(\omega)|^2}{2}\frac{\mu_1 c_1 + \mu_2 c_2}{\mu_s^2\left[\omega^2\left(1 - \omega_0^2/\omega^2\right)^2 + \left(\mu_1 c_1 + \mu_2 c_2\right)^2/\mu_s^2\right]}\\
\mathcal{P}_1 &= \frac{|F_e(\omega)|^2}{2}\frac{\mu_1 c_1}{\mu_s^2\left[\omega^2\left(1 - \omega_0^2/\omega^2\right)^2 + \left(\mu_1 c_1 + \mu_2 c_2\right)^2/\mu_s^2\right]}\\
\mathcal{P}_2 &= \frac{|F_e(\omega)|^2}{2}\frac{\mu_2 c_2}{\mu_s^2\left[\omega^2\left(1 - \omega_0^2/\omega^2\right)^2 + \left(\mu_1 c_1 + \mu_2 c_2\right)^2/\mu_s^2\right]}
\end{aligned} \tag{3.11}$$

Each power has a maximum for $\omega = \omega_0$.

3. *Sound transmission across the piston*

In this case, we now assume that $F_e(\omega) = p_2^i(x,\omega) = 0$ and $p_1^i(x,\omega) = \varpi\mathrm{e}^{\imath k_1 x}$. The reflected pressure $p_1^r(x,\omega)$ and the transmitted pressure $p_2^r(x,\omega)$ have the following form:

$$p_1^r(x,\omega) = R(\omega)\varpi\mathrm{e}^{-\imath k_1 x} \quad , \quad p_2^r(x,\omega) = T(\omega)\varpi\mathrm{e}^{\imath k_2 x}$$

where $R(\omega)$ is called the reflection coefficient and $T(\omega)$ the transmission coefficient. These two coefficients, together with the piston displacement u, satisfy the following system of equations:

$$\begin{pmatrix} -1/\mu_1 c_1 & 0 & \imath\omega \\ 0 & 1/\mu_2 c_2 & \imath\omega \\ -1/\mu_s & 1/\mu_s & \omega_0^2 - \omega^2 \end{pmatrix} \begin{pmatrix} R(\omega) \\ T(\omega) \\ u(\omega)/\varpi \end{pmatrix} = \begin{pmatrix} -1/\mu_1 c_1 \\ 0 \\ 1/\mu_s \end{pmatrix} \tag{3.12}$$

The solution is:

$$\begin{aligned} u(\omega) &= -\frac{2\varpi}{\mu_s\left[\omega^2 - \omega_0^2 + \imath\omega\left(\mu_1 c_1 + \mu_2 c_2\right)/\mu_s\right]} \\ R(\omega) &= \frac{\omega^2 - \omega_0^2 + \imath\omega\left(\mu_2 c_2 - \mu_1 c_1\right)/\mu_s}{\omega^2 - \omega_0^2 + \imath\omega\left(\mu_1 c_1 + \mu_2 c_2\right)/\mu_s} \\ T(\omega) &= \frac{2\imath\omega\mu_2 c_2/\mu_s}{\omega^2 - \omega_0^2 + \imath\omega\left(\mu_1 c_1 + \mu_2 c_2\right)/\mu_s} \end{aligned} \tag{3.13}$$

The particular case in which the two fluids are identical and characterized by a density μ_0 and a sound velocity c_0, is quite interesting. It is, indeed, the simplest model of sound transmission through walls, doors, windows, etc., and which represents one of the basic problems in environmental acoustics. The reflection and transmission coefficients are given by:

$$R(\omega) = \frac{\omega^2 - \omega_0^2}{\omega^2 - \omega_0^2 + 2\imath\omega\mu_0 c_0/\mu_s} \quad , \quad T(\omega) = \frac{2\imath\omega\mu_0 c_0/\mu_s}{\omega^2 - \omega_0^2 + 2\imath\omega\mu_0 c_0/\mu_s} \tag{3.13$'$}$$

Here again, the *in vacuo* resonance frequency ω_0 of the mass–spring system plays a particular role. For $\omega = \omega_0$, the piston displacement has a maximum amplitude, the reflection coefficient $R(\omega_0)$ is zero and the transmission coefficient $T(\omega_0)$ is equal to one.

At this stage, we can introduce two basic notions of architectural acoustics which, in this simple case, can be defined rigorously.

DEFINITION 3.1.– *The ratio $\tau(\omega)$ of the transmitted mean power flow to the incident mean power flow is called the energy transmission rate of the elastically supported piston.*

DEFINITION 3.2.– *The difference $\mathcal{I}(\omega)$ between the incident mean power level and the transmitted mean power level is called the insertion loss index of the elastically supported the piston.*

The transmission rate is equal to $|T(\omega)|^2$. According to equation (3.13′) we have:

$$\tau(\omega) = \frac{\left(2\omega\mu_0 c_0/\mu_s\right)^2}{\left(\omega^2 - \omega_0^2\right)^2 + \left(2\omega\mu_0 c_0/\mu_s\right)^2}$$

$$= \frac{\left(2\mu_0 c_0/\mu_s\omega_0\right)^2}{\left(\omega_0/\omega\right)^2\left(1 - \omega^2/\omega_0^2\right)^2 + \left(2\mu_0 c_0/\mu_s\omega_0\right)^2} \qquad (3.14)$$

$$\mathcal{I}(\omega) = 10\log\bigl(1/\tau(\omega)\bigr)$$

Figures 3.2 and 3.3 show two examples of insertion loss index as functions of the frequency parameter ω/ω_0.

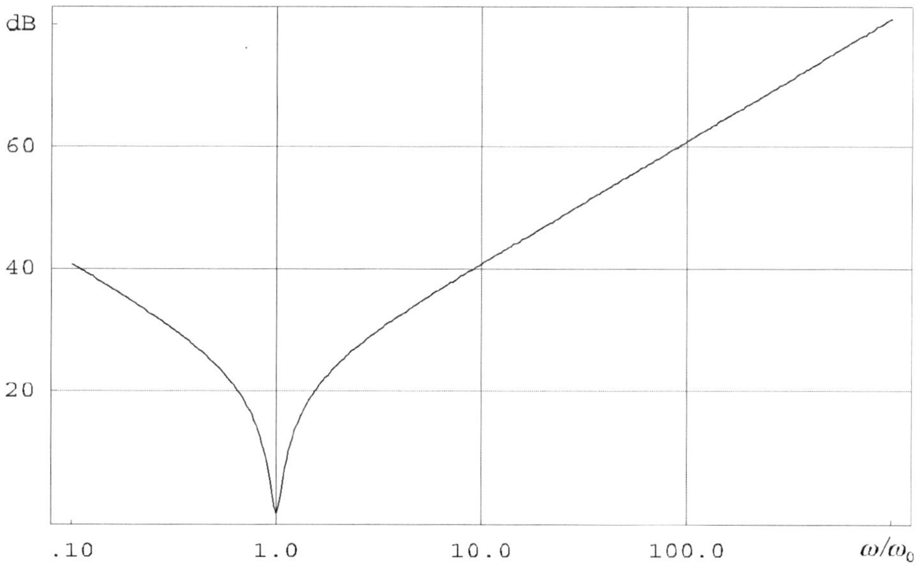

Figure 3.2. *Example of insertion loss index for air:* $\mu_0 c_0/\mu_s\omega_0 = 0.045\ 09\ \text{s}^{-1}$

The mass-spring system is as follows. The piston is a steel plate 0.002 m thick: so, we have $\mu_s = 7800 \times 2 \cdot 10^{-3}$ kg m^{-2}. Instead of choosing the spring stiffness, it is simpler to choose the resonance frequency of the *in vacuo* mass-spring system: it is taken as equal to $\omega_0 = 2\pi \times 100\ \text{s}^{-1}$.

For air, the data are $\mu_0 = 1.3$ kg m^{-3} and $c_0 = 340$ ms^{-2}. Thus, the parameter $\mu_0 c_0/\mu_s\omega_0$ is equal to 0.04509 s^{-1}. The corresponding insertion loss index is shown in Figure 3.2. For water, the data are $\mu_0 = 1000$ kg m^{-3} and $c_0 = 1500$ ms^{-2} which lead to $\mu_0 c_0/\mu_s\omega_0 = 153\ \text{s}^{-1}$. The corresponding insertion loss index is given by Figure 3.3.

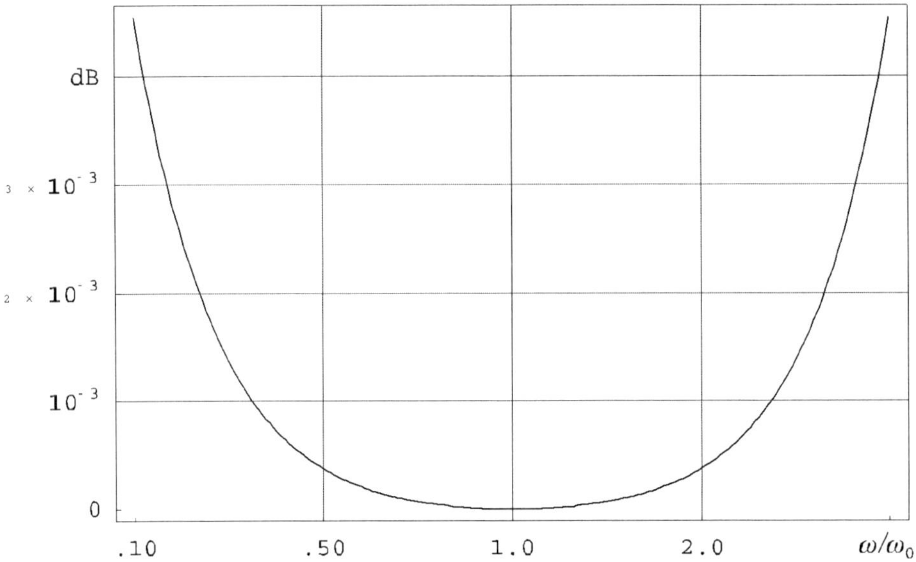

Figure 3.3. *Example of insertion loss index for water:* $\mu_0 c_0/\mu_s \omega_0 = 153 \text{ s}^{-1}$

The important result is that the insertion loss index is large for a gas (away from the mass-spring system resonance frequency, almost all energy is reflected by the piston), while it is much lower for a liquid (the piston ensures a rather good energy transmission between the two waveguides). So, the loading due to a gas is said to be a *weak loading*, while for a liquid it is qualified as a *strong loading*.

In architectural acoustics, walls, doors, windows are characterized by an insertion loss. But, because plane waves never occur in real life, the definition of this parameter is less simple: it corresponds to the mean value of a set of measurements conducted according to a normalized procedure. Nevertheless, for walls made of a single layer of material, the insertion loss index as defined in architectural acoustics always qualitatively follows the curve of Figure 3.2.

In the case of a gas, the ratio $\mu_0 c_0/\mu_s$ is small compared to 1. Thus, at high frequency, that is, for $\omega \gg \omega_0$, the energy transmission rate can be approximated by

$$\tau(\omega) \simeq \frac{4\mu_0^2 c_0^2}{\mu_s \omega^2}$$

$\tau(\omega)$ is approximately proportional to the inverse of the density of the piston: this behavior is known as the *mass law*. It is worth noting that the mass law is valid for actual walls.

3.2.3. *Response of the System to a Transient Excitation of the Piston*

Let us assume that the external excitation force $\tilde{F}_e(t)$ acting on the piston is zero outside a given time interval $[0, T]$ and that $\tilde{F}_e(0) = \tilde{F}_e(T) = 0$. The piston displacement $\tilde{u}(t)$ is given by the inverse Fourier transform of the function $u(\omega)$ defined by the first equation of (3.9) which is the product of $F_e(\omega)$ by the function

$$U(\omega) = \frac{1}{\mu_s\left(\mu_0^2 - \imath\omega\left(\mu_1 c_1 + \mu_2 c_2\right)/\mu_s - \omega^2\right)}$$

The piston displacement is thus the convolution product of $\tilde{F}_e(t)$ by $\tilde{U}(t)$, the inverse Fourier transform of $U(\omega)$. The function $\tilde{U}(t)$ is defined by the integral

$$\tilde{U}(t) = \frac{1}{2\pi}\int_{-\infty}^{+\infty} \frac{\mathrm{e}^{-\imath\omega t}}{\mu_s\left(\omega_0^2 - \imath\omega\left(\mu_1 c_1 + \mu_2 c_2\right)/\mu_s - \omega^2\right)}\,\mathrm{d}\omega \tag{3.15}$$

which can be evaluated by applying the residue theorem. Let us recall that the denominator of the function to be integrated has two roots Ω^+ and Ω^-, given by (3.8). For simplicity, we assume that these roots have opposite real parts, and the same imaginary part, which is negative. Depending on the sign of t, two different integration contours are used. They are defined as the limits, for $R \to \infty$, of the following contours:

- ◇ for $t < 0$, ω runs over the line $[-R, +R]$ and then over the half-circle $R\mathrm{e}^{\imath\theta}$ with $\theta \in [0, +\pi]$;
- ◇ for $t > 0$, ω runs over the line $[-R, +R]$ and then over the half-circle $R\mathrm{e}^{\imath\theta}$ with $\theta \in [0, -\pi]$.

Because the poles Ω^+ and Ω^- are in the half-plane $\Im(\omega) < 0$, $\tilde{U}(t)$ is zero for $t < 0$. For $t > 0$, the following result is easily established:

$$\tilde{U}(t) = Y(t)\frac{\sin\left[\mathring{\Omega} t\right]\mathrm{e}^{-\widehat{\Omega} t}}{\mu_s \mathring{\Omega}}$$

$$\text{with}\quad \mathring{\Omega} = \Re(\Omega^+) = \sqrt{\omega_0^2 - \left(\frac{\mu_1 c_1 + \mu_2 c_2}{2\mu_s}\right)^2} \tag{3.16}$$

$$\widehat{\Omega} = -\Im(\Omega^+) = \frac{\mu_1 c_1 + \mu_2 c_2}{2\mu_s}$$

where $Y(t)$ is the Heaviside step function. The function $\tilde{U}(t)$ is called the *impulse response* of the piston, that is, the response to an impulse unit pressure described by a Dirac measure $\delta(t)$. The piston displacement is deduced from (3.16) by:

$$\tilde{u}(t) = \tilde{F}_e(t) * \tilde{U}(t) = Y(t) \int_0^t \tilde{F}(\tau)\tilde{U}(t-\tau)\,\mathrm{d}\tau \quad \text{for} \quad t < T$$
$$= \int_0^T \tilde{F}(\tau)\tilde{U}(t-\tau)\,\mathrm{d}\tau \quad \text{for} \quad t > T \qquad (3.16')$$

To calculate the radiated acoustic pressures we first rewrite the second and third equations of (3.9) as:

$$p_1(x,\omega) = \imath\omega\mu_1 c_1 u(\omega) \mathrm{e}^{-\imath\omega x/c_1} \quad , \quad p_2(x,\omega) = -\imath\omega\mu_2 c_2 u(\omega)\mathrm{e}^{\imath\omega x/c_2}$$

We use the well known properties of the Fourier transform:

$$\mathcal{F}^{-1}\big(-\imath\omega f(\omega)\big) = \frac{\mathrm{d}\tilde{f}(t)}{\mathrm{d}t} \quad , \quad \mathcal{F}^{-1}\big(f(\omega)\mathrm{e}^{\imath\omega\tau}\big) = \tilde{f}(t-\tau)$$

Thus, the acoustic pressures are expressed in terms of the time derivative of the piston displacement by:

$$\tilde{p}_1(x,t) = -\mu_1 c_1 Y(t + x/c_1) \frac{\mathrm{d}\tilde{u}(t + x/c_1)}{\mathrm{d}t}$$
$$\tilde{p}_2(x,t) = \mu_2 c_2 Y(t - x/c_2) \frac{\mathrm{d}\tilde{u}(t - x/c_2)}{\mathrm{d}t} \qquad (3.17)$$

($\tilde{u}(t)$ being zero for $t = 0$, the derivative of $Y(t)$ does not give any contribution).

Equations (3.16′) and (3.17) show that the vibro-acoustic transient response of the system is governed by the resonance modes of the fluid-loaded structure. This result remains true for any structure of finite dimensions coupled to a fluid (or to several fluids).

3.3. A One-dimensional Example of a Cavity Closed by a Vibrating Boundary

Let us consider a one-dimensional waveguide extending from $x = 0$ to infinity. The extremity $x = 0$ is closed by a perfectly rigid boundary. At $x = L$, there is a mass suspended by a spring, as in the preceding section. The two domains $0 < x < L$ and $L < x < \infty$ are filled with the same fluid (see Figure 3.4). The mechanical properties of the system are the following:

- ◇ the mass-spring system has a mass μ_s and a stiffness r;
- ◇ the fluid is characterized by a density μ_0 and a sound velocity c_0.

Three different configurations are considered: the system oscillates freely or it is excited either by an incident harmonic wave in the domain $x > L$, or by an acoustic point source located at x_0 in the domain $0 < x < L$. First, the free

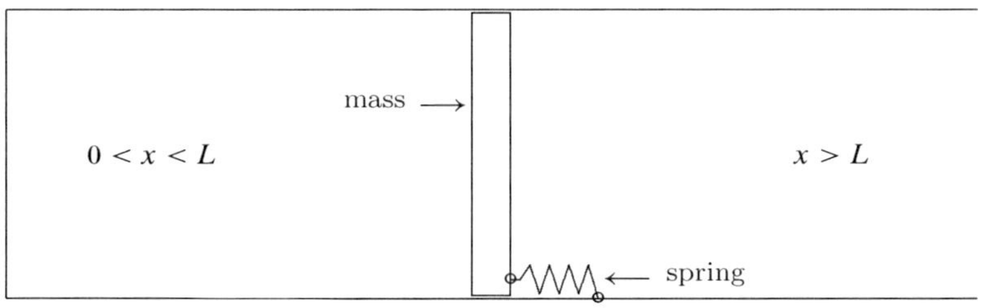

Figure 3.4. *Coupling of a one-dimensional cavity to a semi-infinite waveguide by an elastically supported mass*

oscillations of the system are studied. Then, the forced regime is examined and we focus our attention mainly on the energy transmission across the mass.

3.3.1. *Equations Governing Free Harmonic Oscillations and their Reduced Form*

The mass displacement is denoted by u. The acoustic pressure inside the cavity is denoted by p_1, while the acoustic pressure in the semi-infinite waveguide is p_2. The free harmonic oscillations obey the following equations:

$$
\begin{aligned}
(-r + \mu_s\omega^2)u &= p_1(L) - p_2(L) \\
\frac{d^2 p_1(x)}{dx^2} + \frac{\omega^2}{c_0^2} p_1(x) &= 0 \quad \text{for} \quad x \in]0, L[\\
\frac{d^2 p_2(x)}{dx^2} + \frac{\omega^2}{c_0^2} p_2(x) &= 0 \quad \text{for} \quad x \in]L, \infty[\\
\frac{dp_1(0)}{dx} = 0 \quad , \quad \frac{dp_1(L)}{dx} &= \frac{dp_2(L)}{dx} = \mu_0\omega^2 u
\end{aligned}
\tag{3.18}
$$

This system is completed by an outgoing wave condition on p_2. So, it is obvious that the acoustic pressures have the following form:

$$
p_1(x) = P_1 \frac{e^{\imath\omega x/c_0} + e^{-\imath\omega x/c_0}}{\imath\omega/c_0} \quad , \quad p_2(x) = P_2 \frac{e^{\imath\omega x/c_0}}{\imath\omega/c_0}
$$

The continuity equations and the mass motion equation then lead to:

$$\begin{aligned}
\left[e^{\imath\omega L/c_0} - e^{-\imath\omega L/c_0}\right] P_1 - e^{\imath\omega L/c_0} P_2 &= 0 \\
e^{\imath\omega L/c_0} P_2 - \omega^2 \mu_0 u &= 0 \\
\frac{e^{\imath\omega L/c_0} + e^{-\imath\omega L/c_0}}{\imath\omega/c_0} P_1 - \frac{e^{\imath\omega/c_0}}{\imath\omega/c_0} P_2 - \left(-r + \mu_s\omega^2\right)u &= 0
\end{aligned} \tag{3.19}$$

First, the resonance angular frequency $\omega_0 = \sqrt{r/\mu_s}$ of the mass-spring system is introduced and the angular frequency ω is replaced by $\omega = \Omega\omega_0$ where Ω is the reduced angular frequency. Using $\lambda_0 = 2\pi c_0/\omega_0$ as unit length, the cavity length is expressed with the reduced length Λ defined by $\Lambda = L/\lambda_0 = L\omega_0/2\pi c_0$. Finally, the ratio $\varepsilon = \mu_0/\mu_s$ is introduced. The system of equations (3.19) becomes:

$$\begin{aligned}
\left[e^{2\imath\pi\Omega\Lambda} - e^{-2\imath\pi\Omega\Lambda}\right] P_1 - e^{2\imath\pi\Omega\Lambda} P_2 &= 0 \\
e^{2\imath\pi\Omega\Lambda} P_2 - \varepsilon\mu_s\omega_0^2\Omega^2 u &= 0 \\
c_0\left[e^{2\imath\pi\Omega\Lambda} + e^{-2\imath\pi\Omega\Lambda}\right] P_1 - c_0 e^{2\imath\pi\Omega\Lambda} P_2 - \imath\mu_s\omega_0^3\Omega\left(1 - \Omega^2\right)u &= 0
\end{aligned} \tag{3.19'}$$

This system has a non-zero solution if its determinant is zero, that is, if Ω is a root of the transcendental equation:

$$\mathcal{D}(\Omega) \equiv \imath\omega_0\left(\Omega^2 - 1\right)\left(e^{4\imath\pi\Lambda\Omega} - 1\right) + 2\varepsilon c_0\Omega = 0 \tag{3.20}$$

This equation has a countable sequence of roots Ω_n called *resonance angular frequencies* of the system. Furthermore, it is easily seen that, if Ω_n is such a root, then $-\Omega_n^*$ is also a root. We can thus state the following theorem.

THEOREM 3.1.– *The system cavity–(mass-spring)–semi-infinite waveguide has a countable sequence of pairs of resonance angular frequencies* $\{\Omega_n, (n = 1, 2, \cdots, \infty)\}$ *and* $\Omega_{-n} = -\Omega_n^*$, *with* $\Re(\Omega_n) > 0$ *and* $\Im(\Omega_n) < 0$. *To each resonance frequency, corresponds a free oscillation of the system* $\left(p_1(\Omega_n), p_2(\Omega_n), u(\Omega_n)\right)$ *solution of the system (3.19′); this solution is determined up to an arbitrary multiplicative constant.*

This theorem remains valid for a three-dimensional cavity, closed by a plate (or any other vibrating structure) in contact with an external fluid extending up to infinity.

Equation (3.20) suggests that there could exist a root close to the resonance frequency $\Omega^{\mu_s} = 1$ of the mass-spring system and a sequence of resonance frequencies close to the resonance frequencies Ω_n^c of the cavity with rigid boundaries at each end, that is, the solutions of the equation

$$e^{4\imath\pi\Lambda\Omega} - 1 = 0$$

which are $\Omega_n^c = n/2\Lambda$. Let us write the Taylor series of $\mathcal{D}(\Omega)$ around $\Omega = 1$. We have:

$$\mathcal{D}(\Omega) = 2\varepsilon c_0 + 2\Big[\varepsilon c_0 + \imath\omega_0\Big(\mathrm{e}^{4\imath\pi\Lambda} - 1\Big)\Big](\Omega - 1) + \mathcal{O}\Big((\Omega - 1)^2\Big)$$

(let us recall that the symbol $\mathcal{O}(\epsilon)$ represents a quantity which decreases as fast as ϵ). This relationship provides an approximated root:

$$\Omega \simeq \frac{\imath\omega_0\Big(\mathrm{e}^{4\imath\pi\Lambda} - 1\Big)}{\imath\omega_0\Big(\mathrm{e}^{4\imath\pi\Lambda} - 1\Big) + \varepsilon c_0}$$

which is close to 1 if εc_0 is small compared to the modulus of the first term $\big|\imath\omega_0\big(\mathrm{e}^{4\imath\pi\Lambda} - 1\big)\big|$. Similarly, it can be seen that the other resonance frequencies are close to the resonance frequencies $n/2\Lambda$ of the rigidly closed cavity if εc_0 is small enough.

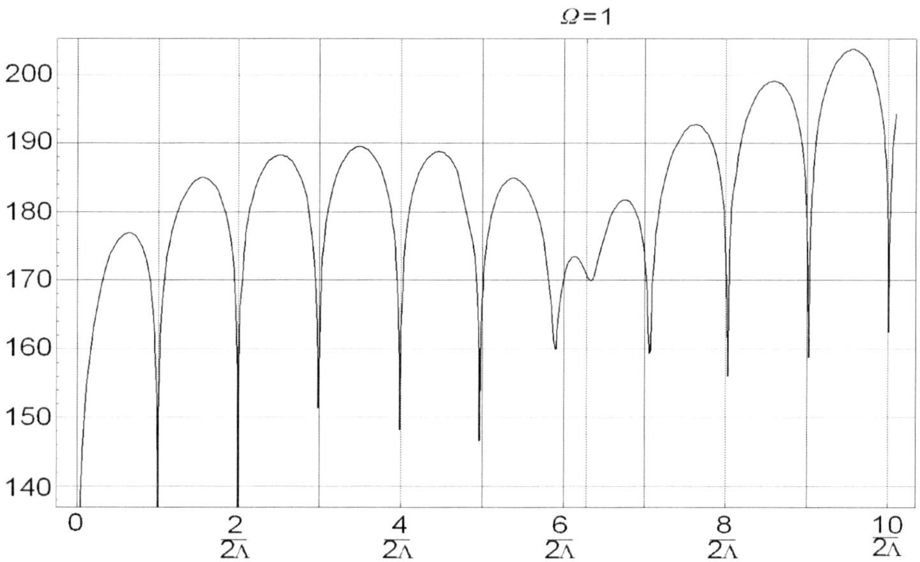

Figure 3.5. *Curve representing* $20 \log |\mathcal{D}(\Omega)|$ *as a function of* Ω

In the following numerical application, the fluid and the mass-spring system have the same characteristics as in the former example. The reduced length of the cavity is $\Lambda = \pi$. Figure 3.5 shows the variations of $20 \log |\mathcal{D}(\Omega)|$ as a function of Ω. It clearly appears that the resonance frequencies of the system are close to the resonance frequencies of the cavity and of the resonance frequency of the mass-spring assembly. Using a Newton's method, an accurate value of

n	Ω_n^c	$\Re(\Omega_n)$	$\Im(\Omega_n)$	$\Re(\Omega_n/\Omega_n^c)$	$\Im(\Omega_n/\Omega_n^c)$
1	0.1591	0.1585	-0.02751×10^{-4}	0.9977	-0.1733×10^{-4}
2	0.3183	0.3175	-0.1280×10^{-4}	0.9975	-0.4020×10^{-4}
3	0.4775	0.4761	-0.3880×10^{-4}	0.9771	-0.8125×10^{-4}
4	0.6366	0.6342	-1.144×10^{-4}	0.9962	-1.7971×10^{-4}
5	0.7758	0.7910	-4.406×10^{-4}	0.9940	-5.537×10^{-4}
6	0.9549	0.9387	-42.827×10^{-4}	0.9830	-44.850×10^{-4}
Ω^{μ_s}	1	1.0104	-147.755×10^{-4}	—	—
7	1.1141	1.1235	-165.410×10^{-4}	1.0084	-148.470×10^{-4}
8	1.2732	1.2783	-4.073×10^{-4}	1.0040	-3.199×10^{-4}

Table 3.1. *Resonance frequencies of the system cavity (mass-spring) semi-infinite waveguide*

the resonance frequencies is easily obtained. The results are presented in Table 3.1: in the first column we give the initial values used, which, in fact, are the cavity resonance frequencies $n/2\Lambda$; in the next two columns the corresponding resonance frequencies of the complete system; and, in the last two columns, the complex ratio of the resonance frequencies of the system cavity (mass-spring) semi-infinite waveguide to the initial values. The resonance frequency of the system corresponding to Ω^{μ_s} is also inserted.

A few comments must be made:

1. as expected, the resonance frequencies of this system have a negative imaginary part, which corresponds to damped resonance modes;

2. below the resonance frequency Ω^{μ_s} of the mass-spring system, the resonance frequencies of the complete system have a real part lower than $n/2\Lambda$, which means that the loading of the cavity by the mass and the external fluid acts like a mass increase of the interior fluid, or as an added length;

3. the resonance frequency corresponding to Ω^{μ_s} is slightly higher than 1, which means that the cavity has a stiffness effect on the mass;

4. finally, for frequencies higher than Ω^{μ_s}, the resonance frequencies of the complete system have a real part higher than $n/2\Lambda$, which means that the loading of the cavity by the mass and the external fluid acts like a stiffness increase of the interior fluid, or a shortening of its length.

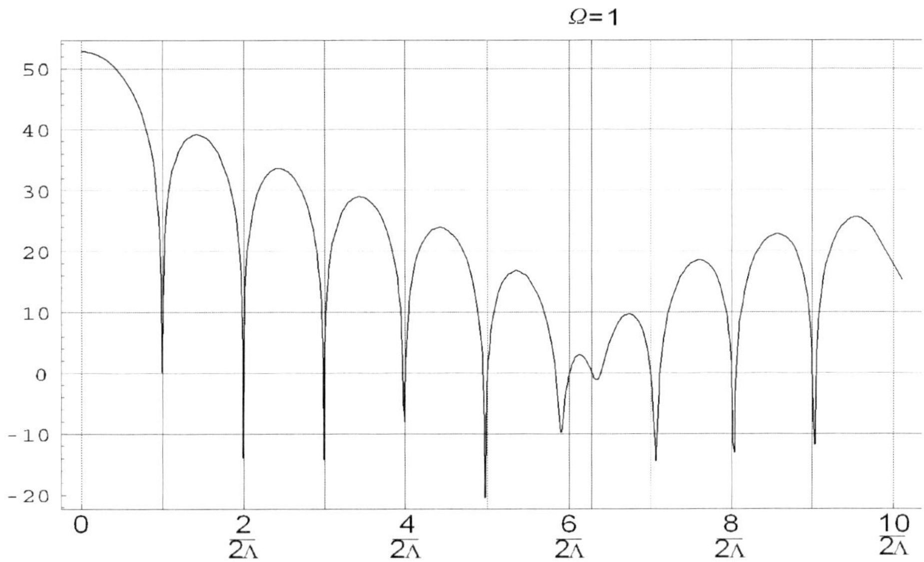

Figure 3.6. *Curve representing* $-20 \log |P_1(\Omega)|$ *as a function of* Ω

3.3.2. *Transmission of Sound Across the Vibrating Boundary*

We consider two different excitations of the system: an incident wave taking place in the semi-infinite waveguide $x > L$; a point source inside the cavity.

1. *Transmission of sound from the exterior to the interior of the cavity*

Let us assume that the system is excited by an incident wave which takes place in the semi-infinite waveguide. The acoustic pressures p_1 and p_2 have the following form:

$$p_1(x) = P_1 \frac{e^{\imath\omega x/c_0} + e^{-\imath\omega x/c_0}}{\imath\omega/c_0} \quad , \quad p_2(x) = \frac{e^{-\imath\omega x/c_0}}{\imath\omega/c_0} + P_2 \frac{e^{\imath\omega x/c_0}}{\imath\omega/c_0}$$

Using the same reduced variables, the unknown parameters u, P_1 and P_2 are the solution of the system:

$$\begin{aligned}
&\left[e^{2\imath\pi\Omega\Lambda} - e^{-2\imath\pi\Omega\Lambda}\right]P_1 - e^{2\imath\pi\Omega\Lambda}P_2 = -e^{-2\imath\pi\Omega\Lambda} \\
&e^{2\imath\Omega\Lambda}P_2 - \varepsilon\mu_s\omega_0^2\Omega^2 u = e^{-2\imath\pi\Omega\Lambda} \\
&c_0\left[e^{2\imath\pi\Omega\Lambda} + e^{2\imath\pi\Omega\Lambda}\right]P_1 - c_0 e^{2\imath\pi\Omega\Lambda}P_2 - \imath\mu_s\omega_0^3\Omega\left(1-\Omega^2\right)u \\
&\qquad = c_0 e^{-2\imath\pi\Omega\Lambda}
\end{aligned} \tag{3.21}$$

We have:

$$
\begin{aligned}
P_1(\Omega) &= \frac{2c_0\varepsilon\Omega}{\mathcal{D}(\Omega)}\\
P_2(\Omega) &= \frac{1}{\mathcal{D}(\Omega)}\Big[2c_0\varepsilon\Omega + \imath\omega_0\big(\Omega^2-1\big)\big(1-\mathrm{e}^{-4\imath\pi\Lambda\Omega}\big)\Big]\\
u(\Omega) &= \frac{4\imath c_0 \sin 2\pi\Lambda\Omega}{\mu_s\omega_0^2\Omega\mathcal{D}(\Omega)}\\
\text{with}\quad \mathcal{D}(\Omega) &= 2c_0\varepsilon\Omega - \imath\omega_0\big(\Omega^2-1\big)\big(1-\mathrm{e}^{4\imath\pi\Lambda\Omega}\big)
\end{aligned}
\tag{3.22}
$$

Furthermore, it is easily seen that $|P_2| = 1$ for any value of Ω. Finally, it should be remarked that the mass displacement is zero at $\Omega = n/2\Lambda$, where n is an integer.

The transmission of sound is characterized by P_1. Figure 3.6 represents $-20\log|P_1(\Omega)|$ as a function of Ω, which is equivalent to the insulation index defined in the preceding section. It appears that the transmitted amplitude is maximum at frequencies equal to the real part of the resonance frequencies of the complete system, that is, at resonance frequencies close to the mass-spring resonance frequency and close to the cavity resonance frequencies.

2. *Transmission of sound from the interior to the exterior of the cavity*

We now consider the case of a sound source inside the cavity, at $x = x_0$ or, equivalently in reduced coordinates, at $\xi_0 = x_0/\lambda_0$. The acoustic pressures have the form:

$$
\begin{aligned}
p_1(x) &= \frac{\mathrm{e}^{\imath\omega|x-x_0|/c_0}}{\imath\omega/c_0} + \frac{\mathrm{e}^{\imath\omega(x+x_0)/c_0}}{\imath\omega/c_0} + P_1\frac{\mathrm{e}^{\imath\omega x/c_0}+\mathrm{e}^{-\imath\omega x/c_0}}{\imath\omega/c_0}\\
p_2(x) &= P_2\frac{\mathrm{e}^{\imath\omega x/c_0}}{\imath\omega/c_0}
\end{aligned}
$$

The unknown parameters satisfy the following system:

$$
\begin{aligned}
\Big[\mathrm{e}^{2\imath\pi\Omega\Lambda} - \mathrm{e}^{-2\imath\pi\Omega\Lambda}\Big]P_1 - \mathrm{e}^{2\imath\pi\Omega\Lambda}P_2 &= -2\mathrm{e}^{2\imath\pi\Omega\Lambda}\cos(2\pi\Omega\xi_0)\\
\mathrm{e}^{2\imath\Omega\Lambda}P_2 - \varepsilon\mu_s\omega_0^2\Omega^2 u &= 0\\
c_0\Big[\mathrm{e}^{2\imath\pi\Omega\Lambda} + \mathrm{e}^{2\imath\pi\Omega\Lambda}\Big]P_1 - c_0\mathrm{e}^{2\imath\pi\Omega\Lambda}P_2 - \imath\mu_s\omega_0^3\Omega\big(1-\Omega^2\big)u&\\
&= -2c_0\mathrm{e}^{2\imath\pi\Omega\Lambda}\cos(2\pi\Omega\xi_0)
\end{aligned}
\tag{3.21$'$}
$$

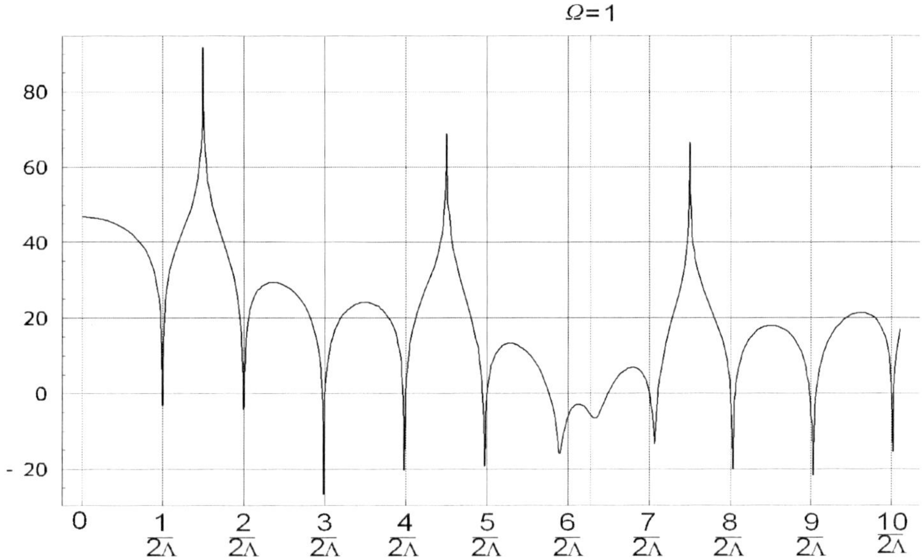

Figure 3.7. *Curve representing* $-20 \log |P_2(\Omega)|$ *as a function of* Ω

The solution is:

$$P_1(\Omega) = -2\imath\pi\omega_0(\Omega^2 - 1)\mathrm{e}^{4\imath\pi\Lambda\Omega}\frac{\cos(2\pi\Omega\xi_0)}{\mathcal{D}(\Omega)}$$

$$P_2(\Omega) = 4c_0\varepsilon\Omega\frac{\cos(2\pi\Omega\xi_0)}{\mathcal{D}(\Omega)} \qquad (3.22')$$

$$u(\Omega) = \frac{4c_0\mathrm{e}^{2\imath\pi\Lambda\Omega}}{\mu_s\omega_0^2\Omega}\frac{\cos(2\pi\Omega\xi_0)}{\mathcal{D}(\Omega)}$$

with $\mathcal{D}(\Omega)$ as before. We first remark that this solution is identically zero for $\Omega = (n + 1/2)/2\xi_0$ with n integer. For these values of Ω, the acoustic pressure component

$$\frac{\mathrm{e}^{\imath\omega|x-x_0|/c_0}}{\imath\omega/c_0} + \frac{\mathrm{e}^{\imath\omega(x+x_0)/c_0}}{\imath\omega/c_0}$$

which satisfies the Neumann condition at $x = 0$, is zero at $x > x_0$. Thus, it is obvious that the mass is not excited and, thus, there is no acoustic pressure generated by the mass motion. In this case again, the sound transmission is characterized by P_2. Figure 3.7 represents $-20 \log |P_2(\Omega)|$ as a function of Ω for $\xi_0 = \Lambda/3$. Once more, the resonance frequencies of the cavity and that of the mass-spring assembly have an important role: they correspond to the maxima of transmitted amplitude of the sound pressure.

3.4. A Little Acoustics

Before going further into the study of acoustic radiation by vibrating structures, it is necessary to recall a few basic notions in theoretical acoustics. The reader can find a complete and detailed presentation in the following textbooks [MOR 53, MOR 68, PIE 81, FIL 99, BRU 06]. The acoustics equations are obtained by linearizing the general fluid-mechanics equations which can be found in many textbooks (for example, [GER 73]).

This section starts with the variational form of the wave and the Helmholtz equations.

Then we recall the expressions of the free-field Green's function of the Helmholtz equation, in two and three dimensions, and their expansions into series of special functions.

The section ends with the establishment of the Green's formula for the acoustics equations and a few properties of the layer potentials involved in the Green's formula.

3.4.1. *Variational Form of the Wave Equation and of the Helmholtz Equation*

We consider a fluid characterized by a density μ_0 and a sound speed c_0. It occupies a bounded domain Ω, with boundary $\partial\Omega$. We denote by $\vec{n}$ the unit vector normal to $\partial\Omega$ and pointing out to the exterior of Ω. Let P_0 be the static pressure in the fluid, $\widetilde{P}$ the time-dependent acoustic pressure and $\vec{\widetilde{U}}$ the time-dependent particle displacement vector. The total energy of the fluid is:

$$\mathcal{E} = \int_{\Omega} \left[-(\widetilde{P} + P_0)\vec{\nabla}.\vec{\widetilde{U}} + \frac{\mu_0 \dot{\vec{\widetilde{U}}}^2}{2} \right] \mathrm{d}\Omega$$

The variation $\delta\mathcal{E}$ of the total energy due to an elementary displacement $\delta\vec{\widetilde{U}}$ has the following expression:

$$\delta\mathcal{E} = \int_{\Omega} \left[-(\widetilde{P} + P_0)\vec{\nabla}.\delta\vec{\widetilde{U}} + \mu_0 \ddot{\vec{\widetilde{U}}}.\delta\vec{\widetilde{U}} \right] \mathrm{d}\Omega \tag{3.23}$$

An acoustic motion has a mean time value which is zero. Thus, the term $P_0\vec{\nabla}.\delta\vec{\widetilde{U}}$, having a mean time value equal to zero, does not concern the acoustic wave and can be omitted. In an acoustic motion, the rotational of the displacement is zero. This allows us to express the elementary displacement as the gradient of a scalar function, that is $\delta\vec{\vec{U}} = \vec{\nabla}\widetilde{\varphi}$. So, the variation of the total energy of the fluid takes the form:

$$\delta\mathcal{E} = \int_{\Omega} \left[-\widetilde{P}\,\Delta\widetilde{\varphi} + \mu_0 \ddot{\vec{\vec{U}}}.\vec{\nabla}\widetilde{\varphi} \right] \mathrm{d}\Omega \tag{3.24}$$

To express $\delta\mathcal{E}$ in terms of P only, we use the equations of mass and momentum conservation

$$\frac{\dot{\widetilde{P}}}{{c_0}^2} + \mu_0 \vec{\nabla}.\dot{\vec{\vec{U}}} = 0 \qquad , \qquad \mu_0 \ddot{\vec{\vec{U}}} + \vec{\nabla}\widetilde{P} = 0$$

and we perform integrations by parts:

$$-\int_{\Omega} \widetilde{P}\,\Delta\widetilde{\varphi}\,\mathrm{d}\Omega = \int_{\Omega} \vec{\nabla}\widetilde{P}.\vec{\nabla}\widetilde{\varphi}\,\mathrm{d}\Omega - \int_{\partial\Omega} \mathrm{Tr}\,\widetilde{P}\;\mathrm{Tr}\big(\vec{n}.\vec{\nabla}\widetilde{\varphi}\big)\,\mathrm{d}\sigma$$

$$\int_{\Omega} \mu_0 \ddot{\vec{\vec{U}}}.\vec{\nabla}\widetilde{\varphi}\,\mathrm{d}\Omega = -\int_{\Omega} \mu_0 \vec{\nabla}.\ddot{\vec{\vec{U}}}\widetilde{\varphi}\,\mathrm{d}\Omega + \int_{\partial\Omega} \mu_0\,\mathrm{Tr}\big(\vec{n}.\ddot{\vec{\vec{U}}}\big)\,\mathrm{Tr}\,\widetilde{\varphi}\,\mathrm{d}\sigma$$

$$= \int_{\Omega} \frac{\ddot{\widetilde{P}}}{{c_0}^2}\widetilde{\varphi}\,\mathrm{d}\Omega - \int_{\partial\Omega} \mathrm{Tr}\big(\vec{n}.\vec{\nabla}\widetilde{P}\big)\,\mathrm{Tr}\,\varphi\,\mathrm{d}\sigma$$

(the meaning of the symbol "Tr" is as in Chapter 1). The energy variation is due to the work of external forces with volume density $\vec{\vec{F}}$ and of the boundary forces $-\,\mathrm{Tr}\,\big(\widetilde{P} + P_0\big)\vec{n}$, that is:

$$\delta\mathcal{E} = \int_{\Omega^i} \vec{\vec{F}}.\delta\vec{\vec{U}}\,\mathrm{d}\Omega - \int_{\partial\Omega} \mathrm{Tr}\big(\widetilde{P} + P_0\big)\,\mathrm{Tr}\big(\vec{n}.\delta\vec{\vec{U}}\big)\,\mathrm{d}\sigma$$

$$= -\int_{\Omega} \vec{\nabla}.\vec{\vec{F}}\widetilde{\varphi}\,\mathrm{d}\Omega - \int_{\partial\Omega} \mathrm{Tr}\big(\widetilde{P} + P_0\big)\,\mathrm{Tr}\big(\vec{n}.\vec{\nabla}\widetilde{\varphi}\big)\,\mathrm{d}\sigma$$

Here again, the term with P_0 can be omitted. Defining $\widetilde{f} = \vec{\nabla}.\vec{\vec{F}}$, the final result writes:

$$\int_{\Omega} \left[\vec{\nabla}\widetilde{P}.\vec{\nabla}\widetilde{\varphi} + \frac{\ddot{\widetilde{P}}}{{c_0}^2}\widetilde{\varphi} \right] \mathrm{d}\Omega = -\int_{\Omega} \widetilde{f}\widetilde{\varphi}\,\mathrm{d}\Omega + \int_{\partial\Omega} \mathrm{Tr}\big(\vec{n}.\vec{\nabla}\widetilde{P}\big)\,\mathrm{Tr}\,\widetilde{\varphi}\,\mathrm{d}\sigma \tag{3.25}$$

which is the variational form of the wave equation. For harmonic regime, this equation becomes:

$$\int_{\Omega} \left[\vec{\nabla}P.\vec{\nabla}\varphi - \frac{\omega^2}{c_0^2} P\varphi \right] \mathrm{d}\Omega = -\int_{\Omega} f\varphi\,\mathrm{d}\Omega + \int_{\partial\Omega} \mathrm{Tr}\,\partial_n P\;\mathrm{Tr}\,\varphi\,\mathrm{d}\sigma \tag{3.26}$$

where the time independent functions are denoted by the same symbol without the over-sign "$\sim$" and ∂_n stands for $\vec{n}.\vec{\nabla}$. Equation (3.26) is the variational form of the Helmholtz equation.

These last two equations show that any linear relationship between the pressure and its normal derivative – which is proportional to the normal component of the particle velocity – is an admissible boundary condition.

3.4.2. *Free-field Green's Function of the Helmholtz Equation*

Let us consider the propagation of a harmonic wave due to a point unit isotropic source δ_O, located at the origin of the infinite space $\mathbf{R}^2$ or $\mathbf{R}^3$. The acoustic pressure $G(O, M)$ satisfies the Helmholtz equation

$$(\Delta + k^2)G(O, M) = \delta_O(M)$$

where $k = \omega/c_0$ is the wavenumber. To ensure the uniqueness of the solution, we must add a condition at infinity, such as, for example the Sommerfeld condition, which, for the time dependence $e^{-\imath\omega t}$ chosen in this book, writes:

$$\lim_{r\to\infty} G = \mathcal{O}\left(r^{(1-n)/2}\right)$$
$$\lim_{r\to\infty} \left(\partial_r G - \imath k G\right) = o\left(r^{(1-n)/2}\right)$$

where r is the radial coordinate of the point M, and n is the dimension of the space. The symbol $\mathcal{O}(\epsilon)$ means that the quantity concerned decreases as fast as ϵ, while the symbol $o(\epsilon)$ indicates a decrease faster than ϵ. The result is:

$$\begin{aligned} G(O, M) &= -\frac{\imath}{4} H_0^{(1)}(kr) \quad \text{in} \quad \mathbf{R}^2 \\ &= -\frac{e^{\imath kr}}{4\pi r} \quad \text{in} \quad \mathbf{R}^3 \end{aligned} \tag{3.27}$$

where $H_0^{(1)}(kr)$ is the Hankel function of the first kind and order 0 (it is generally denoted by $H_0(kr)$ if there is no ambiguity).

If the point source is located at another point S, r will be replaced by $r(S, M)$, the distance between M and the source point S.

3.4.3. *Series Expansions of the Free Field Green's Function of the Helmholtz Equation*

It is often useful to express the Green's function $G(S, M)$ in cylindrical or spherical coordinates with origin O different from the source point.

1. *In dimension 2*

Let (ρ_S, θ_S) be the cylindrical coordinates of S and (ρ, θ) those of M. The Green's function can be expanded into two different series of Bessel functions, depending on the relative values of ρ_S and ρ:

$$\begin{aligned} G(O,M) &= -\frac{\imath}{4} H_0(kr) \\ &= -\frac{\imath}{4} \sum_{n=-\infty}^{+\infty} H_n(k\rho_S) J_n(k\rho) \mathrm{e}^{\imath n(\theta_S - \theta)} \quad \text{for} \quad \rho < \rho_S \\ &= -\frac{\imath}{4} \sum_{n=-\infty}^{+\infty} H_n(k\rho) J_n(k\rho_S) \mathrm{e}^{\imath n(\theta - \theta_S)} \quad \text{for} \quad \rho > \rho_S \end{aligned} \tag{3.28}$$

where $H_n(z)$ denotes the Hankel function of order n and of the first kind, and $J_n(z)$ is the Bessel function of order n.

2. *In dimension 3*

Let $(\rho_S, \theta_S, \varphi_S)$ be the spherical coordinates of S and (ρ, θ, φ) those of M. The Green's function can be expanded into two different series of spherical Bessel functions and Legendre functions, depending on the relative values of ρ_S and ρ:

$$\begin{aligned} G(O,M) &= -\frac{\mathrm{e}^{\imath kr}}{4\pi r} \\ &= -\frac{\imath k}{2\pi} \sum_{n=0}^{+\infty} (2n+1) \sum_{m=0}^{n} \varepsilon_m \frac{(n-m)!}{(n+m)!} \cos m(\varphi - \varphi_S) \\ &\quad P_n^m(\cos\theta) P_n^m(\cos\theta_S) j_n(k\rho) h_n(k\rho_S) \quad \text{for} \quad \rho < \rho_S \\ &= -\frac{\imath k}{2\pi} \sum_{n=0}^{+\infty} (2n+1) \sum_{m=0}^{n} \varepsilon_m \frac{(n-m)!}{(n+m)!} \cos m(\varphi - \varphi_S) \\ &\quad P_n^m(\cos\theta) P_n^m(\cos\theta_S) j_n(k\rho_S) h_n(k\rho) \quad \text{for} \quad \rho > \rho_S \end{aligned} \tag{3.29}$$

where $\varepsilon_0 = 1$ and $\varepsilon_{m \neq 0} = 2$; $h_n(z)$ is the spherical Hankel function of order n and of the first kind, $j_n(z)$ is the spherical Bessel function of order n, and $P_n^m(z)$ is the Legendre function of degree n and order m.

3.4.4. *Green's Formula for the Helmholtz Operator and Green's Representation of the Solution of the Helmholtz Equation*

From the variational equation (3.26), integrations by parts lead to the following relationships:

$$\int_{\Omega} P\left(\Delta\varphi + k^2\varphi\right) \mathrm{d}\Omega = \int_{\Omega} f\varphi \,\mathrm{d}\Omega + \int_{\partial\Omega} \left(\mathrm{Tr}\, P \,\mathrm{Tr}\, \partial_n \varphi - \mathrm{Tr}\, \partial_n P \,\mathrm{Tr}\, \varphi\right) \mathrm{d}\sigma$$

$$\int_{\Omega} \varphi\left(\Delta P + k^2 P\right) \mathrm{d}\Omega = \int_{\Omega} f\varphi \,\mathrm{d}\Omega$$

By difference, we obtain the Green's formula for the Laplace operator (which is also the Green's formula for the Helmholtz operator):

$$\int_{\Omega} \left(P\Delta\varphi - \varphi\Delta P\right) \mathrm{d}\Omega = \int_{\partial\Omega} \left(\mathrm{Tr}\, P \,\mathrm{Tr}\, \partial_n \varphi - \mathrm{Tr}\, \varphi \,\mathrm{Tr}\, \partial_n P\right) \mathrm{d}\sigma \tag{3.30}$$

We now look for the solution of the Helmholtz equation

$$\left(\Delta + k^2\right) P(M) = f(M) \quad \text{for} \quad M \in \Omega$$

By writing the Green's formula for $P(M)$ and for the Green's function $G(M, M')$, one obtains:

$$\int_{\Omega} \left(P(M')\Delta_{M'} G(M, M') - G(M, M')\Delta_{M'} P(M')\right) \mathrm{d}\Omega(M')$$
$$= \int_{\partial\Omega} \left(\mathrm{Tr}\, P(Q') \,\mathrm{Tr}\, \partial_{n'} G(M, Q') - \mathrm{Tr}\, G(M, Q') \,\mathrm{Tr}\, \partial_{n'} P(Q')\right) \mathrm{d}\sigma(Q')$$

where the symbol $\partial_{n'}$ means that the normal derivative is taken with respect to the integration point Q'. In the following, the symbol " Tr " in front of the Green's function and its normal derivative will be omitted. Using the equations satisfied by $P(M)$ and $G(M, M')$, we obtain the Green's representation of the solution as:

$$P(M) = \int_{\Omega} f(M') G(M, M') \,\mathrm{d}\Omega(M') + \int_{\partial\Omega} \Big(\mathrm{Tr}\, P(Q') \partial_{n'} G(M, Q')$$
$$- G(M, Q') \,\mathrm{Tr}\, \partial_{n'} P(Q')\Big) \mathrm{d}\sigma(Q') \tag{3.31}$$

The pressure $P(M)$ is determined when $\operatorname{Tr} P(M')$ and $\operatorname{Tr} \partial_{n'} P(M')$ are known. To do so, a boundary condition must be given and a boundary integral equation has to be solved.

Expression (3.31) involves a simple layer potential and a double layer potential, respectively given by:

$$P_1(M) = \int_{\partial\Omega} \operatorname{Tr} \partial_{n'} P(Q') G(M, Q') \, d\sigma(Q')$$

$$P_2(M) = -\int_{\partial\Omega} \operatorname{Tr} P(Q') \partial_{n'} G(M, Q') \, d\sigma(Q')$$

The simple layer potential is a continuous function with a discontinuous normal derivative, while the double layer potential is a discontinuous function with a continuous normal derivative. More precisely, if a point $M \in \Omega$ tends to a point $Q \in \partial\Omega$, we have:

$$\begin{aligned} \operatorname{Tr} P_1(Q) &= \int_{\partial\Omega} \operatorname{Tr} \partial_{n'} P(Q') G(Q, Q') \, d\sigma(Q') \\ \operatorname{Tr} \partial_n P_1(Q) &= -\frac{\operatorname{Tr} \partial_n P(Q)}{2} + \int_{\partial\Omega} \operatorname{Tr} \partial_{n'} P(Q') \partial_n G(Q, Q') \, d\sigma(Q') \end{aligned} \tag{3.32}$$

and

$$\begin{aligned} \operatorname{Tr} P_2(Q) &= -\frac{\operatorname{Tr} P(Q)}{2} - \int_{\partial\Omega} \operatorname{Tr} P(Q') \partial_{n'} G(Q, Q') \, d\sigma(Q') \\ \operatorname{Tr} \partial_n P_2(Q) &= -\text{Pf.} \int_{\partial\Omega} \operatorname{Tr} P(Q') \partial_n \partial_{n'} G(Q, Q') \, d\sigma(Q') \end{aligned} \tag{3.32'}$$

where the symbol "Pf." designates the finite part of the integral which is divergent in the Riemann sense (see, for example, [SCH 61]).

All these classical results, which can be found in every basic books on acoustics or on applied mathematics, can be easily established using elementary results of the theory of distributions.

To do so, we first recall the expression of the Laplace operator in the distribution sense. Let P be a continuous function, with continuous first and second order derivatives in Ω and in its complement $\overline{\Omega}$. We assume that P and its first order normal derivative can have discontinuities – respectively $(\operatorname{Tr}^+ P - \operatorname{Tr}^- P)$ and $(\operatorname{Tr}^+ \partial_n P - \operatorname{Tr}^- \partial_n P)$ – across $\partial\Omega$. The Laplace operator applied to P in the distribution sense writes:

$$\Delta P = \{\Delta P\} + (\mathrm{Tr}^+ P - \mathrm{Tr}^- P) \otimes \delta'_{\partial\Omega} + (\mathrm{Tr}^+ \partial_n P - \mathrm{Tr}^- \partial_n P) \otimes \delta_{\partial\Omega}$$

In this relationship, $\{\Delta P\}$ is the Laplacian of P in the classical sense, which is defined everywhere but on $\partial\Omega$. The distribution $\delta_{\partial\Omega}$ is the Dirac measure with support $\partial\Omega$ and $\delta'_{\partial\Omega}$ is its normal derivative; more precisely, these distributions are given by

$$\langle \mu \otimes \delta_{\partial\Omega}, \psi \rangle = \int_{\partial\Omega} \mu(Q)\psi(Q)\,\mathrm{d}\sigma(Q)\ ,\ \langle \mu \otimes \delta'_{\partial\Omega}, \psi \rangle = -\int_{\partial\Omega} \mu(Q)\partial_n \psi(Q)\,\mathrm{d}\sigma(Q)$$

where ψ is an indefinitely derivable function with compact support.

Let us now consider, for example, a simple layer potential defined by:

$$P(M) = \int_{\partial\Omega} \mu(Q')G(M, Q')\,\mathrm{d}\sigma(Q')$$

It is defined everywhere and it satisfies, in the distribution sense, the following Helmholtz equation:

$$\begin{aligned}(\Delta + k^2)P &= (\mathrm{Tr}^+ P - \mathrm{Tr}^- P) \otimes \delta'_{\partial\Omega} + (\mathrm{Tr}^+ \partial_n P - \mathrm{Tr}^- \partial_n P) \otimes \delta_{\partial\Omega} \\ &= \mu \otimes \delta_{\partial\Omega}\end{aligned}$$

with

$$\mathrm{Tr}^- P(Q) = \lim_{M \in \Omega \to Q \in \partial\Omega} P(M) \quad , \quad \mathrm{Tr}^+ P(Q) = \lim_{M \in \overline{\Omega} \to Q \in \partial\Omega} P(M)$$

$$\mathrm{Tr}^- \partial_n P(Q) = \lim_{M \in \Omega \to Q \in \partial\Omega} \vec{n} \cdot \vec{\nabla} P(M)$$

$$\mathrm{Tr}^+ \partial_n P(Q) = \lim_{M \in \overline{\Omega} \to Q \in \partial\Omega} \vec{n} \cdot \vec{\nabla} P(M)$$

We immediately have that P is a continuous function (the coefficient of $\delta'_{\partial\Omega}$ is zero), while its normal derivative has a discontinuity equal to μ.

Let us define a couple of points M_1 and M_2, which are on the line normal to $\partial\Omega$ at the point Q and in symmetric positions with respect to Q. The function $P(M_1) + P(M_2)$ is continuous with a continuous normal derivative when M_1 crosses $\partial\Omega$. Thus, we have:

$$\begin{aligned}\lim_{M_1 \in \Omega \to Q \in \partial\Omega} \partial_n\big(P(M_1) + P(M_2)\big) &= \mathrm{Tr}^+ \partial_n P(Q) + \mathrm{Tr}^- \partial_n P(Q) \\ &= 2\int_{\partial\Omega} \mu(Q')\partial_n G(Q, Q')\,\mathrm{d}\sigma(Q')\end{aligned}$$

From this last result, we deduce that:

$$\mathrm{Tr}^{+}\partial_n P(Q) = \frac{\mu(Q)}{2} + \int_{\partial\Omega} \mu(Q')\partial_n G(Q, Q')\,\mathrm{d}\sigma(Q')$$

$$\mathrm{Tr}^{-}\partial_n P(Q) = -\frac{\mu(Q)}{2} + \int_{\partial\Omega} \mu(Q')\partial_n G(Q, Q')\,\mathrm{d}\sigma(Q')$$

The same method gives the result concerning the double layer potential.

3.4.5. *Numerical Difficulties*

The integrals in expressions (3.32) and (3.32′) involve a singular term. Let us examine these singularities in dimension 3 (in dimension 2, the conclusions are identical).

In the first integral in (3.32), the singularity of the kernel $G(Q, Q')$ is $1/r(Q, Q')$ which is integrable in the Riemann sense. Thus, the integral can be evaluated by any method, such as, for instance, a Gauss algorithm.

The second integral in (3.32) and the first one in (3.32′) involve singularities which are respectively of the form

$$\frac{1}{r(Q, Q')}\frac{\partial r(Q, Q')}{\partial \vec{n}(Q')} \quad \text{or} \quad \frac{1}{r(Q, Q')}\frac{\partial r(Q, Q')}{\partial \vec{n}(Q)}$$

Such a singularity is integrable as a Cauchy principal value. That is, a small domain $\delta\partial\Omega$ defined by $r(Q, Q') < \rho$ is isolated around the point Q. The value of the integral is the limit, for $\rho \to 0$ of the integral on the domain $(\partial\Omega - \delta\partial\Omega)$. For numerical applications, the following approximation is commonly used: a/around the point Q, the surface $\partial\Omega$ is replaced by its tangent plane and the domain $\delta\partial\Omega$ is replaced by a disk with radius ρ in this plane; b/ the quantity $\mathrm{Tr}\,\partial_{n'} P(Q')$ (resp. $\mathrm{Tr}\,P(Q')$) is replaced by the constant value $\mathrm{Tr}\,\partial_{n'} P(Q)$ (resp. $\mathrm{Tr}\,(Q)$); c/ the contribution of the small domain $\delta\partial\Omega$ is then evaluated analytically.

The second integral in (3.32′) involves a non-integrable singularity. However, the existence and uniqueness of the following limit

$$\mathrm{Tr}\,\partial_n P_2(Q) = \lim_{M\in\Omega\to Q\in\partial\Omega} \frac{\partial}{\partial n(Q)} \int_{\partial\Omega} \mathrm{Tr}\,P(Q')\frac{\partial G(M, Q')}{\partial n(Q')}\,\mathrm{d}\sigma(Q')$$

can be proved. The only way to evaluate such an integral numerically is to use a limit procedure corresponding to the above expression. A classical method is to proceed exactly as in the previous case. Around the point Q, the surface $\partial\Omega$ is replaced by is tangent plane and a small disk $\delta\partial\Omega$ in this plane, with radius ρ, is isolated. The function $\operatorname{Tr} P(Q')$ is replaced by the constant value $\operatorname{Tr} P(Q)$. Then the integral

$$\Psi(M) = \operatorname{Tr} P(Q) \int_{\delta\partial\Omega} \frac{\partial G(M, Q')}{\partial n(Q')} \, d\sigma(Q')$$

over the disk $\delta\partial\Omega$ is approximated analytically by taking into account the first terms of the Taylor series of $\partial_{n'} G(M, Q')$. The result is derivated with respect to the normal $n(Q)$ and the limit, for $M \to Q$, is taken (this is left as an exercise for the reader).

3.5. Infinite Structures

We now consider simple three-dimensional examples. The first structure considered is an infinite plate immersed in two different fluids extending up to infinity. The second structure is a circular cylindrical shell of infinite length, immersed in a fluid extending up to infinity and containing either a *vacuum* or a second fluid.

3.5.1. *Infinite Plate in Contact with a Single Fluid or Two Different Fluids*

1. *General equations*

Let us consider a plate of infinite extent occupying the plane $\Sigma \equiv (z = 0)$. Its mechanical characteristics are denoted as in the previous chapters

$$\mu_s \ , \ h \ , \ \bar{\mu} = \mu_s h \ , \ E \ , \ \nu \ , \ D = Eh^3/12(1 - \nu^2)$$

The half-space $\Omega^+ \equiv (z > 0)$ is occupied by a fluid with density μ_0^+ and sound velocity c_0^+; the half-space $\Omega^- \equiv (z < 0)$ is occupied by a fluid with mechanical characteristics μ_0^- and c_0^-. If there is a unique fluid on both sides of the plate, its characteristics are denoted by μ_0 and c_0. A point on the plate will, in general, be denoted by M, or by its coordinates (x, y). A point in Ω^+ or Ω^- will, in general, be denoted by Q, or by its coordinates (x, y, z).

The system can be excited by three different sources which start at $t \geq 0$: an external force $\tilde{F}(M;t)$ acting on the plate and acoustic sources $\tilde{S}^+(Q;t)$ and $\tilde{S}^-(Q;t)$ respectively in Ω^+ and Ω^-. The plate displacement is denoted by $\tilde{w}(M;t)$ and the acoustic pressures by $\tilde{p}^+(Q;t)$ and $\tilde{p}^-(Q;t)$. These functions satisfy the following equations:

$$\left(D\Delta^2 + \bar{\mu}\frac{\partial^2}{\partial t^2}\right)\tilde{w}(M;t) + \operatorname{Tr}\tilde{p}^+(M;t) - \operatorname{Tr}\tilde{p}^-(M;t) = \tilde{F}(M;t) \quad \text{for} \quad M \in \Sigma \quad (3.33)$$

$$\begin{aligned}&\left(\Delta - \frac{1}{c_0^{+2}}\frac{\partial^2}{\partial t^2}\right)\tilde{p}^+(Q;t) = \tilde{S}^+(Q;t) \quad \text{for} \quad Q \in \Omega^+ \\ &\left(\Delta - \frac{1}{c_0^{-2}}\frac{\partial^2}{\partial t^2}\right)\tilde{p}^-(Q;t) = \tilde{S}^-(Q;t) \quad \text{for} \quad Q \in \Omega^-\end{aligned} \quad (3.33')$$

The coupling between the plate and the fluids is expressed by the continuity of the plate acceleration and the normal component of the fluid particle acceleration:

$$\frac{1}{\mu_0^+}\operatorname{Tr}\frac{\partial\tilde{p}^+}{\partial z}(M;t) = \frac{1}{\mu_0^-}\operatorname{Tr}\frac{\partial\tilde{p}^-}{\partial z}(M;t) = -\frac{\partial^2\tilde{w}}{\partial t^2}(M;t) \quad M \in \Sigma \quad (3.33'')$$

To ensure the uniqueness of the solution of these equations, we assume that the fluid and the plate are at rest for $t < 0$ and that, at infinity in the plate and in the fluids, only outgoing waves are present.

We will restrict this section to the study of time harmonic regimes. Thus, the equations are:

$$D\left(\Delta^2 - \lambda^4\right)w(M;\omega) + \operatorname{Tr}p^+(M;\omega) - \operatorname{Tr}p^-(M;\omega) = F(M;\omega) \text{ for } M \in \Sigma \quad (3.34)$$

$$\begin{cases}\left(\Delta - k_0^{+2}\right)p^+(Q;\omega) = S^+(Q;\omega) \quad \text{for} \quad Q \in \Omega^+ \\ \left(\Delta - k_0^{-2}\right)p^-(Q;\omega) = S^-(Q;\omega) \quad \text{for} \quad Q \in \Omega^-\end{cases} \quad (3.34')$$

$$\frac{1}{\mu_0^+}\operatorname{Tr}\frac{\partial p^+}{\partial z}(M;\omega) = \frac{1}{\mu_0^-}\operatorname{Tr}\frac{\partial p^-}{\partial z}(M;\omega) = \omega^2 w(M;\omega) \quad M \in \Sigma \quad (3.34'')$$

$$\text{with} \quad \lambda^4 = \frac{\bar{\mu}\omega^2}{D} \quad \text{and} \quad k_0^{\pm} = \frac{\omega}{c_0^{\pm}}$$

The uniqueness of the solution is ensured by adding a condition at infinity for the three unknowns $w(M;\omega)$, $p^+(Q;\omega)$ and $p^-(Q;\omega)$, such as, for example, a Sommerfeld condition, or by using the limit absorption principle. In what follows, we will omit the variable ω, unless it is useful.

2. *Free waves in the plate immersed in a unique fluid*

We first look at the possibility of solutions for the homogenous equations (3.34) to (3.34″). More precisely, we examine the condition under which a solution exists for the plate displacement of the form $w(M) = e^{\imath \Lambda x}$. Boundary conditions (3.34″) together with equations (3.34′) imply that the acoustic pressures have the following form

$$p^+(Q) = \omega^2 \mu_0 \frac{e^{(\imath \Lambda x + \alpha z)}}{\alpha} \quad , \quad p^-(Q) = -\omega^2 \mu_0 \frac{e^{(\imath \Lambda x - \alpha z)}}{\alpha} \qquad (3.35)$$
$$\text{with} \quad \alpha^2 = \Lambda^2 - k_0^2$$

There is a particular angular frequency Ω_{cr} – called *the critical frequency* – for which $k_0 = \lambda$, and which is given by:

$$\Omega_{\text{cr}} = c_0^2 \sqrt{\frac{\bar{\mu}}{D}}$$

We will see later that it plays a particular role.

Up to now, the parameter Λ is undetermined and, thus the parameter α is unknown. Let us introduce expressions (3.35) of the acoustic pressures into equation (3.34). We obtain the so-called *dispersion equation of the fluid-loaded plate*:

$$\alpha\left(\Lambda^4 - \lambda^4\right) + 2\frac{\mu_0}{\bar{\mu}}\lambda^4 = 0 \qquad (3.36)$$

which depends on the parameter ω that is supposed to have a given positive value. If Λ is a solution of this equation, then $-\Lambda$ is a solution too. In terms of the parameter α, the dispersion equation writes:

$$\mathcal{Y}(\alpha) \equiv \frac{\alpha\left[\left(\alpha^2 + k_0^2\right)^2 - \lambda^4\right]}{\lambda^4} + \frac{2\mu_0}{\bar{\mu}} = 0 \qquad (3.36')$$

Equation (3.36′) has five roots $\alpha_j (j = 1, \cdots, 5)$. The study of the function $\mathcal{Y}(\alpha)$, for α real, shows that:

- ◇ for $k_0 > \lambda$, there is only one real root which is negative;
- ◇ for $k_0 < \lambda$, there is one real negative root, and, depending on the parameters k_0/λ and $2\mu_0/\bar{\mu}$, zero or two positive real roots;
- ◇ the other 4 or 2 roots are pairs of complex conjugate numbers.

With each root α_j, we associate two wavenumbers $\Lambda_j^+ = \sqrt{k_0^2 + \alpha_j^2}$ and $\Lambda_j^- = -\sqrt{k_0^2 + \alpha_j^2}$.

Figure 3.8 shows two curves $\mathcal{Y}(\alpha)$ for $k_0 < \lambda$ (the value $k_0 = \Omega_{\text{cr}}/(15\ c_0)$ has been chosen). The fluid is either air ($\mu_0 = 1.3$ kg m^{-3}, $c_0 = 340$ m s^{-1}), or

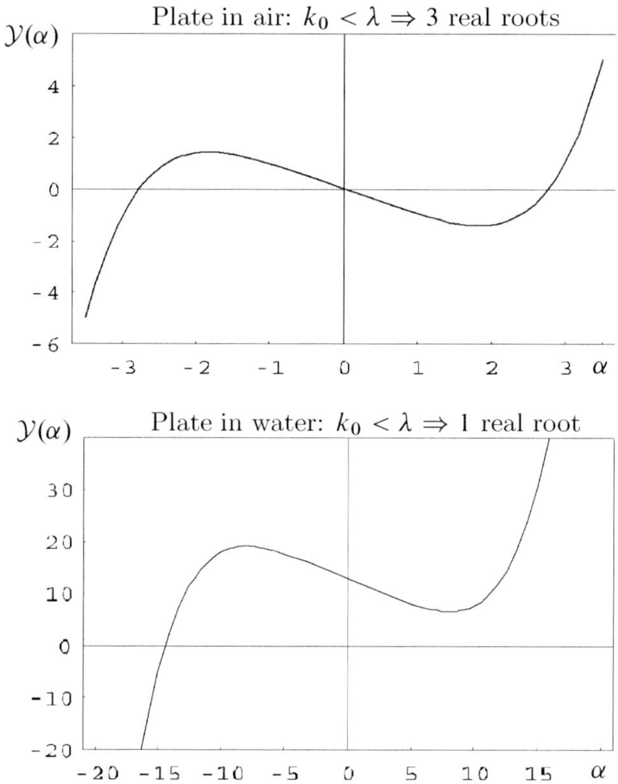

Figure 3.8. *Dispersion curves $\mathcal{Y}(\alpha)$ for $k_0 < \lambda$ ($\omega = \Omega_{\rm cr}/15$), corresponding to a steel plate in air or in water*

water ($\mu_0 = 1000$ kg m^{-3}, $c_0 = 1500$ m s^{-1}). The plate is 2 cm thick and made of steel ($\mu_s = 7800$ kg m^{-3}, $E = 2\ 10^{11}$ Pa, $\nu = 1/3$). The critical frequency is equal to 600 Hz in air and to 11700 Hz in water. Table 3.2 presents the roots α_j of the dispersion equation (3.36′) for the plate in air or in water, and for both cases $k_0 < \lambda$ and $k_0 > \lambda$.

A few comments can be made on these free oscillations of the fluid-loaded plate:

◇ For α_j real and negative, the acoustic waves are evanescent. The wavenumber Λ_j^+ (resp. Λ_j^-) is real and positive (resp. negative) and the plate displacement – which behaves like a wave propagating in the positive (resp. negative) x direction – has a constant amplitude. No energy is transmitted from the plate to the fluid.

◇ For α_j real and positive, the acoustic waves increase exponentially with

steel plate in air				
$k_0 < \lambda$				
−2.772	−(0.00391 +ı2.958)	−(0.00391 −ı2.958)	0.0167	2.763
$k_0 > \lambda$				
−0.00555	−(0.00139 +ı27.175)	−(0.00139 −ı27.175)	(0.00417 −ı15.689)	(0.00417 +ı15.689)
steel plate in water				
$k_0 < \lambda$				
−14.448	−(2.225 +ı14.046)	−(2.225 −ı14.046)	−(9.449 +ı4.757)	−(9.449 −ı4.757)
$k_0 > \lambda$				
−3.938	−(1.0191 −ı27.346)	−(1.0191 +ı27.346)	(2.988 −ı15.959)	(2.988 +ı15.959)

Table 3.2. *Roots α_j of the dispersion equation (3.36′) for a plate in air or in water: for $k_0 < \lambda$ the angular frequency is $\Omega_{cr}/15$; for $k_0 > \lambda$, it is equal to $2\Omega_{cr}$*

$|z|$. The wavenumbers Λ_j^+ and Λ_j^- are real. The physical interpretation of such a situation is not clear.

◇ For α_j complex, the acoustic waves are evanescent if $\Re(\alpha_j) < 0$, or increase exponentially with $|z|$ if $\Re(\alpha_j) > 0$. The wavenumbers Λ_j^+ and Λ_j^- are complex: their imaginary parts have a sign which depends on α_j. The plate displacement decreases exponentially in one direction, depending on the sign of the imaginary part of the wavenumber: this can be interpreted as a damped wave in the plate.

If the fluid is a gas, the parameter $\varepsilon = \mu_0/\bar{\mu}$ is small and, thus, the roots of the dispersion equation (3.36′) are in general close to those of the equation

$$\mathcal{Y}^0(\alpha) \equiv \alpha\left[\left(\alpha^2 + k_0^2\right)^2 - \lambda^4\right] = 0 \qquad (3.36'')$$

which are:

$$\alpha_1^0 = 0 \quad , \quad \alpha_2^0 = -\imath\sqrt{\lambda^2 + k_0^2} \quad , \quad \alpha_3^0 = \imath\sqrt{\lambda^2 + k_0^2}$$

$$\alpha_4^0 = -\sqrt{\lambda^2 - k_0^2} \quad , \quad \alpha_5^0 = \sqrt{\lambda^2 - k_0^2}$$

We look for the roots of the dispersion equation as a formal Taylor series in terms of the successive powers of ε

$$\alpha_j = \alpha_j^0 + \alpha_j^1\varepsilon + \alpha_j^2\varepsilon^2 + \cdots$$

If we limit these series to the first order correcting term, we find:

$$\alpha_1 \simeq \frac{2\lambda^4}{\lambda^4 - k_0^4}\varepsilon$$

$$\alpha_2 \simeq -\imath\sqrt{\lambda^2 + k_0^2} - \frac{\lambda^2}{2(\lambda^2 + k_0^2)}\varepsilon \,,\; \alpha_3 \simeq \imath\sqrt{\lambda^2 + k_0^2} - \frac{\lambda^2}{2(\lambda^2 + k_0^2)}\varepsilon$$

$$\alpha_4 \simeq -\sqrt{\lambda^2 - k_0^2} - \frac{\lambda^2}{2(\lambda^2 - k_0^2)}\varepsilon \,,\; \alpha_5 \simeq \sqrt{\lambda^2 - k_0^2} - \frac{\lambda^2}{2(\lambda^2 - k_0^2)}\varepsilon$$

The approximations of the α_j-s are valid if the correcting term is small. This implies that the hypothesis $\varepsilon \ll 1$ is not sufficient: we must assume that $\varepsilon/(\lambda^2 - k_0^2)$ is small compared to 1 to ensure that the expressions given for α_1, α_4 and α_5 are approximations of the exact roots. In other words, the roots of the dispersion equation can be sought as Taylor series of the successive powers of $\mu_0/\bar{\mu}$ if this parameter is small and if the angular frequency is far enough from the critical angular frequency.

REMARK.– The critical frequency defines a critical wavelength in the fluid, $\Lambda_{\mathrm{cr}} = c_0\Omega_{\mathrm{cr}}/2\pi$, which is sometimes used as the unit length to establish a system of reduced equations.

3. *Transmission and reflection of a plane harmonic incident wave: two identical fluids*

Let us consider a plate in contact with two identical fluids, and excited by an acoustic plane wave in the domain $z > 0$. The total acoustic pressure $p^+(M)$ has the form

$$p^+(x, y, z) = \mathrm{e}^{\imath k_0(x \sin\theta - z\cos\theta)} + p^r(x, y, z) \tag{3.37}$$

where θ is the angle of incidence of the incident wave.

The plate displacement $w(x, y)$, the reflected acoustic pressure $p^r(x, y, z)$ and the transmitted acoustic pressure $p^t(x, y, z)$ satisfy the system of equations:

$$D\left(\Delta^2 - \lambda^4\right)w(x, y) + p^r(x, y, 0) - p^t(x, y, 0) = -\mathrm{e}^{\imath k_0 x \sin\theta} \tag{3.38}$$

$$\begin{cases} \left(\Delta - k_0^2\right)p^r(x, y, z) = 0 \quad \text{for} \quad z > 0 \\ \left(\Delta - k_0^2\right)p^t(x, y, z) = 0 \quad \text{for} \quad z < 0 \end{cases} \tag{3.38'}$$

$$\frac{1}{\mu_0}\frac{\partial p^r}{\partial z}(x, y, 0) - \frac{\imath k_0\cos\theta}{\mu_0}\mathrm{e}^{\imath k_0 x \sin\theta} = \frac{1}{\mu_0}\frac{\partial p^t}{\partial z}(x, y, 0) = \omega^2 w(x, y) \tag{3.38''}$$

As usual, an outgoing wave condition for $p^r(x,z)$ and $p^t(x,z)$ is used to ensure the uniqueness of the solution. The data – the incident pressure in equation (3.38) and its derivative with respect to z in equation (3.38″) – do not depend on the variable y. Thus, the unknown functions are also independent of this variable, which will be omitted in what follows. Furthermore, the data are proportional to $e^{\imath k_0 x \sin\theta}$, and so are the unknown functions, that is:

$$w(x) = W e^{\imath k_0 x \sin\theta}$$

$$p^r(x,z) = R\, e^{\imath(k_0 x \sin\theta + \alpha^r z)} \quad , \quad p^t(x,z) = T\, e^{\imath(k_0 x \sin\theta - \alpha^t z)}$$

The problem reduces to the determination of the five constants W, R, α^r, T and α^t. Equations (3.38′) together with the outgoing wave condition imply that

$$\alpha^r = k_0 \cos\theta \quad \Rightarrow \quad p^r(x,z) = R\, e^{\imath k_0 (x\sin\theta + z\cos\theta)}$$
$$\alpha^t = k_0 \cos\theta \quad \Rightarrow \quad p^t(x,z) = T\, e^{\imath k_0 (x\sin\theta - z\cos\theta)}$$

By introducing these expressions into equations (3.38) and (3.38″), we obtain:

$$D\left(k_0^4 \sin^4\theta - \lambda^4\right)W + R - T = -1$$
$$R + T = 1$$
$$\mu_0 \omega^2 W + \imath k_0 \cos\theta\; T = 0$$

Solving this system leads to the following solution of the problem:

$$\begin{aligned} w(x) &= -\frac{2\imath k_0 \cos\theta}{\imath k_0 D \cos\theta\left(k_0^4 \sin^4\theta - \lambda^4\right) + 2\mu_0\omega^2} e^{\imath k_0 x \sin\theta} \\ p^r(x,z) &= \frac{\imath k_0 D \cos\theta\left(k_0^4 \sin^4\theta - \lambda^4\right)}{\imath k_0 D \cos\theta\left(k_0^4 \sin^4\theta - \lambda^4\right) + 2\mu_0\omega^2} e^{\imath k_0 (x\sin\theta + z\cos\theta)} \\ p^t(x,z) &= \frac{2\mu_0\omega^2}{\imath k_0 D \cos\theta\left(k_0^4 \sin^4\theta - \lambda^4\right) + 2\mu_0\omega^2} e^{\imath k_0 (x\sin\theta - z\cos\theta)} \end{aligned} \tag{3.39}$$

The denominator that appears in these three expressions is no more than the first member of the dispersion equation (3.36) with $\Lambda = k_0 \sin\theta$. Because all the parameters have real values, this denominator can never be zero. Its modulus has a minimum for $k_0 \sin\theta - \lambda = 0$. For a given angle of incidence, this equality defines an angular frequency Ω_c, called *coincidence angular frequency* and given by:

$$\Omega_c(\theta) = \frac{c_0^2}{\sin^2\theta}\sqrt{\frac{\mu}{D}} = \frac{\Omega_{cr}}{\sin^2\theta}$$

The coincidence frequency is always larger than the critical frequency and decreases as the angle of incidence is increased; it tends to the critical frequency as θ tends to $\pi/2$. At the coincidence frequency, we have

$$\Delta^2 w = k_0^4 \sin^4\theta\; w = \lambda^4 w$$

So, the plate displacement satisfies the homogenous *in vacuo* plate equation, that is it corresponds to a wave which can propagate freely in the plate. The reflected acoustic wave is identically zero, while the transmitted wave is equal to the incident wave: in other words, the plate is totally transparent to the acoustic field. We have already seen such a phenomenon in the one-dimensional previous example.

The energy transmission rate $\tau(\theta, \omega)$ of the plate is defined as the ratio of the mean power of the transmitted wave to the mean power of the incident wave. Here we have

$$\tau(\theta,\omega) = |T(\theta,\omega)|^2 = \left| \frac{2\mu_0\omega^2}{\imath k_0 D \cos\theta \big(k_0^4 \sin^4\theta - \lambda^4\big) + 2\mu_0\omega^2} \right|^2$$

The insertion loss index, which characterizes the acoustic insulation due to the presence of the plate, is defined by:

$$\begin{aligned} \mathcal{I}(\theta,\omega) = 10\log\frac{1}{\tau(\theta,\omega)} &= -20\log\left[\frac{2\mu_0\omega^2}{\left|\imath k_0 D\cos\theta\big(k_0^4\sin^4\theta - \lambda^4\big) + 2\mu_0\omega^2\right|}\right] \\ &= 20\log\left|\imath\frac{\omega D\cos\theta}{2c_0\mu_0}\left(\frac{\omega^2}{c_0^4}\sin^4\theta - \frac{\bar{\mu}}{D}\right) + 1\right| \qquad (3.40) \end{aligned}$$

and it takes the following simple form for $\theta = 0$:

$$\mathcal{I}(0,\omega) = 10\log\left[1 + \frac{\omega^2\bar{\mu}^2}{4c_0^2\mu_0^2}\right]$$

For any angle of incidence, the insertion loss index is zero for $\omega = 0$. As already mentioned, the insertion loss index is zero at the coincidence frequency, and it tends to infinity as the frequency increases.

For $\theta \neq 0$ the following approximations can be established:

- at low frequency, and more precisely if $\omega \ll c_0^2\sqrt{\bar{\mu}/D}/\sin^2\theta$, the insertion loss index has the asymptotic form

  $$\mathcal{I}(\theta,\omega) \simeq 10\log\left[1 + \left(\omega\frac{\bar{\mu}\cos\theta}{2\mu_0 c_0}\right)^2\right]$$

  and if we add the assumption $\omega\bar{\mu}\cos\theta/2\mu_0 c_0 \gg 1$, then we have

  $$\mathcal{I}(\theta,\omega) \simeq 20\log\left[\omega\frac{\bar{\mu}\cos\theta}{2\mu_0 c_0}\right]$$

◇ at high frequency and for $\theta \neq 0$, the insertion loss index has the asymptotic form

$$\mathcal{I}(\theta,\omega) \simeq 20 \log \frac{\omega^3 D \cos\theta \sin^4\theta}{2\mu_0 c_0^5} \quad \text{for} \quad \omega \gg 1$$

We can conclude that, for a given fluid, the insertion loss index is governed at low frequency by the density of the plate, while, at high frequency, the governing parameter is the plate rigidity.

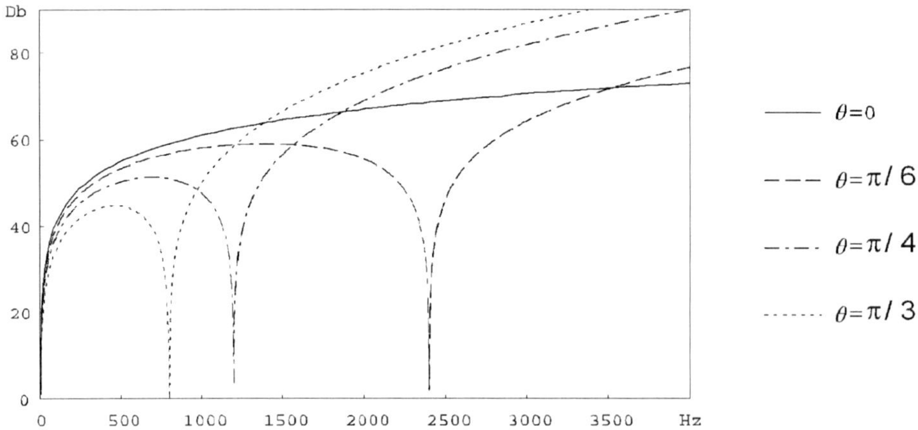

Figure 3.9. *Insertion loss index for an infinite plate for four angles of incidence*

Figure 3.9 shows the insertion loss index, as a function of the frequency $\nu = \omega/2\pi$, for an infinite plate in air, with the same mechanical characteristics as previously. Four angles of incidence are considered: 0, $\pi/6$, $\pi/4$ and $\pi/3$.

4. *Transmission and reflection of a plane harmonic incident wave: two different fluids*

We now consider a plate that is in contact with two different fluids and excited by an incident plane wave. The fluids are respectively characterized by (μ_0^+, c_0^+) and (μ_0^-, c_0^-). As before, the total acoustic pressure in the half-space $z > 0$ has the form:

$$p^+(x,y,z) = e^{\imath k_0^+ (x \sin\theta - z\cos\theta)} + p^r(x,y,z) \tag{3.41}$$

The plate displacement $w(x, y)$, the reflected acoustic pressure $p^r(x, y, z)$ and the transmitted acoustic pressure $p^t(x, y, z)$ satisfy the system of equations:

$$D\left(\Delta^2 - \lambda^4\right)w(x, y) + p^r(x, y, 0) - p^t(x, y, 0) = -e^{\imath k_0^+ x \sin\theta} \quad \text{with} \quad \lambda^4 = \frac{\bar{\mu}\omega^2}{D} \tag{3.42}$$

$$\begin{cases} \left(\Delta - k_0^{+2}\right)p^r(x, y, z) = 0 & \text{for} \quad z > 0 \\ \left(\Delta - k_0^{-2}\right)p^t(x, y, z) = 0 & \text{for} \quad z < 0 \end{cases} \tag{3.42'}$$

$$\frac{1}{\mu_0^+}\left[\frac{\partial p^r}{\partial z}(x, y, 0) - \imath k_0^+ \cos\theta \; e^{\imath k_0^+ x \sin\theta}\right] = \frac{1}{\mu_0^-}\frac{\partial p^t}{\partial z}(x, y, 0) = \omega^2 w(x, y) \tag{3.42''}$$

Equation (3.42″) implies that the reflected pressure and the transmitted pressure do not depend on y. Furthermore, we must have:

$$w(x) = W e^{\imath k_0^+ x \sin\theta}$$

$$p^r(x, z) = R \; e^{\imath (k_0^+ x \sin\theta + \omega z \alpha^r)} \quad , \quad p^t(x, z) = T \; e^{\imath (k_0^+ x \sin\theta - \omega z \alpha^t)}$$

From equations (3.42′) and the outgoing wave condition for p^r, we obtain:

$$\omega\alpha^r = k_0^+ \cos\theta \quad \Rightarrow \quad p^r(x, z) = R \; e^{\imath k_0^+ (x \sin\theta + z \cos\theta)}$$

The coefficient $(\alpha^t)^2$ is equal to $1/c_0^{-2} - \sin^2\theta/c_0^{+2}$, which can be positive or negative:

1. if $1/c_0^{-2} - \sin^2\theta/c_0^{+2} > 0$, then $\alpha^t = \sqrt{1/c_0^{-2} - \sin^2\theta/c_0^{+2}}$ is real and positive and the acoustic wave is a propagating wave;

2. if $1/c_0^{-2} - \sin^2\theta/c_0^{+2} < 0$, then $\alpha^t = -\imath\sqrt{1/c_0^{-2} - \sin^2\theta/c_0^{+2}}$ is imaginary and negative and the acoustic wave is an evanescent wave.

This can be summarized as:

$$\alpha^t = \operatorname{sgn}\left[\frac{1}{c_0^{-2}} - \frac{\sin^2\theta}{c_0^{+2}}\right]\sqrt{\frac{1}{c_0^{-2}} - \frac{\sin^2\theta}{c_0^{+2}}} \quad \Rightarrow \quad p^t(x, z) = T \; e^{\imath\omega(x \sin\theta / c_0^+ - z\alpha^t)}$$

REMARK.– If $c_0^+ > c_0^-$, we can always define a refraction angle $\theta' < \theta$ by $(\sin\theta')/c_0^- = (\sin\theta)/c_0^+$. We then have $\alpha^t = k_0^- \cos\theta'$ and p^t is a propagating wave. For $c_0^+ < c_0^-$, we can define in the same way a refraction angle for $\theta < \theta_{\max} = \arcsin(c_0^+/c_0^-)$. The transmitted pressure is a propagating wave for $\theta < \theta_{\max}$ and an evanescent wave for $\theta > \theta_{\max}$.

The three unknown constants W, R and T are obtained by introducing the expressions of the plate displacement and the reflected and transmitted pressures into equations (3.42), (3.42′) and (3.42″). An easy calculation leads to the following result:

$$W(\theta,\omega) = -\frac{2}{\mathcal{D}} \frac{\alpha^t \cos\theta}{\omega c_0^+ \mu_0^+ \mu_0^-}$$

$$R(\theta,\omega) = \frac{1}{\mathcal{D}}\left\{\left[D\frac{\omega^2 \sin^4\theta}{c_0^{+4}} - \bar{\mu}\right]\frac{\omega\alpha^t\cos\theta}{c_0^+\mu_0^+\mu_0^-} - \imath\left[\frac{\cos\theta}{c_0^+\mu_0^+} - \frac{\alpha^t}{\mu_0^-}\right]\right\} \tag{3.43}$$

$$T(\theta,\omega) = -\frac{2\imath}{\mathcal{D}}\frac{\cos\theta}{c_0^+\mu_0^+}$$

with $$\mathcal{D} = \left[D\frac{\omega^2\sin^4\theta}{c_0^{+4}} - \bar{\mu}\right]\frac{\omega\alpha^t\cos\theta}{c_0^+\mu_0^+\mu_0^-} - \imath\left[\frac{\cos\theta}{c_0^+\mu_0^+} + \frac{\alpha^t}{\mu_0^-}\right]$$

The insertion loss introduced by the presence of the plate must be defined in a slightly different way. The wave transmitted through the plate must be compared to the wave transmitted in the second fluid in the absence of the plate.

Let us consider the transmission of acoustic energy between the two fluids in the absence of the plate. The reflected wave ϖ^r and the transmitted wave ϖ^t satisfy the homogenous Helmholtz equations (3.42′) and the following continuity conditions:

$$\varpi^r(x,y,0) - \varpi^t(x,y,0) = -e^{\imath k_0^+ x\sin\theta}$$

$$\frac{1}{\mu_0^+}\frac{\partial\varpi^r}{\partial z}(x,y,0) - \frac{1}{\mu_0^-}\frac{\partial\varpi^t}{\partial z}(x,y,0) = \frac{\imath k_0^+\cos\theta}{\mu_0^+}e^{\imath k_0^+ x\sin\theta}$$

We immediately find that the reflected and transmitted pressures are independent of y and they are given by:

$$\varpi^r(x,z) = R_0\ \mathrm{e}^{\imath\omega(x\sin\theta + z\cos\theta)/c_0^+}$$

$$\varpi^t(x,z) = T_0\ \mathrm{e}^{\imath\omega(x\sin\theta/c_0^+ - z\alpha^t)}$$

$$R_0(\theta) = \frac{1}{\mathcal{D}_0}\left[\frac{\cos\theta}{\mu_0^+ c_0^+} - \frac{\alpha^t}{\mu_0^-}\right] \ , \quad T_0(\theta) = \frac{1}{\mathcal{D}_0}\frac{2\cos\theta}{\mu_0^+ c_0^+} \tag{3.44}$$

$$\mathcal{D}_0 = \frac{\cos\theta}{\mu_0^+ c_0^+} + \frac{\alpha^t}{\mu_0^-}$$

Thus, we can define an insertion loss index by

$$\widehat{\mathcal{I}}(\theta,\omega) = -20\log\left|\frac{T(\theta,\omega)}{T_0(\theta)}\right|$$

At the coincidence angular frequency $\Omega_c(\theta) = c_0^2\sqrt{\bar{\mu}/D}/\sin^2\theta$, the insertion loss index is again equal to zero.

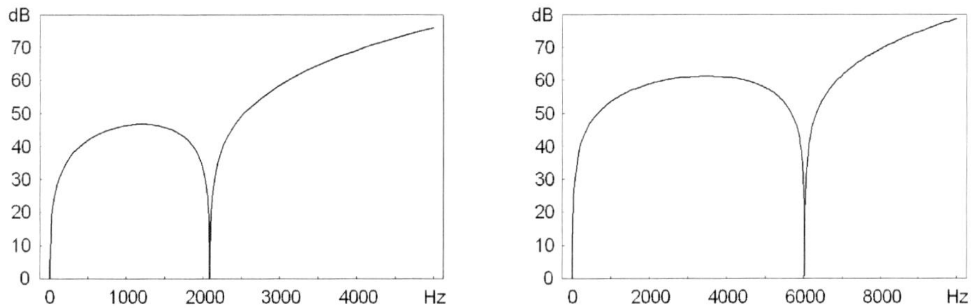

Figure 3.10. *Insertion loss index: from air to a fictitious fluid (left); from a fictitious fluid to air (right)*

Numerical examples of this insertion loss are presented in Figures 3.10 and 3.11, and the plate characteristics are the same as in paragraph 2 of the present section (steel plate):

◇ Figure 3.10 corresponds to air and a fictitious fluid with a density equal to 1.7 times the air density and a sound speed equal to 1.7 times the sound speed in air. For the transmission from air to the fictitious fluid, the limit angle is $\theta_{\max} = 0.4\ \pi/2$. Both curves are calculated for $\theta = 0.9\ \theta_{\max}$. The minima of the insertion loss index occur at 2000 Hz and at 6000 Hz. The left hand side curve corresponds to the transmission from air to the fictitious fluid, while the right hand side one to the transmission from the fictitious fluid to air. These curves are very similar to the those in Figure 3.9.

◇ Figure 3.11 corresponds to air and water. For the transmission from air to water, the limit angle is $\theta_{\max} = 0.146\,\pi/2$. Both curves are calculated for $\theta = 0.9\,\theta_{\max}$. The minima of the insertion loss index occur around 15 Hz and 380 Hz. The left hand side curve corresponds to the transmission from air to water, while the right hand side one to the transmission from water to air. The curve corresponding to the transmission from air to water is different from the curves in Figure 3.9: the minimum is much less sharp.

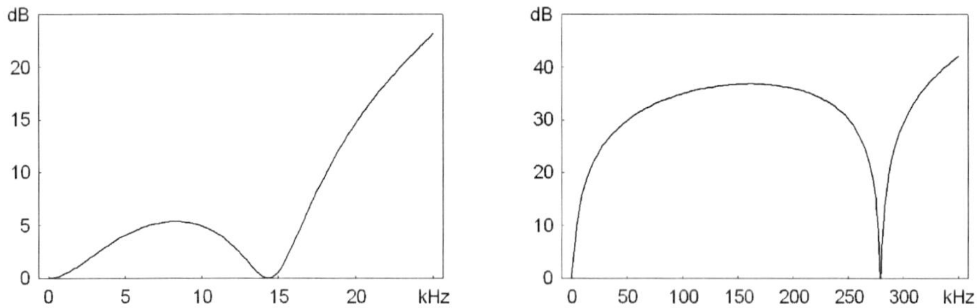

Figure 3.11. *Insertion loss index: from air to water (left); from water to air (right)*

The main difference between the previous system – a plate separating two identical fluids – and this one is that a limit angle of incidence $\theta_{\max}$ appears, which plays a fundamental role. For incidence angles less than this limit angle, both systems behave in a similar way, but for incidence angles larger than $\theta_{\max}$ – that is when α^t is purely imaginary – we observe a phenomenon of total reflection: the modulus of $R(\theta,\omega)$ is equal to one, and no energy is transmitted through the plate into the second fluid.

5. *Infinite plate immersed in a unique fluid and excited by a point harmonic force*

Let $F\delta_O$ be a point harmonic force acting on the plate at the coordinate origin O. Equations (3.34-3.34′-3.34″) reduce to:

$$D\left(\Delta^2 - \lambda^4\right)w(M) + \operatorname{Tr} p^+(M) - \operatorname{Tr} p^-(M) = F\delta_O \quad M \in \Sigma \tag{3.45}$$

$$\left(\Delta - k_0^2\right)p^\pm(Q) = 0 \quad \text{for} \quad z > 0 \quad \text{or} \quad z < 0 \tag{3.45'}$$

$$\frac{1}{\mu_0}\operatorname{Tr}\frac{\partial p^+}{\partial z}(M)=\frac{1}{\mu_0}\operatorname{Tr}\frac{\partial p^-}{\partial z}(M)=\omega^2 w(M)\quad M\in\Sigma \tag{3.45''}$$

The acoustic pressures can be explicitly expressed in terms of the plate velocity. Let $\mathcal{G}(M,M')$ be the Green's function for the Helmholtz equation in the half-space $z>0$ which satisfies the homogenous Neumann condition on $z=0$, that is, the solution of

$$\left(\Delta-k_0^2\right)\mathcal{G}(Q,Q')=\delta_{Q'}(Q)\quad\text{for}\quad Q\text{ and }Q'\in z>0$$
$$\partial_z\mathcal{G}(M,Q')=0\quad\text{for}\quad M\in z=0$$

which satisfies the Sommerfeld condition at infinity. It is obvious that this function is given by:

$$\mathcal{G}(Q,Q')=G(Q,Q')+G(Q,Q'')$$
$$\text{with } G(Q,Q')=-\frac{e^{ik_0r(Q,Q')}}{4\pi r(Q,Q')}\,,\ r(Q,Q')=\text{distance between } Q \text{ and } Q'$$
$$\text{and } Q'' \text{ symmetric of } Q' \text{ with respect of the plane } z=0$$

The Green's function which satisfies the Helmholtz equation in the half-space $z<0$ and the Neumann condition on $z=0$ has the same expression. By applying the Green's formula to $p^{\pm}(Q)$ and $\mathcal{G}(Q,Q')$, the Green's representation of the acoustic pressures in terms of the plate displacement is obtained:

$$p^{\pm}(Q)=\pm 2\omega^2\mu_0\int_{\Sigma}G(Q,M')w(M')\,\mathrm{d}\sigma(M') \tag{3.46}$$

Introducing this expression into equation (3.45), we obtain an integro-differential equation for w only:

$$D\left(\Delta^2-\lambda^4\right)w(M)$$
$$+\,4\omega^2\mu_0\int_{\Sigma}G(M,M')w(M')\,\mathrm{d}\sigma(M')=F\delta_O\,,\ M\in\Sigma \tag{3.47}$$

No analytical solution can be obtained for this equation. Nevertheless, the use of the two-dimensional space Fourier transform leads to an integral expression of the plate displacement which can be computed numerically and which provides an analytic expression of the far field acoustic pressures.

Fourier transform of the plate displacement.– Because the excitation is a point isotropic force at the coordinate origin, the unknowns have a cylindrical symmetry. In particular, the plate displacement will depend on the radial cylindrical coordinate ρ only. It is well known [SCH 61, ERD 54-2] that if a function f defined in a plane depends on the radial coordinate only, its Fourier transform $\hat{f}$ depends of the radial coordinate ξ in the Fourier variables. The functions f and $\hat{f}$ are related by:

$$\hat{f}(\xi) = \mathcal{F}f = 2\pi \int_0^\infty f(\rho) J_0(2\pi\xi\rho)\rho\, \mathrm{d}\rho$$

$$f(\rho) = 2\pi \int_0^\infty \hat{f}(\xi) J_0(2\pi\xi\rho)\xi\, \mathrm{d}\xi = \pi \int_{\infty e^{\imath\pi}}^\infty \hat{f}(\xi) H_0(2\pi\xi)\xi\, \mathrm{d}\xi$$

The Fourier transform of the squared Laplace operator is $16\pi^4\xi^4$. The Fourier transform of the integral in (3.47) is less straightforward to obtain. We first remark that the acoustic pressures $p^+(M)$ and $p^-(M)$ given by (3.46) are convolution products

$$p^\pm(Q) = \pm 2\omega^2\mu_0 G * w \otimes \delta_\Sigma(Q)$$

where $w \otimes \delta_\Sigma$ is the distribution supported by Σ with density w, and the symbol "$*$" stands for the three-dimensional space convolution product. The two-dimensional space Fourier transform of $p^\pm$ is readily obtained:

$$\mathcal{F}\left[-\frac{e^{\imath k_0 r}}{4\pi r}\right] = \frac{e^{\imath K|z|}}{2\imath K} \quad \text{with} \quad K^2 = k_0^2 - 4\pi^2\xi^2 \quad , \quad \mathcal{F}(w \otimes \delta_\Sigma) = \hat{w}(\xi) \otimes \delta_z$$

$$\Longrightarrow \mathcal{F}p^\pm = \pm 2\omega^2\mu_0 \mathcal{F}\big[G * w \otimes \delta_\Sigma\big] = \pm 2\omega^2\mu_0 \frac{\hat{w}(\xi) e^{\imath K|z|}}{2\imath K}$$

If K^2 is positive, we choose $K > 0$; for K^2 negative or complex, we choose $\Im(K) > 0$. From the expression of $\mathcal{F}p^\pm$ we deduce that the Fourier transform of the integral term in equation (3.47) is:

$$\mathcal{F}\left[4\omega^2\mu_0 \int_\Sigma G(M, M') w(M')\, \mathrm{d}\sigma(M')\right] = 2\omega^2\mu_0 \lim_{z\to 0}\left[\frac{\hat{w}(\xi) e^{\imath K|z|}}{\imath K}\right]$$

$$= 2\omega^2\mu_0 \frac{\hat{w}(\xi)}{\imath K}$$

Then equation (3.47) leads to:

$$D\left[\left(16\pi^4\xi^4 - \lambda^4\right) + \frac{2\lambda^4\mu_0}{\imath K\bar{\mu}}\right]\hat{w}(\xi) = F$$

which gives

$$\hat{w}(\xi) = \frac{F}{D}\frac{\imath K}{\imath K\left(16\pi^4\xi^4 - \lambda^4\right) + 2\lambda^4\mu_0/\bar{\mu}} \tag{3.48}$$

where ω^2 has been replaced by $\lambda^4 D/\bar{\mu}$.

Expression of the plate displacement.– An exhaustive analytical study of the plate displacement is a difficult task which is beyond the scope of this textbook. Nevertheless, it is useful to have an overview of the method which can be used.

The plate displacement is given by the following integral

$$w(\rho) = \frac{F\pi}{D} \int_{\infty e^{\imath\pi}}^{\infty} \frac{\imath K}{\imath K\left(16\pi^4\xi^4 - \lambda^4\right) + 2\lambda^4\mu_0/\bar{\mu}} H_0(2\pi\xi)\xi \,\mathrm{d}\xi \tag{3.49}$$

which can be evaluated by a contour integral, but some caution must first be taken to determine which poles of the integrand must be accounted for. First, let us remark that if ξ_j is a pole of the integrand, then $-\xi_j$ is a pole too. The integrand has two kinds of poles. Some of them, say $\pm\xi_j^c$, are complex and do not cause any trouble because only half of them will lie inside the integration contour. The other ones, say $\pm\xi_j^r$, are real and it is not obvious, at first sight, to determine which ones must be accounted for. One way is to use the limit absorption principle, which ensures the uniqueness of the plate displacement.

Let us first assume that the system plate/fluid is slightly absorbing. A small damping is equivalent to replacing the real angular frequency ω with a complex angular frequency $\omega_\varepsilon = \omega(1 + \imath\varepsilon)$, with $\varepsilon > 0$ and $\varepsilon \ll 1$. We look for a displacement $w_\varepsilon(\rho)$ defined by

$$w_\varepsilon(\rho) = \frac{F\pi}{D} \int_{\infty e^{\imath\pi}}^{\infty} \frac{\imath K_\varepsilon}{\imath K_\varepsilon\left(16\pi^4\xi^4 - \lambda_\varepsilon^4\right) + 2\,\lambda_\varepsilon^4\mu_0/\bar{\mu}} H_0(2\pi\xi)\xi \,\mathrm{d}\xi$$

$$\text{with} \quad \lambda_\varepsilon^4 = \frac{\bar{\mu}\omega_\varepsilon^2}{D} \quad , \quad K_\varepsilon^2 = \frac{\omega_\varepsilon^2}{c_0^2} - 4\pi^2\xi^2 \quad \text{and} \quad \Im(K_\varepsilon) > 0$$

which is finite everywhere. Because the angular frequency has a non-zero imaginary part, we are ensured that the integrand has no pole along the real ξ-axis. Let R and ε' be positive numbers (R will be allowed to grow up to infinity and ε' will tend to 0), and let us define an integration contour C by (see Figure 3.12):

1. the half-circle $|\xi| = R$ in the half-plane $\Im(\xi) > 0$;

2. the parts of the real axis $-R \leq \xi \leq -\varepsilon'$ and $+\varepsilon' \leq \xi \leq R$;

3. any contour starting at $\xi = \varepsilon'$, passing around the point $\xi = k_0(1+\imath\varepsilon)/2\pi$ and ending at $\xi = \varepsilon'$.

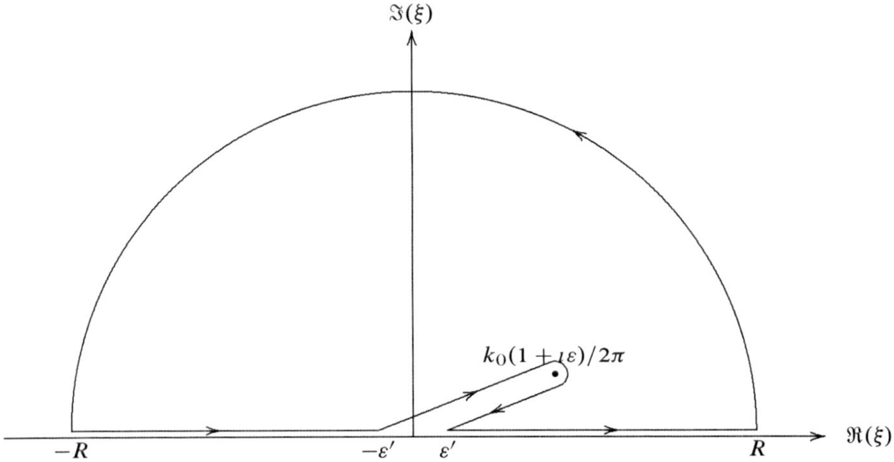

Figure 3.12. *Integration contour for the evaluation of the fluid-loaded plate displacement*

It is easily shown that, when $R \to \infty$ and $\varepsilon' \to 0$, the limit of the integral

$$w_\varepsilon^C(\rho) = \frac{F\pi}{D} \int_C \frac{\imath K_\varepsilon}{\imath K_\varepsilon\left(16\pi^4\xi^4 - \lambda_\varepsilon^4\right) + 2\lambda_\varepsilon^4 \mu_0/\bar{\mu}} H_0(2\pi\xi)\xi \, d\xi$$

is equal to the sum of $w_\varepsilon(\rho)$ and the integral along the branch-cut $[0, k_0(1+\imath\varepsilon)/2\pi]$. Indeed, when $R \to \infty$, the function to be integrated tends exponentially to zero because of the asymptotic behavior of the Hankel function: thus, the integral along the half-circle tends to zero. When $R \to \infty$ and $\varepsilon' \to 0$, the integral along the second part of the contour tends to $w_\varepsilon(\rho)$. We can conclude that the plate displacement is given by a double limit

$$w(\rho) = \lim_{\varepsilon \to 0} w_\varepsilon(\rho) = \lim_{\varepsilon \to 0}\left[\lim_{R \to \infty,\ \varepsilon' \to 0}\left(w_\varepsilon^C(\rho) - \text{branch cut integral}\right)\right]$$

The branch-cut integral cannot be evaluated analytically: analytical or numerical approximations can be found in the literature. The limit of the contour integral $w_\varepsilon^C(\rho)$ is easy to obtain. Inside contour C, the integrand is a meromorphic function. Thus, $w_\varepsilon^C(\rho)$ is the sum of the residues due to the poles of the integrand, that is, the roots of the dispersion equation

$$\imath K_\varepsilon\left(\Lambda^4 - \lambda_\varepsilon^4\right) + 2\lambda_\varepsilon^4 \frac{\mu_0}{\bar{\mu}} = 0 \quad \text{with} \quad \Lambda = 2\pi\xi \tag{3.50}$$

which have a positive imaginary part. The roots of equation (3.50) are of two kinds: the roots $\pm\xi_j^c(\varepsilon)$ that remain complex and tend to $\pm\xi_j^c$ when ε tends to zero; and the roots $\pm\xi_j^r(\varepsilon)$ that become real and tend to $\pm\xi_j^r$ when ε tends to zero. This limit process enables us to choose which of the real roots must be used to evaluate the limit, for $\varepsilon \to 0$, of the contour integral $w_\varepsilon^C(\rho)$.

The far field acoustic pressure.– The acoustic pressure $p^+(Q)$, as given by expression (3.46), that is

$$p^+(Q) = -\frac{1}{2\pi}\omega^2\mu_0 \int_\Sigma \frac{e^{\imath k_0 r(Q,M')}}{r(Q,M')} w(M')\,\mathrm{d}\sigma(M') \tag{3.51}$$

cannot be calculated analytically. However, its asymptotic behavior can easily be expressed in terms of the Fourier transform of the plate displacement, as will be shown.

Let (R,θ,φ) be the spherical coordinates of the point Q, and $(R',\theta' = \pi/2,\varphi')$ those of M'. We look for the asymptotic form of $p^+(Q) = p^+(R,\theta,\varphi)$ for $R \to \infty$. We first introduce the expansion of the Green's function in terms of spherical harmonics (see [MOR 53]):

$$\frac{e^{\imath k_0 r(Q,M')}}{r(Q,M')} = \imath k_0 \sum_{n=0}^{\infty}(2n+1)\sum_{m=0}^{n}\varepsilon_m \frac{(n-m)!}{(n+m)!}\cos\big[m(\varphi-\varphi')\big]$$
$$P_n^m(\cos\theta)P_n^m(\cos\theta')h_n(k_0R)j_n(k_0R') \quad , \quad \text{for } R > R' \tag{3.52}$$

where:

- ◇ $\varepsilon_m = 1$ for $m = 0$ and $\varepsilon_m = 2$ for $m > 0$;
- ◇ $h_n(z)$ are the spherical Hankel functions of the first kind and order n, and $j_n(z)$ are the spherical Bessel functions of order n;
- ◇ $P_n^m(z)$ are the associated Legendre functions of the first kind of degree n and order m.

The definitions and properties of these functions can be found in many classical books as [ABR 65, ERD 54-1, MOR 53]. Expansion (3.52) is convergent for $R > R'$, only. Thus, it cannot be used to compute the acoustic pressure at a finite distance because the integral extends up to infinity. But, as far as an asymptotic behavior is concerned, the convergence restriction $R > R'$ can be ignored (the proof is beyond the scope of this monograph). We then introduce the asymptotic behavior of the spherical Hankel function

$$h_n(k_0R) = \imath^{-n}\frac{e^{\imath k_0 R}}{\imath k_0 R} + \mathcal{O}\big(k_0^{-2}R^{-2}\big)$$

which leads to

$$\frac{\mathrm{e}^{\imath k_0 r(Q,M')}}{r(Q,M')} \simeq \frac{\mathrm{e}^{\imath k_0 R}}{R} \sum_{n=0}^{\infty} \imath^{-n}(2n+1) \sum_{m=0}^{n} \varepsilon_m \frac{(n-m)!}{(n+m)!} \cos\big[m(\varphi-\varphi')\big]$$
$$P_n^m(\cos\theta) P_n^m(\cos\theta') j_n(k_0 R') \quad , \quad \text{for } R > R' \quad (3.52')$$

Let us now define the following two vectors $\vec{X}$ and $\vec{\Xi}$, with coordinates:

$$\vec{X} = \begin{pmatrix} x_1' = R'\sin\theta' \, \cos\varphi' \\ x_2' = R'\sin\theta' \, \sin\varphi' \\ x_3' = R'\cos\theta' \end{pmatrix} \quad , \quad \vec{\Xi} = \begin{pmatrix} \xi_1 = k_0 \sin\theta \, \cos\varphi \,/2\pi \\ \xi_2 = k_0 \sin\theta \, \sin\varphi \,/2\pi \\ \xi_3 = k_0 \cos\theta \,/2\pi \end{pmatrix}$$

The function $\mathrm{e}^{-\imath\vec{\Xi}\cdot\vec{X}}$ defines a plane wave which propagates in the direction θ, φ and which can be expanded into the following series (see, for example, [MOR 53]):

$$\mathrm{e}^{-\imath\vec{\Xi}\cdot\vec{X}} = \sum_{n=0}^{\infty} \imath^{-n}(2n+1) \sum_{m=0}^{n} \varepsilon_m \frac{(n-m)!}{(n+m)!} \cos\big[m(\varphi-\varphi')\big]$$
$$P_n^m(\cos\theta) P_n^m(\cos\theta') j_n(k_0 R') \quad (3.53)$$

By introducing (3.53) into (3.52′), we have

$$\frac{\mathrm{e}^{\imath k_0 r(Q,M')}}{r(Q,M')} \simeq \frac{\mathrm{e}^{\imath k_0 R}}{R} \mathrm{e}^{-\imath\vec{\Xi}\cdot\vec{X}} \quad (3.54)$$

The asymptotic expression of integral (3.51) is thus:

$$p^+(Q) \simeq -\frac{\omega^2\mu_0}{2\pi}\frac{\mathrm{e}^{\imath k_0 R}}{R} \int_{-\infty}^{\infty}\int_{-\infty}^{\infty} w(x_1', x_2') \mathrm{e}^{-\imath k_0 (x_1' sin\theta \cos\varphi + x_2' \sin\theta \sin\varphi)/2\pi} \, \mathrm{d}x_1' \, \mathrm{d}x_2'$$
$$= -\frac{\omega^2\mu_0}{2\pi}\frac{\mathrm{e}^{\imath k_0 R}}{R} \hat{w}\Big(\frac{k_0 \sin\theta\cos\varphi}{2\pi}, \frac{k_0\sin\theta\sin\varphi}{2\pi}\Big)$$

As we mentioned earlier, the Fourier transform of the plate displacement depends on the radial variable only, that is on $\xi = k_0 \sin\theta \sqrt{\cos\theta^2 + \sin\theta^2}/2\pi = k_0 \sin\theta/2\pi$. The final result is:

$$p^+(Q) = -\frac{F}{D}\frac{\omega^2\mu_0}{2\pi} \frac{\imath k_0\cos\theta}{\imath k_0 \cos\theta\big(k_0^4\sin\theta^4 - \lambda^4\big) + 2\lambda^4\mu_0/\bar{\mu}} \frac{\mathrm{e}^{\imath k_0 R}}{R}$$
$$+ \mathcal{O}(k_0^{-2}R^{-2}) \quad (3.55)$$

In acoustics, it is classical to compare the radiation of a source at infinity to that of an isotropic point source of unit amplitude. This is a means to characterize the directivity of the source radiation. The diagram thus obtained is called the *directivity pattern* or *radiation pattern* of the source. The directivity pattern of the infinite plate is the curve which represents the function

$$20\ \log_{10}\left|\frac{F}{D}\,\frac{\omega^2\mu_0}{2\pi}\,\frac{\imath k_0\cos\theta}{\imath k_0\cos\theta\big(k_0^4\sin\theta^4-\lambda^4\big)+2\lambda^4\mu_0/\bar{\mu}}\right|$$

for a given angular frequency. The first remark is that the plate radiates no energy in the direction $\theta=\pi/2$. If $\omega<\Omega_c$, the modulus of the denominator is always larger than $2\lambda^4\mu_0/\bar{\mu}$. If $\omega>\Omega_c$, the term $k_0^4\sin\theta^4-\lambda^4$ is zero for the incidence angle $\bar{\theta}=\arcsin(\lambda/k_0)$: we can expect that the directivity pattern is maximum in that direction.

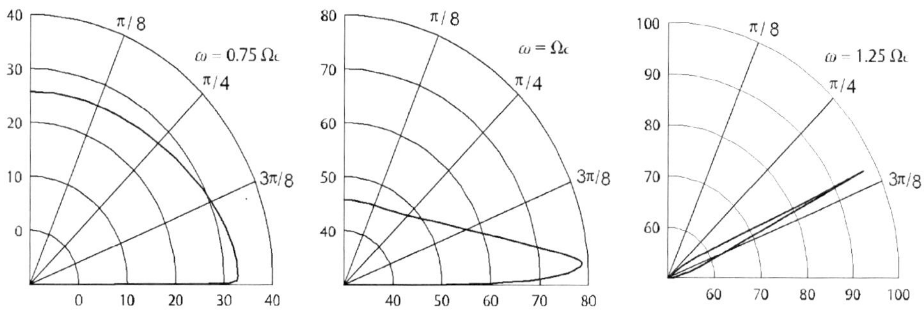

Figure 3.13. *Directivity pattern of the infinite plate in air*

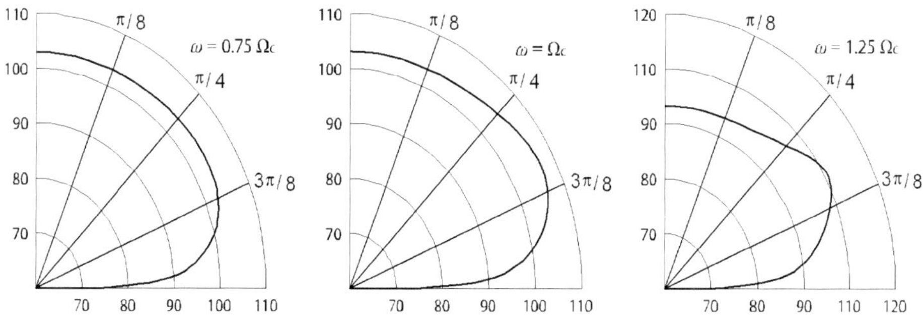

Figure 3.14. *Directivity pattern of the infinite plate in water*

Figures 3.13 and 3.14 show examples of directivity patterns of a plate in air or in water, for $F/D=1$. The mechanical data are the same as in the preceding examples. For each fluid, the angular frequencies are $0.75\,\Omega_c$, Ω_c and $1.25\,\Omega_c$. For this last frequency, we have $\bar{\theta}=0.35\,\pi$: the directivity pattern has a maximum in this direction, which is sharper for the air-loaded plate than for the water-loaded plate.

3.5.2. *Free Oscillations of an Infinite Circular Cylindrical Shell Filled with a vacuum and Immersed in a Fluid of Infinite Extent*

As in section 2.4, we consider a thin elastic shell, with radius R and thickness h. It occupies the domain $\Sigma \equiv \{\rho = R, 0 \leq \varphi < 2\pi, -\infty < z < +\infty\}$. The mechanical parameters of the shell are μ_s, E and ν. It is surrounded by a fluid, characterized by a density μ_0 and a sound velocity c_0, which occupies the domain $\Omega \equiv \{\rho > R, 0 \leq \varphi < 2\pi, -\infty < z < +\infty\}$.

The system is excited by a harmonic force with components (f_z, f_φ, f_ρ) and by a harmonic incident acoustic wave p_0. The angular frequency of the excitation is ω. Let (u, v, w) be the components of the shell displacement. The total acoustic pressure is denoted by $p_0 + p$. The component p corresponds to two different phenomena: the diffraction of the incident pressure by the shell and the acoustic radiation induced by the external force exciting the shell. The governing equations are:

$$\begin{gathered}
\frac{Eh}{1-\nu^2}\mathcal{M}^c\begin{pmatrix} u \\ v \\ w \end{pmatrix} - \omega^2\mu_s h\begin{pmatrix} u \\ v \\ w \end{pmatrix} + \begin{pmatrix} 0 \\ 0 \\ p \end{pmatrix} = \begin{pmatrix} f_z \\ f_\varphi \\ f_\rho \end{pmatrix} - \begin{pmatrix} 0 \\ 0 \\ p_0 \end{pmatrix} \\
\left(\Delta + k_0^2\right)p = 0 \quad \text{in} \quad \Omega \quad , \quad \text{with} \quad k_0 = \frac{\omega}{c_0} \\
\frac{\partial_\rho p + \partial_\rho p_0}{\mu_0} = \omega^2 w \quad \text{on} \quad \Sigma
\end{gathered} \tag{3.56}$$

where the matrix operator $\mathcal{M}^c$ is given by:

$$\begin{pmatrix}
-\dfrac{\partial^2}{\partial z^2} - \dfrac{1-\nu}{2R^2}\dfrac{\partial^2}{\partial\varphi^2} & -\dfrac{1+\nu}{2R}\dfrac{\partial^2}{\partial z\partial\varphi} & -\dfrac{\nu}{R}\dfrac{\partial}{\partial z} \\
-\dfrac{1+\nu}{2R}\dfrac{\partial^2}{\partial z\partial\varphi} & -\dfrac{1-\nu}{2}\dfrac{\partial^2}{\partial z^2} - \dfrac{1}{R^2}\dfrac{\partial^2}{\partial\varphi^2} & -\dfrac{1}{R^2}\dfrac{\partial}{\partial\varphi} \\
\dfrac{\nu}{R}\dfrac{\partial}{\partial z} & \dfrac{1}{R^2}\dfrac{\partial}{\partial\varphi} & \begin{aligned}&\dfrac{1}{R^2} + \dfrac{h^2}{12}\Big[\dfrac{\partial^4}{\partial z^4} \\ &+\dfrac{2}{R^2}\dfrac{\partial^4}{\partial z^2\partial\varphi^2} + \dfrac{1}{R^4}\dfrac{\partial^4}{\partial\varphi^4}\Big]\end{aligned}
\end{pmatrix}$$

These equations are completed by convenient conditions at infinity for both the shell displacement and the acoustic pressure.

As done for the infinite plate, we can use the Green's function for the exterior Neumann problem to express the acoustic pressure p in term of w. Let

$\mathcal{G}(M, M')$ be the Green's function for the Helmholtz equation which satisfies the homogenous Neumann condition on Σ and the condition at infinity. The acoustic pressure p is thus given by:

$$p(M) = \omega^2 \mu_0 \int_\Sigma w(M')\mathcal{G}(M, M')\,\mathrm{d}\sigma(M') - \int_\Sigma \partial_\rho p_0(M')\mathcal{G}(M, M')\,\mathrm{d}\sigma(M') \qquad (3.57)$$

This expression is introduced into the first equation (3.56) leading to a system of integro-differential equations for the shell displacement only:

$$\frac{Eh}{1-\nu^2}\mathcal{M}^c\begin{pmatrix} u \\ v \\ w \end{pmatrix} - \omega^2 \mu_s h \begin{pmatrix} u \\ v \\ w \end{pmatrix} + \begin{pmatrix} 0 \\ 0 \\ \omega^2 \mu_0 \int_\Sigma w\mathcal{G} \end{pmatrix} = \begin{pmatrix} f_z \\ f_\varphi \\ f_\rho \end{pmatrix} - \begin{pmatrix} 0 \\ 0 \\ p_0 - \int_\Sigma \partial_\rho p_0 \mathcal{G} \end{pmatrix} \qquad (3.58)$$

The difficulty is that the Green's function $\mathcal{G}(M, M')$ cannot be expressed in terms of analytical functions: only its Fourier transform in z is known. Thus, the response of the fluid-loaded shell to any excitation is not at all easy to compute numerically. Nevertheless, as we will see, it is straightforward to establish the system of equations which govern the free oscillations. From these equations, it is rather easy to understand how the resonance frequencies of the shell vary as the fluid loading increases.

1. *Green's function of the Helmholtz equation for the exterior Neumann problem*

Let us introduce the cylindrical coordinates (ρ, φ, z) of M, and (ρ', φ', z') of M'. The z-Fourier transform of $\mathcal{G}(M, M')$, defined by

$$\widehat{\mathcal{G}}(\rho, \rho'; \varphi, \varphi'; \zeta, z') = \int_{-\infty}^{+\infty} \mathcal{G}(\rho, \rho'; \varphi, \varphi'; z, z')\mathrm{e}^{-2\imath\pi\zeta z}\,\mathrm{d}z$$

satisfies the following boundary value problem:

$$\left[\Delta_2 + \left(\frac{\omega^2}{c_0^2} - 4\pi^2\zeta^2\right)\right]\widehat{\mathcal{G}}(\rho, \rho'; \varphi, \varphi'; \zeta, z') = \delta_{(\rho', \varphi')}\mathrm{e}^{-2\imath\pi\zeta z'} \quad \text{for} \quad \rho > R$$

$$\frac{\partial\widehat{\mathcal{G}}(\rho, \rho'; \varphi, \varphi'; \zeta, z')}{\partial\rho} = 0 \quad \text{for} \quad \rho = R$$

Here Δ_2 is the Laplace operator in two dimensions, and $\delta_{(\rho',\varphi')}$ is the Dirac distribution at point (ρ', φ'). The solution of this equation is developed in all classical books dealing with the Helmholtz equation (see, for example, [MOR 53]). We only recall its expression:

$$\widehat{\mathcal{G}}(\rho, \rho'; \varphi, \varphi'; \zeta, z') = -\frac{\imath}{4}\Bigg[H_0(Kd) - \sum_{n=-\infty}^{+\infty} \frac{J_n'(KR)}{H_n'(KR)} H_n(K\rho)H_n(K\rho')\mathrm{e}^{\imath n(\varphi-\varphi')}\Bigg]\mathrm{e}^{-2\imath\pi\zeta z'}$$

$$\text{with} \quad K^2 = \frac{\omega^2}{c_0^2} - 4\pi^2\zeta^2 \quad \text{and} \quad \Im(K) < 0 \qquad (3.59)$$

where d is the distance between the point with coordinates (ρ, φ) and that with coordinates (ρ', φ'). $J_n(z)$ is the Bessel function of order n, $H_n(z)$ is the Hankel function of the first kind and order n; and $J_n'(z)$ and $H_n'(z)$ are their derivatives. This expression enables us to express the z-Fourier transform of the pressure p on the shell in terms of the Fourier transform of the displacement component w.

2. *z-Fourier transform of the acoustic pressure p*

Let us recall that the Hankel function $H_0(Kd)$ can be expanded into series

$$H_0(Kd) = \sum_{n=-\infty}^{+\infty} H_n(K\rho)J_n(K\rho')\mathrm{e}^{\imath n(\varphi-\varphi')} \quad \text{for} \quad \rho > \rho'$$

$$= \sum_{n=-\infty}^{+\infty} J_n(K\rho)H_n(K\rho')\mathrm{e}^{\imath n(\varphi-\varphi')} \quad \text{for} \quad \rho < \rho'$$

Thus, for $\rho' < \rho$, $\widehat{\mathcal{G}}(\rho, \rho'; \varphi, \varphi'; \zeta, z')$ takes the form

$$\widehat{\mathcal{G}}(\rho, \rho'; \varphi, \varphi'; \zeta, z') = -\frac{\imath}{4}\mathrm{e}^{-2\imath\pi\zeta z'} \sum_{n=-\infty}^{+\infty} \widehat{\mathcal{G}}_n(\rho, \rho')\mathrm{e}^{\imath n(\varphi-\varphi')}$$

$$\text{with} \quad \widehat{\mathcal{G}}_n(\rho, \rho') = H_n(K\rho)\left[J_n(K\rho') - \frac{J_n'(KR)}{H_n'(KR)}H_n(K\rho')\right] \qquad (3.60)$$

The Fourier transform of $p(\rho, \varphi, z)$ with respect to z is given by the following integral:

$$\hat{p}(\rho, \varphi, \zeta) = \omega^2\mu_0 \int_{-\infty}^{+\infty} \mathrm{e}^{-2\imath\pi\zeta z}\,\mathrm{d}z \int_0^{2\pi} R\,\mathrm{d}\varphi' \int_{z'=-\infty}^{+\infty} w(z', \varphi')\mathcal{G}(\rho, R; \varphi, \varphi'; z, z')\,\mathrm{d}z'$$

$$= \omega^2 \mu_0 \int_0^{2\pi} R \, \mathrm{d}\varphi' \int_{z'=-\infty}^{+\infty} w(z', \varphi') \, \mathrm{d}z' \int_{-\infty}^{+\infty} \mathrm{e}^{-2\imath\pi\zeta z} \mathcal{G}(\rho, R; \varphi, \varphi'; z, z') \, \mathrm{d}z$$

$$= \omega^2 \mu_0 \int_0^{2\pi} R \, \mathrm{d}\varphi' \int_{z'=-\infty}^{+\infty} w(z', \varphi') \widehat{\mathcal{G}}(\rho, R; \varphi, \varphi'; \zeta, z') \, \mathrm{d}z' \quad (3.61)$$

By introducing (3.60) into (3.61), we obtain:

$$\hat{p}(\rho, \varphi, \zeta) = -\frac{\imath}{4} \omega^2 \mu_0 \int_0^{2\pi} R \int_{z'=-\infty}^{+\infty} w(z', \varphi') \mathrm{e}^{-2\imath\pi\zeta z'} \, \mathrm{d}z' \sum_{n=-\infty}^{+\infty} \widehat{\mathcal{G}}_n(\rho, R) \mathrm{e}^{\imath n(\varphi - \varphi')} \, \mathrm{d}\varphi'$$

$$= -\frac{\imath}{4} \omega^2 \mu_0 \int_0^{2\pi} R \hat{w}(\zeta, \varphi') \sum_{n=-\infty}^{+\infty} \widehat{\mathcal{G}}_n(\rho, R) \mathrm{e}^{\imath n(\varphi - \varphi')} \, \mathrm{d}\varphi' \quad (3.62)$$

where $\hat{w}(\zeta, \varphi')$ is the z-Fourier transform of $w(z, \varphi')$. Because of the geometry of a shell, the function $\hat{w}(\zeta, \varphi')$ can be expanded into a Fourier series with respect to the variable φ'

$$\hat{w}(\zeta, \varphi') = \sum_{n=-\infty}^{+\infty} \hat{w}_n(\zeta) \mathrm{e}^{\imath n \varphi'}$$

and finally the z-Fourier transform of $p(\rho, \varphi, z)$ is expressed as the following series:

$$\hat{p}(\rho, \varphi, \zeta) = -\frac{\imath \pi R}{2} \omega^2 \mu_0 \sum_{n=-\infty}^{+\infty} \hat{w}_n(\zeta) \widehat{\mathcal{G}}_n(\rho, R) \mathrm{e}^{\imath n \varphi} \quad (3.63)$$

3. *Dispersion equation for the fluid-loaded shell*

We can now obtain the system of equations satisfied by the resonance frequencies of the fluid-loaded shell. Let $\hat{u}(\varphi, \zeta)$ and $\hat{v}(\varphi, \zeta)$ be the z-Fourier transforms of $u(\varphi, z)$ and $v(\varphi, z)$ which are expanded into Fourier series

$$\hat{u}(\varphi, \zeta) = \sum_{n=-\infty}^{+\infty} \hat{u}_n(\zeta) \mathrm{e}^{\imath n \varphi} \quad , \quad \hat{v}(\varphi, \zeta) = \sum_{n=-\infty}^{+\infty} \hat{v}_n(\zeta) \mathrm{e}^{\imath n \varphi}$$

The z-Fourier transform of equations (3.58) reduces to a sequence of linear systems of equations:

$$\frac{Eh}{1-\nu^2}\widehat{\mathcal{M}}^{cn}\begin{pmatrix}\hat{u}_n\\ \hat{v}_n\\ \hat{w}_n\end{pmatrix}-\omega^2\mu_s h\begin{pmatrix}\hat{u}_n\\ \hat{v}_n\\ \hat{w}_n\end{pmatrix}-\frac{\imath\pi R}{2}\begin{pmatrix}0\\ 0\\ \omega^2\mu_0\hat{w}_n\hat{g}_n\end{pmatrix}=\begin{pmatrix}0\\0\\0\end{pmatrix}$$
$$n=0,1,\ldots,\infty \quad (3.64)$$

where $\hat{g}_n$ is defined as

$$\hat{g}_n=H_n(KR)\left[J_n(KR)-\frac{J_n'(KR)}{H_n'(KR)}H_n(KR)\right]=\frac{2\imath}{\pi KR}\frac{H_n(KR)}{H_n'(KR)}$$

(to obtain the last form of $\hat{g}_n$ use is made of the Wronskian of $J_n(KR)$ and $H_n(KR)$ – see [ABR 65]). The matrix operator $\widehat{\mathcal{M}}^{cn}$ is given by

$$\widehat{\mathcal{M}}^{cn}=\begin{pmatrix}4\pi^2\zeta^2+n^2\frac{1-\nu}{2R^2} & 2n\pi\zeta\frac{1+\nu}{2R} & -2\imath\pi\zeta\frac{\nu}{R}\\ 2n\pi\zeta\frac{1+\nu}{2R} & 4\pi^2\zeta^2\frac{1-\nu}{2}+\frac{n^2}{R^2} & -\frac{\imath n}{R^2}\\ 2\imath\pi\zeta\frac{\nu}{R} & \frac{\imath n}{R^2} & \frac{1}{R^2}+\frac{h^2}{12}\left[16\pi^4\zeta^4+\frac{8\pi^2n^2\zeta^2}{R^2}+\frac{n^4}{R^4}\right]\end{pmatrix}$$

A solution of equation (3.64) corresponds to a free wave in the fluid-loaded shell which is an exponential function of z of the form $\mathrm{e}^{\imath\zeta z}$.

As we did for the shell *in vacuo*, we introduce the following reduced variables $\Lambda=2\pi\zeta R$ and $\Omega=\omega R\sqrt{\mu_s(1-\nu^2)/E}$. The dispersion equation then becomes:

$$\left(\widehat{\mathcal{M}}^{cn}_{\mathrm{red}}-\Omega^2\mathbf{I}\right)\begin{pmatrix}\hat{u}_n\\ \hat{v}_n\\ \hat{w}_n\end{pmatrix}-\Omega^2\frac{\imath\pi R}{\mu_s h}\begin{pmatrix}0\\ 0\\ \hat{g}_n\hat{w}_n\end{pmatrix}=\begin{pmatrix}0\\0\\0\end{pmatrix}$$
$$\widehat{\mathcal{M}}^{cn}_{\mathrm{red}}=\begin{pmatrix}\Lambda^2+n^2\frac{1-\nu}{2} & \Lambda n\frac{1+\nu}{2} & -\imath\Lambda\nu\\ \Lambda n\frac{1+\nu}{2} & \Lambda^2\frac{1-\nu}{2}+n^2 & -\imath n\\ \imath\Lambda\nu & \imath n & 1+\frac{h^2}{12R^2}\left(\Lambda^2+n^2\right)^2\end{pmatrix} \quad (3.65)$$

This system has a non-zero solution if its determinant is zero, that is, if the couple (Λ,Ω) satisfies the equation

$$\mathrm{Det}\left[\widehat{\mathcal{M}}^{cn}_{\mathrm{red}}-\Omega^2\mathbf{I}-\begin{pmatrix}0&0&0\\0&0&0\\0&0&\Omega^2\frac{\imath\pi R}{\mu_s h}\hat{g}_n\end{pmatrix}\right]=0 \quad (3.66)$$

We can either fix the wavenumber Λ and look for the values of the angular frequency Ω, or choose Ω and search for the possible values of Λ. But, because of the term $\hat{g}_n$, equation (3.66) is non-linear. To find a solution, it is necessary to begin with a rather good approximation.

For the *in vacuo* shell, we showed that the resonance frequencies Ω can be determined easily for a given Λ (and *vice versa*). When the fluid is a gas, the *in vacuo* resonance frequencies provide a good approximation. This is not the case for a heavy fluid. But, an interesting task is to start from an *in vacuo* resonance frequency and to follow it as the fluid loading increases. For example, if the fluid is water, we will follow a resonance frequency by increasing the fluid density and the sound velocity step by step, starting from air density μ_a and sound velocity c_a up to their values μ_w and c_w in water. We can choose a linear function

$$c_0 = \tau c_a + (1-\tau)c_w \quad , \quad \mu_0 = \tau\mu_a + (1-\tau)\mu_w \quad , \quad 0 \leq \tau \leq 1$$

At each step, the initial value of the resonance frequency chosen to solve the equation is its value at the previous step. If τ varies slowly enough, the procedure is stable.

As an example, let us consider the variation of Ω when $\Lambda = 0$, that is, when the shell displacement is independent of ζ. The matrix $(\widehat{\mathcal{M}}^{cn}_{\text{red}} - \Omega^2\mathbf{I})$ takes the simplified form:

$$\widehat{\mathcal{M}}^{cn}_{\text{red}} - \Omega^2\mathbf{I} = \begin{pmatrix} n^2\dfrac{1-\nu}{2} - \Omega^2 & 0 & 0 \\ 0 & n^2 - \Omega^2 & -\imath n \\ 0 & \imath n & 1 + \dfrac{h^2}{12R^2}n^4 - \Omega^2 \end{pmatrix}$$

and the *in vacuo* dispersion equation is:

$$\left(n^2\frac{1-\nu}{2} - \Omega^2\right)\left[\left(n^2 - \Omega^2\right)\left(1 + \frac{h^2}{12R^2}n^4 - \Omega^2\right) - n^2\right] = 0$$

Obviously, the roots $\pm\Omega = n\sqrt{(1-\nu)/2}$, which appear for the *in vacuo* shell, will also be present for the shell surrounded by any fluid. But these roots correspond to a purely tangential displacement of the shell in the z direction ($u \not\equiv 0, v = 0, w = 0$): such a displacement is not affected by the presence of the fluid. We thus ignore it. For the *in vacuo* shell, the dispersion equation which provides the other roots is given by:

$$\left(n^2 - \Omega^2\right)\left(1 + \frac{h^2}{12R^2}n^4 - \Omega^2\right) - n^2 = 0$$

For $n = 0$, there is one pair of real roots with opposite signs. Otherwise, there are two pairs of real roots with opposite signs. It is sufficient to pay attention to the positive roots. We use the same shell data as in Chapter 2. For $n = 0$, the unique *in vacuo* root is equal to 1. For $n = 2$, there are two *in vacuo* roots equal to 0.0206 and 2.236. Figure 3.15 presents the variations of these roots as the fluid loading increases. We observe that:

1. the real part of the angular resonance frequencies decreases: this is classically interpreted as an effect of "added mass", that is, the shell behaves as if it had a larger density;
2. the angular resonance frequencies have a negative imaginary part which corresponds to a time damping due to a loss of energy in the fluid.

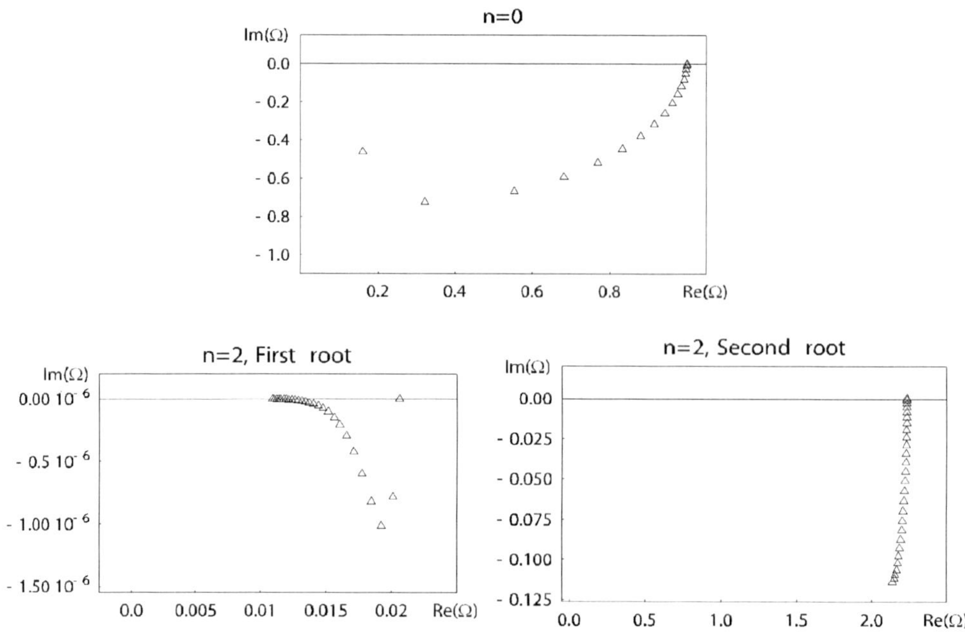

Figure 3.15. *Variation of the angular frequency Ω from vacuum to water loading: for $n = 0$ the in vacuo value is 1; for $n = 2$, the in vacuo first root is 0.0206 and the second one is 2.236*

3.5.3. *A Few Remarks on the Free Oscillations of an Infinite Circular Cylindrical Shell containing a Fluid and Immersed in a Second Fluid of Infinite Extent*

We can proceed exactly as in the previous case. The Green's function of the interior Neumann problem enables us to represent the interior acoustic pressure. By taking the z-Fourier transform of the interior pressure, we obtain a dispersion equation very similar to equation (3.66), where $\hat{g}_n$ is replaced by the sum of the nth component of the exterior and the interior Green's functions.

Nevertheless, the interior Green's function is not defined for any real angular frequency. Indeed, it is well known that a waveguide has a set of cut-off frequencies (real for non-absorbing boundaries) for which the homogenous boundary value problem (homogenous Helmholtz equation and boundary condition) has a non-zero solution of arbitrary amplitude. This implies that the Green's representation of the interior Neumann problem cannot be used for all real frequencies. In a forthcoming section, we solve the problem of sound radiation from a baffled shell excited by an internal turbulent flow. We will show how to overcome the theoretical and numerical difficulties induced by the cut-off frequencies.

3.6. Baffled Rectangular Plate

This section is devoted to the study of a thin plate, extended by a perfectly rigid infinite plane surface and immersed in a single fluid which extends up to infinity. The study of a plate in contact with two different fluids does not present any additional difficulty. With such a simple geometry, general theoretical results are easy to establish and powerful numerical methods can be developed.

We first begin with some general considerations on the eigenmodes and the resonance modes of the fluid-loaded plate. We show that the response of the plate to any excitation (periodic, transient, random, etc.) can be expanded into a series of resonance modes.

The main problem is to compute analytical or numerical approximations of the resonance modes. We propose several methods and compare their results. We also compare numerical results to experiments.

3.6.1. *General Theory: Eigenmodes, Resonance Modes, Series Expansion of the Response of the System*

Let us consider a finite dimension plate, occupying a domain Σ, with boundary $\partial\Sigma$, of the plane $z = 0$. It is extended by a perfectly rigid baffle which occupies the plane complement $\overline{\Sigma}$ of Σ (see Figure 3.16). The domain Ω is composed of the two half-spaces $z < 0$ and $z > 0$ which are occupied by the same perfect fluid (for two different fluids, the modifications to be introduced are quite obvious). The plate thickness is defined by a function $h(x, y) = h_0(1 + \varepsilon(x, y))$

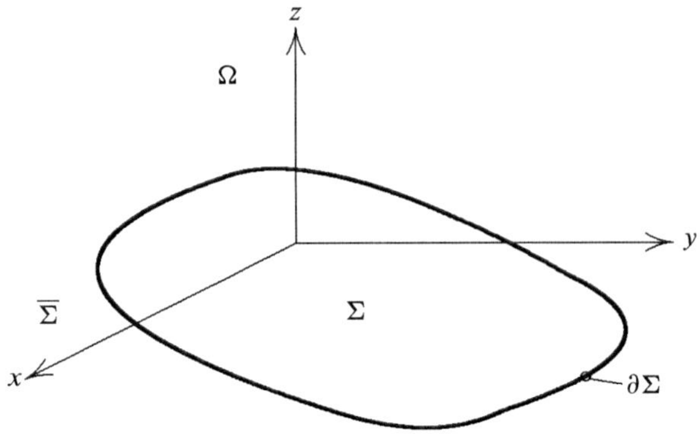

Figure 3.16. *Sketch of the baffled fluid-loaded plate*

which describes its space variations around the value h_0. The mechanical characteristics of the plate material are E, ν and μ_s. The fluid characteristics are μ_0 and c_0. The boundary conditions satisfied by the plate displacement along $\partial\Sigma$ can be of any kind and will be specified when necessary. For simplicity, we assume that there is no energy source in the fluid and that the plate is excited by a mechanical force $\widetilde{F}(M,t) = \widetilde{\Psi}(t)f(M)$, which is identically zero for $t < 0$. A final assumption is that the system is at rest for $t < 0$.

1. *Basic equations*

Let $\widetilde{W}(M,t)$ be the plate displacement and $\widetilde{P}(Q,t)$ stand for the acoustic pressure. These functions satisfy the following system of equations:

$$\begin{aligned} &\left(\Delta - \frac{1}{c_0^2}\frac{\partial^2}{\partial t^2}\right)\widetilde{P} = 0 \quad \text{in} \quad \Omega \\ &\left(\mathcal{D} + \mu_s h\frac{\partial^2}{\partial t^2}\right)\widetilde{W} + \widetilde{\overline{P}} = \widetilde{F} \quad \text{on} \quad \Sigma \\ \text{with} \quad &\mathcal{D}\widetilde{W} := \Delta(D\Delta\widetilde{W}) + 2\frac{\partial^2}{\partial x\partial y}\left((1-\nu)D\frac{\partial^2}{\partial x\partial y}\widetilde{W}\right) \\ &- \frac{\partial^2}{\partial y^2}\left((1-\nu)D\frac{\partial^2}{\partial x^2}\widetilde{W}\right) - \frac{\partial^2}{\partial x^2}\left((1-\nu)D\frac{\partial^2}{\partial y^2}\widetilde{W}\right) \end{aligned} \tag{3.67}$$

where $\widetilde{\overline{P}}$ is the step of the acoustic pressure across the plane $z = 0$, that is, $\widetilde{\overline{P}} = \lim_{z>0,z\to 0}(\widetilde{P}(x,y,z;t) - \widetilde{P}(x,y,-z;t))$. Remember that, for a plate of constant thickness, the operator $\mathcal{D}$ reduces to $D\Delta^2$. On the plate, the fluid particle acceleration is equal to the plate acceleration; and on the baffle, it is zero. This is expressed by:

$$\begin{aligned}\frac{\partial \widetilde{P}}{\partial z} &= -\mu_0 \frac{\partial^2 \widetilde{W}}{\partial t^2} \quad \text{on} \quad \Sigma \\ &= 0 \quad \text{on} \quad \overline{\Sigma}\end{aligned} \tag{3.68}$$

To complete these equations, an out-going wave condition for the pressure is added (no energy is coming back from infinity). Thus, the solution is unique.

2. *Integro-differential equation for the plate displacement*

Because the geometry of the system is simple, the acoustic pressure can be expressed in terms of the plate displacement. Let $\widetilde{\mathcal{G}}(Q;t)$ be the Green's function of the wave equation which satisfies the homogenous Neumann condition on the plane $z = 0$ and the out-going wave condition. $\widetilde{\mathcal{G}}(Q;t)$ is the acoustic field emitted by an impulsive point source $\delta_S \otimes \delta_{t'}$:

$$\widetilde{\mathcal{G}}(Q, S; t, t') = -\frac{\delta(t - t' - R/c_0)}{4\pi R} - \frac{\delta(t - t' - R'/c_0)}{4\pi R'}$$

where R is the distance between Q and S and R' is the distance between Q and the image S' of S with respect to the plane $z = 0$. Using this Green's function, the acoustic pressure writes:

$$\begin{aligned}\widetilde{P}(Q,t) &= -\mu_0 \operatorname{sgn}(z)\, \widetilde{\mathcal{G}} \underset{(Q,t)}{*} \left(\frac{\partial^2 \widetilde{W}}{\partial t^2} \otimes \delta_\Sigma \right) \\ &= -\mu_0 \operatorname{sgn}(z) \int_0^\infty \mathrm{d}t' \int_\Sigma \widetilde{\mathcal{G}}(Q, M'; t - t') \frac{\partial^2 \widetilde{W}(M', t')}{\partial t'^2} \,\mathrm{d}M'\end{aligned} \tag{3.69}$$

where the symbol $\underset{(Q,t)}{*}$ stands for the convolution product over the space and time variables; δ_Σ is the Dirac measure on Σ. Introducing this relationship in the second equation of (3.67) leads to an integro-differential equation for the plate displacement only:

$$\left(\mathcal{D} + \mu_s h \frac{\partial^2}{\partial t^2} \right) \widetilde{W} - 2\mu_0\, \widetilde{\mathcal{G}} \underset{(Q,t)}{*} \left(\frac{\partial^2 \widetilde{W}}{\partial t^2} \otimes \delta_\Sigma \right) = \widetilde{F} \tag{3.70}$$

We now show that the solution of (3.70) can be expressed as a series of resonance modes of the fluid-loaded plate.

3. *Eigenmodes of the fluid-loaded plate – Eigenmodes series expansion of the response to a harmonic excitation*

Let us introduce the time Fourier transform defined by

$$\phi(\omega) = \int_{-\infty}^{+\infty} \tilde{\phi}(t)\mathrm{e}^{-\imath\omega t}\,\mathrm{d}t$$

The time Fourier transform of equation (3.70) is:

$$\begin{aligned}
&\mathcal{D}W - \mu_s h\omega^2 W + 2\mu_0\omega^2\, \mathcal{G}_\omega \underset{M}{*} (W \otimes \delta_\Sigma) = f\psi \\
&\text{with}\quad \mathcal{G}_\omega(S, Q) = -\frac{\mathrm{e}^{ik_0 R}}{4\pi R} - \frac{\mathrm{e}^{ik_0 R'}}{4\pi R'} \quad , \quad k_0 = \frac{\omega}{c_0} \\
&\text{and}\quad \mathcal{G}_\omega \underset{Q}{*} (W \otimes \delta_\Sigma) = \int_\Sigma \mathcal{G}_\omega(Q, M')W(M')\,\mathrm{d}M'
\end{aligned} \tag{3.71}$$

The symbol $\underset{Q}{*}$ represents the space convolution product. It is now necessary to use the variational form of equation (3.71). The solution W is sought in the functional space $\mathbf{H}(\Sigma)$ of functions that are square integrable together with their derivatives up to order 2, and which satisfy the given boundary conditions. Thus, for any function $U \in \mathbf{H}(\Sigma)$, the plate displacement satisfies the equality

$$\mathcal{A}(W, U) - \mu_s h_0\omega^2\left[\int_\Sigma (1+\varepsilon)WU^* - \frac{\mu_0}{\mu_s h_0}\beta_\omega(W, U)\right] = \psi(\omega)\langle f, U\rangle$$

with :

$$\begin{aligned}
\mathcal{A}(W, U) = \int_\Sigma D\Bigg[&\left(\frac{\partial^2 W}{\partial x^2} + \frac{\partial^2 W}{\partial y^2}\right)\left(\frac{\partial^2 U^*}{\partial x^2} + \frac{\partial^2 U^*}{\partial y^2}\right) \\
&+(1-\nu)\left(2\frac{\partial^2 W}{\partial x\partial y}\frac{\partial^2 U^*}{\partial x\partial y} - \frac{\partial^2 W}{\partial x^2}\frac{\partial^2 U^*}{\partial y^2} - \frac{\partial^2 W}{\partial y^2}\frac{\partial^2 U^*}{\partial x^2}\right)\Bigg] \\
\beta_\omega(W, U) = 2&\iint_\Sigma W(M)\mathcal{G}_\omega(M, M')U^*(M')\,\mathrm{d}M\,\mathrm{d}M' \\
&\text{and}\quad \langle f, U\rangle = \int_\Sigma fU^*
\end{aligned} \tag{3.72}$$

U^* is the complex conjugate of U. The operator β_ω describes the coupling between the plate and the fluid.

DEFINITION 3.3 (EIGENMODES AND EIGENVALUES).– *The* eigenmodes $\widehat{W}_n$ *of the fluid-loaded plate are the non-zero solutions of the homogenous variational equation*

$$\mathcal{A}(\widehat{W}_n, U) - \Lambda_n\left[\int_\Sigma (1+\varepsilon)\widehat{W}_n U^* - \frac{\mu_0}{\mu_s h_0}\beta_\omega(\widehat{W}_n, U)\right] = 0 \qquad n = 1, 2, \cdots, \infty \quad (3.73)$$

The corresponding values Λ_n *of the parameter* $\Lambda = \mu_s h_0 \omega^2$ *are called* eigenvalues.

The eigenmodes and eigenvalues depend on the excitation frequency ω because the coupling operator β_ω depends on ω. The eigenmodes of the fluid-loaded plate – like the eigenmodes of the *in vacuo* plate which correspond to $\mu_0 = 0$ – satisfy the following orthogonality property:

$$\mathcal{A}(\widehat{W}_n, \widehat{W}_m^*) = \Lambda_n\left[\int_\Sigma (1+\varepsilon)\widehat{W}_n \widehat{W}_m - \frac{\mu_0}{\mu_s h_0}\beta_\omega(\widehat{W}_n, \widehat{W}_m^*)\right] = 0 \qquad \text{for} \quad m \neq n$$

$$\mathcal{A}(\widehat{W}_n, \widehat{W}_n^*) = \Lambda_n\left[\int_\Sigma (1+\varepsilon)\widehat{W}_n \widehat{W}_n - \frac{\mu_0}{\mu_s h_0}\beta_\omega(\widehat{W}_n, \widehat{W}_n^*)\right] = \Lambda_n N_n(\omega) \quad (3.74)$$

Because the functions $\widehat{W}_n$ are defined *modulo* a multiplicative constant, we can choose $N_n(\omega) = 1$.

The solution of equation (3.72) can be expanded into a series of the form $W = \sum_{n=1}^{\infty} a_n \widehat{W}_n$. The coefficients a_n are determined by introducing this expansion into (3.72) in which U is successively replaced by each of the $\widehat{W}_m^*$. This leads to the result:

$$W(M,\omega) = \Psi(\omega)\sum_{n=1}^{\infty}\frac{\langle f, \widehat{W}_n^*\rangle}{\Lambda_n - \mu_s h_0 \omega^2}\widehat{W}_n(M)$$

$$P(M,\omega) = -\mu_0\omega^2 \operatorname{sgn}(z)\Psi(\omega)\sum_{n=1}^{\infty}\frac{\langle f, \widehat{W}_n^*\rangle}{\Lambda_n - \mu_s h_0 \omega^2}\times \times\int_\Sigma \mathcal{G}_\omega(Q, M')\widehat{W}_n(M')\,\mathrm{d}M' \quad (3.75)$$

4. *Resonance modes of the fluid-loaded plate – Resonance modes series expansion of the response to a harmonic and a transient excitation*

The resonance modes describe the free oscillations of the fluid-loaded plate. They are defined as follows:

DEFINITION 3.4 (RESONANCE MODES AND RESONANCE FREQUENCIES).– *The* resonance modes W_n *of the fluid-loaded plate are the non-zero solutions of the homogenous variational equation*

$$\mathcal{A}(W_n, U) - \mu_s h_0 \omega_n^2 \left[\int_\Sigma (1+\varepsilon) W_n U^* - \frac{\mu_0}{\mu_s h_0} \beta_{\omega_n}(W_n, U) \right] = 0$$
$$n = 1, 2, \cdots, \infty \quad (3.76)$$

The corresponding values ω_n *of the angular frequency* ω *are called* angular resonance frequencies.

The angular resonance frequencies are solutions of the equations

$$\Lambda_n(\omega_n) = \mu_s h_0 \omega_n^2$$

It is easily seen that if $\Omega_n - \imath\tau_n$ (with $\Omega_n > 0$ and $\tau_n > 0$) is a resonance frequency, then $-\Omega_n - \imath\tau_n$ is a resonance frequency too. The corresponding resonance modes are complex conjugate. So, it is convenient to associate the resonance modes by pairs with symmetric indices

$$\left\{W_n, \omega_n = \Omega_n - \imath\tau_n\right\}, \left\{W_{-n} = W_n^*, \omega_{-n} = -\omega_n^* = -\Omega_n - \imath\tau_n\right\}, n = 1, 2, \cdots, \infty$$

The resonance modes are related to the eigenmodes by

$$W_n(M) = \widehat{W}_n(M, \omega_n) \quad W_{-n}(M) = \widehat{W}_n(M, -\omega_n^*) = W_n^*$$

They do not satisfy an orthogonality relationship. They depend on the mechanical and geometrical properties of the plate and of the surrounding fluid only, while the eigenmodes and eigenfrequencies also depend on the excitation frequency.

By taking the inverse Fourier transform of the eigenmodes series expansion of harmonic regime (3.75), the transient response of the fluid-loaded plate is obtained. In its expression, the resonance modes and the resonance frequencies appear in a straightforward way. The transient plate displacement is the time convolution product of $\tilde{\Psi}(t)$ by the inverse Fourier transform of the eigenmodes series which can easily be obtained by applying the residue theorem to the integrals

$$I_n = \frac{1}{2\pi} \int_{-\infty}^{+\infty} \frac{\langle f, \widehat{W}_n^* \rangle}{\Lambda_n - \mu_s h_0 \omega^2} \widehat{W}_n(M) \mathrm{e}^{-\imath\omega t} \, \mathrm{d}\omega$$

For each n, the function to be integrated has two poles, which are the values of the resonance angular frequencies ω_n and $-\omega_n^*$. The result is:

$$\begin{aligned}\widetilde{W}(M,t) = -\imath\, \widetilde{\Psi}(t) \underset{t}{*} Y(t) \sum_{n=1}^{\infty} \Bigg[& \frac{\langle f, W_n^* \rangle}{\Lambda_n'(\omega_n) - 2\mu_s h_0 \omega_n} W_n(M) \mathrm{e}^{-\imath \Omega_n t} \\ & - \frac{\langle f, W_n^* \rangle^*}{\Lambda_n'^{\,*}(\omega_n) - 2\mu_s h_0 \omega_n^*} W_n^*(M) \mathrm{e}^{+\imath \Omega_n t} \Bigg] \mathrm{e}^{-\tau_n t}\end{aligned} \tag{3.77}$$

where $\underset{t}{*}$ stands for the time convolution product and $Y(t)$ is the Heaviside step function. The acoustic pressure is directly deduced from expression (3.69) in which $\widetilde{W}(M,t)$ is replaced by the series (3.77):

$$\begin{aligned}\widetilde{P}(M,t) = \operatorname{sgn}(z)\, \mu_0\, \widetilde{\mathcal{G}}(M,t) \underset{(M,t)}{*} \frac{\partial^2}{\partial t^2} \Bigg\{ & -\imath\, \widetilde{\Psi}(t) \underset{t}{*} Y(t) \\ & \sum_{n=1}^{\infty} \Bigg[\frac{\langle f, W_n^* \rangle}{\Lambda_n'(\omega_n) - 2\mu_s h_0 \omega_n} W_n(M) \mathrm{e}^{-\imath \Omega_n t} \\ & - \frac{\langle f, W_n^* \rangle^*}{\Lambda_n'^{\,*}(\omega_n) - 2\mu_s h_0 \omega_n^*} W_n^*(M) \mathrm{e}^{+\imath \Omega_n t} \Bigg] \mathrm{e}^{-\tau_n t} \Bigg\}\end{aligned} \tag{3.78}$$

Finally, the Fourier transform of expressions (3.77) and (3.78) gives the representations of W and P in terms of the fluid-loaded plate resonance modes:

$$\begin{aligned}W(M,\omega) &= \Psi(\omega) \sum_{n=1}^{\infty} \Bigg[\frac{\langle f, W_n^* \rangle}{\Lambda_n'(\omega_n) - 2\mu_s h_0 \omega_n} \frac{W_n(M)}{\omega - \omega_n} \\ &\qquad - \frac{\langle f, W_n^* \rangle^*}{\Lambda_n'^{\,*}(\omega_n) - 2\mu_s h_0 \omega_n^*} \frac{W_n^*(M)}{\omega + \omega_n^*} \Bigg] \\ P(M,\omega) &= \operatorname{sgn}(z) \omega^2 \mu_0 \Psi(\omega) \\ &\sum_{n=1}^{\infty} \Bigg[\frac{\langle f, W_n^* \rangle}{\Lambda_n'(\omega_n) - 2\mu_s h_0 \omega_n} \int_{\Sigma} \frac{W_n(M')}{\omega - \omega_n} \mathcal{G}_\omega(M,M')\, \mathrm{d}\sigma(M') \\ &\qquad - \frac{\langle f, W_n^* \rangle^*}{\Lambda_n'^{\,*}(\omega_n) - 2\mu_s h_0 \omega_n^*} \int_{\Sigma} \frac{W_n^*(M')}{\omega + \omega_n^*} \mathcal{G}_\omega(M,M')\, \mathrm{d}\sigma(M') \Bigg]\end{aligned} \tag{3.79}$$

3.6.2. *Rectangular Plate Clamped along its Boundary: Numerical Approximation of the Resonance Modes*

The plate domain is defined by $\Sigma \equiv \{-L_1 < x_1 < L_1, -L_2 < x_2 < L_2\}$. Let us recall that for a clamped plate, the boundary conditions write:

$$\begin{aligned} w(\pm L_1, x_2) = 0 \quad &, \quad w(x_1, \pm L_2) = 0 \\ \partial_{x_1} w(\pm L_1, x_2) = 0 \quad &, \quad \partial_{x_2} w(x_1, \pm L_2) = 0 \end{aligned}$$

The resonance modes belong to the space $\mathbf{H}(\Sigma)$ of functions which are square integrable together with their partial derivatives up to order 2 and which satisfy the boundary conditions. Let us denote by $(U_m, m = 1, 2, ...)$ a basis of $\mathbf{H}(\Sigma)$. Each resonance mode is sought in the following form:

$$W_n(M) = \sum_{m=1}^{\infty} u_n^m U_m(M) \tag{3.80}$$

Equation (3.76) leads to the infinite system of equations:

$$\sum_{m=1}^{\infty} u_n^m \left\{ \mathcal{A}(U_m, U_q^*) - \mu_s h_0 \omega_n^2 \left[\int_{\Sigma} (1+\varepsilon) U_m U_q - \frac{\mu_0}{\mu_s h_0} \mathcal{B}_{\omega_n}(U_m, U_q^*) \right] \right\} = 0$$
$$q = 1, 2, \cdots, +\infty \tag{3.81}$$

As done in the previous chapter, we choose the basis

$$U_m(x_1, x_2) = \frac{V_{m_1}(x_1/2L_1) V_{m_2}(x_2/2L_2)}{2\sqrt{L_1 L_2}} \tag{3.82}$$

where $V_q(\xi)$ is the set of normalized resonance modes of a beam extending from $\xi = -1/2$ to $\xi = +1/2$. The functions $U_m(x_1, x_2)$ can also be numbered with two indices $U_{m_1 m_2}(x_1, x_2)$, the correspondence between m and the couple (m_1, m_2) and with the total number of nodal lines is given in Table 3.3 (identical to Table 2.2).

m_1	0	0	1	0	1	2	0	1	2	3	...
m_2	0	1	0	2	1	0	3	2	1	0	...
m	1	2	3	4	5	6	7	8	9	10	...
number of nodal lines	0	1			2			3			...

Table 3.3. *Correspondence between the single index numbering of the functions $U_m(x_1, x_2)$ and the double index numbering $U_{m_1 m_2}(x_1, x_2)$*

When this basis is used for computing the *in vacuo* resonance modes, a mode which has, in the interior of the domain Σ, m_1 nodal lines parallel to the x_2-direction and m_2 nodal lines parallel to the x_1-direction, has a strong component on the function $V_{m_1}(x_1/2L_1)V_{m_2}(x_2/2L_2)$. This property remains true when a surrounding fluid is present. Thus, the corresponding resonance modes can still be numbered by a pair of indices which indicate the number of nodal lines in each direction.

For numerical applications, the number of basis functions must be finite, i.e. we choose $0 < m_1 \leq N_1$ and $0 < m_2 \leq N_2$ which corresponds to a total number of functions $N = (N_1 + 1)(N_2 + 1)$. The choice of the upper bounds is, of course, arbitrary but it is reasonable to choose the ratio N_1/N_2 as close as possible to L_1/L_2 so that the function $V_{N_1}(x_1/2L_1)V_{N_2}(x_2/2L_2)$ has almost the same density of nodal lines in both directions (as a consequence, in the highest order group, some functions can be missing and the numbering of the functions in that group must be modified). The infinite system of equations (3.81) is replaced by:

$$\sum_{m=1}^{N} u_n^m \left\{ \mathcal{A}(U_m, U_q^*) - \mu_s h_0 \omega_n^2 \left[\int_\Sigma (1+\varepsilon) U_m U_q - \frac{\mu_0}{\mu_s h_0} \beta_{\omega_n}(U_m, U_q^*) \right] \right\} = 0$$

$$q = 1, 2, \cdots, N \quad (3.83)$$

Although this system has an infinite number of resonance angular frequencies, only the first N can be considered as approximations of the exact resonance angular frequencies.

Equation (3.83) is not a classical eigenvalue problem. Indeed, the resonance frequency appears in a non-linear way in the coupling term:

$$\beta_{\omega_n}(U_m, U_q^*) = -\int_\Sigma U_m(x_1, x_2) \int_\Sigma U_q(x_1', x_2') g(|x_1 - x_1'|, |x_2 - x_2'|)\, \mathrm{d}x_1'\, \mathrm{d}x_2'\, \mathrm{d}x_1\, \mathrm{d}x_2 \quad (3.84)$$

$$\text{where} \quad g(|x_1 - x_1'|, |x_2 - x_2'|) = \frac{\mathrm{e}^{\imath \omega_n r(P,P')/c_0}}{\pi r(P, P')}$$

$$\text{and} \quad r(P, P') = \left[(x_1 - x_1')^2 + (x_2 - x_2')^2 \right]^{1/2}$$

and a specific method must be used.

If the fluid is a gas, its influence is small and the coupling term $\mu_0/\mu_s h_0$ $\beta_{\omega_n}(U_m, U_q^*)$ is a small perturbation of the plate operator. So, it is natural to make use of an iterative method. For a heavy fluid like water, the coupling term has a larger influence on the behavior of the plate. Nevertheless, the coupling operator remains "small" compared to the plate operator in the sense that the plate displacement is mainly governed by the partial differential operator. Thus, an iterative method can still be used.

1. *Iterative procedure*

We make the following hypothesis: to each *in vacuo* mode corresponds a fluid-loaded mode which has the same number of nodal lines in the x_1- and in the x_2-directions. To our knowledge, this assumption is confirmed experimentally. Nevertheless, we do not know if it can be proved. A subsequent assumption is that the order of multiplicity of an *in vacuo* mode and of the corresponding fluid-loaded mode are the same. For simplicity, in what follows we consider modes with a multiplicity order equal to 1.

The direct way of evaluating the first N resonance frequencies is to look for the values of ω for which the system (3.83) has non-zero solutions. A Marquardt's algorithm is well adapted to this problem and provides very accurate results, but requires a large amount of calculation.

The iterative procedure that we propose here is as follows. In a first step, the *in vacuo* resonance angular frequencies $\omega_n^{(0)}$ and modes $W_n^{(0)}$ are approximated by solving the classical eigenvalue problem:

$$\sum_{m=1}^{N} u_n^{m\,(0)} \left\{ \mathcal{A}(U_m, U_q^*) - \mu_s h_0 \omega_n^{(0)\,2} \int_\Sigma (1+\varepsilon) U_m U_q \right\} = 0$$

$$\text{for} \quad q = 1, 2, \cdots, N \qquad (3.85)$$

where the *in vacuo* modes are approximated by finite sums:

$$W_n^{(0)}(M) \simeq \sum_{m=1}^{N} u_n^{m\,(0)} U_m(M) \quad \text{with} \quad \| W_n^{(0)} \| = 1$$

the functions $U_m(M)$ still being defined by (3.82). The system of equations (3.85) provides approximations of N *in vacuo* resonance frequencies $\omega_n^{(0)}$ and resonance modes $W_n^{(0)}$. The main component of $W_n^{(0)}$ has indexes (n_1, n_2). Then, in a second step, for each *in vacuo* resonance frequency, the operator β_{ω_n} in (3.83) is replaced by $\beta_{\omega_n^{(0)}}$. A sequence of N classical eigenvalue problems is obtained:

$$\sum_{m=1}^{N} u_\nu^{m\,(1)} \left\{ \mathcal{A}(U_m, U_q^*) - \Lambda_\nu^{(1)} \left[\int_\Sigma (1+\varepsilon) U_m U_q - \frac{\mu_0}{\mu_s h_0} \beta_{\omega_n^{(0)}} (U_m, U_q^*) \right] \right\} = 0$$

$$q = 1, 2, \cdots, N \qquad (3.86)$$

For a fixed n, this system has N eigenfunctions $W_\nu^{(1)}$ with a main component of indexes (ν_1, ν_2). The approximation $W_n^{(1)}$ of the fluid-loaded mode that corresponds to $W_n^{(0)}$ is the eigenfunction for which the couple (ν_1, ν_2) is equal to (n_1, n_2).

Let $\Lambda_n^{(1)}$ be the corresponding eigenvalue. The corresponding approximation $\omega_n^{(1)}$ of the angular frequency, which is used to restart the iterative process, is chosen as follows:

$$\omega_n^{(1)} = \sqrt{\frac{1}{\mu_s h_0}} \sqrt{\Lambda_n^{(1)}} \quad \text{if} \quad \Im(\Lambda_n^{(1)}) < 0$$
$$= \sqrt{\frac{1}{\mu_s h_0}} \sqrt{\Lambda_n^{(1)*}} \quad \text{if} \quad \Im(\Lambda_n^{(1)}) > 0 \tag{3.87}$$

Let us remember that, because of our choice of time dependence, the resonance frequencies must have a negative imaginary part (the resonance modes are damped functions of time). So, for $\Im(\Lambda_n^{(1)}) < 0$, the former choice is quite obvious. For $\Im(\Lambda_n^{(1)}) > 0$, we could choose $-\sqrt{1/\mu_s h_0}\sqrt{\Lambda_n^{(1)}}$, which gives a resonance frequency with a negative real part. But we know that if ω_n is a resonance frequency, then $-\omega_n^*$ is also a resonance frequency. Thus, the choice made is admissible. Furthermore, it ensures that we obtain approximations of the resonance frequencies which all have a positive real part.

Let us now evaluate the accuracy of this first order approximation. To do so, let us define the vectors $V^{(0)}$ and $V^{(1)}$ by:

$$V^{(0)} = \sum_q \gamma_q^{(0)} U_q \quad , \quad V^{(1)} = \sum_q \gamma_q^{(1)} U_q \tag{3.88}$$

with:

$$\gamma_q^{(0)} = \sum_{m=1}^{R} u_n^{m\,(0)} \Bigg\{ \mathcal{A}(U_m, U_q^*) - \mu_s h_0 \omega_n^{(0)2} \Bigg[\int_\Sigma (1+\varepsilon) U_m U_q - \frac{\mu_0}{\mu_s h_0} \beta_{\omega_n^{(0)}}(U_m, U_q^*) \Bigg] \Bigg\} \tag{3.88'}$$

and

$$\gamma_q^{(1)} = \sum_{m=1}^{R} u_n^{m\,(1)} \Bigg\{ \mathcal{A}(U_m, U_q^*) - \mu_s h_0 \omega_n^{(1)2} \Bigg[\int_\Sigma (1+\varepsilon) U_m U_q - \frac{\mu_0}{\mu_s h_0} \beta_{\omega_n^{(1)}}(U_m, U_q^*) \Bigg] \Bigg\} \tag{3.88''}$$

The norm of $V^{(0)}$ (resp. $V^{(1)}$) measures the error made when, in equations (3.83), the exact solution (u_n^m, ω_n) is replaced by $(u_n^{m\,(0)}, \omega_n^{(0)})$, (resp. $(u_n^{m\,(1)}, \omega_n^{(1)})$). A good indicator of the accuracy of the first approximation is given by the ratio $\|V^{(1)}\| / \|V^{(0)}\|$.

As indicated before, $u_n^{m(0)}$ are the components of the *in vacuo* resonance mode, and $u_n^{m(1)}$ those of the first approximation of the fluid-loaded resonance mode. Because of its construction, $V^{(0)}$ is different from zero.

If the ratio $\|V^{(1)}\|/\|V^{(0)}\|$ is not small enough, the procedure is repeated and a sequence of approximate resonance angular frequencies $\omega_n^{(r)}$ and modes is obtained. We do not have any proof that the iterative procedure is convergent. We just assume that this is the case if the successive vectors $V^{(r)}$ tend to zero. The accuracy of the r-th approximation is estimated by the ratio $\|V^{(r)}\|/\|V^{(0)}\|$. Examples will show that an experimental convergence is observed even for a strong coupling.

Expressions (3.77) to (3.79) of the transient and harmonic plate displacement and pressure require the value $\Lambda_n'(\omega_n)$ of the derivative of Λ_n with respect to the angular frequency. An approximation of the derivative Λ_n' is obtained by a finite difference formula from the successive approximations of Λ_n.

2. *The Warburton's approximation*

This approximation has proved to be powerful for *in vacuo* plates. It is reasonable to make use of it for a fluid-loaded plate. Equation (3.83) will be replaced by:

$$\mathcal{A}(U_m, U_m^*) - \mu_s h_0 \omega_m^2 \left[\int_\Sigma (1+\varepsilon) U_m U_m - \frac{\mu_0}{\mu_s h_0} \beta_{\omega_m}(U_m, U_m^*) \right] = 0$$

$$m = 1, 2, \cdots, N \qquad (3.89)$$

Here again the iterative procedure previously described enables us to obtain, for each mode, a sequence of angular frequencies $\omega_n^{(r)}$ which converges to the exact solution of (3.89), but, of course, not to the exact resonance frequency. Nevertheless, the advantage of this approximation is to divide the computation time in half, or so, while providing quite an accurate result.

The accuracy of this approximation can be greatly improved by constructing the Warburton's approximation with the *in vacuo* modes of the plate instead of the basis functions U_m. In this case, equation (3.83) becomes:

$$\sum_{m=1}^{N} u_n^m \left\{ \left[\omega_m^{(0)2} - \omega_n^2 \right] \int_\Sigma (1+\varepsilon) W_m^{(0)} W_q^{(0)} - \omega_n^2 \frac{\mu_0}{\mu_s h_0} \beta_{\omega_n^{(0)}}(W_m^{(0)}, W_q^{(0)*}) \right\} = 0$$

$$q = 1, 2, \cdots, N \qquad (3.90)$$

Taking the first iteration of the Warburton's approximation of this equation leads to:

$$\omega_m^2 \simeq \omega_m^{(0)2}\left[1 - \frac{\mu_0}{\mu_s h_0}\frac{\beta_{\omega_n^{(0)}}(W_m^{(0)}, W_m^{(0)*})}{\int_\Sigma (1+\varepsilon) W_m^{(0)} W_m^{(0)}}\right]^{-1} \tag{3.91}$$

3. *The light-fluid approximation*

Let us recall the resonance mode equation:

$$\mathcal{A}(W_m, U) - \mu_s h_0 \omega_m^2 \left[\int_\Sigma (1+\varepsilon) W_m U^* - \eta \beta_{\omega_n}(W_m, U)\right] = 0 \quad \text{with} \quad \eta = \frac{\mu_0}{\mu_s h_0}$$

where U is any function in $\mathbf{H}(\Sigma)$. Assuming a weak fluid-loading, that is $\eta \ll 1$, the resonance modes and the squared resonance angular frequencies are sought as formal Taylor series of this parameter:

$$\begin{aligned} W_m &= W_m^{(0)} + \eta W_m^{(1)} + \eta^2 W_m^{(2)} + \cdots \\ \omega_m^2 &= \omega_m^{(0)2} + \eta \varpi_m^{(1)} + \eta^2 \varpi_m^{(2)} + \cdots \end{aligned}$$

where $W_m^{(0)}$ and $\omega_m^{(0)}$ are the resonance mode and resonance angular frequency of the *in vacuo* plate. These series are introduced into the resonance mode equation and, by making the successive coefficients of η^p $(p = 0, 1, \ldots)$ equal to zero, a sequence of equations is obtained. The first two are:

$$\mathcal{A}(W_m^{(0)}, U) - \mu_s h_0 \omega_m^{(0)2} \int_\Sigma (1+\varepsilon) W_m^{(0)} U^* = 0$$

$$\mathcal{A}(W_m^{(1)}, U) - \mu_s h_0 \omega_m^{(0)2} \int_\Sigma (1+\varepsilon) W_m^{(1)} U^*$$
$$-\mu_s h_0 \varpi_m^{(1)} \int_\Sigma (1+\varepsilon) W_m^{(0)} U^* + \mu_s h_0 \omega_m^2 \beta_{\omega_n^{(0)}}(W_m^{(0)}, U) = 0$$

The first equation, which is no more than the *in vacuo* resonance mode equation, is satisfied. The second equation must be satisfied for any function U and, in particular, for $U = W_m^{(0)*}$. This leads immediately to the value of $\varpi_m^{(1)}$

$$\varpi_m^{(1)} = \frac{\beta_{\omega_n^{(0)}}(W_m^{(0)}, W_m^{(0)*})}{\int_\Sigma (1+\varepsilon) W_m^{(0)} W_m^{(0)}}$$

Thus, the first-order light-fluid approximation of the resonance frequency is given by:

$$\omega_m^2 \simeq \omega_m^{(0)^2}\left[1 + \frac{\mu_0}{\mu_s h_0}\frac{\beta_{\omega_n^{(0)}}(W_m^{(0)}, W_m^{(0)^*})}{\int_\Sigma (1+\varepsilon)W_m^{(0)} W_m^{(0)}}\right] \tag{3.92}$$

Comparing equations (3.91) and (3.92), it is easily seen that (3.92) is an approximation of (3.91) if the ratio within brackets is smaller than 1, which is so for a weak coupling.

The classical approximation of "added mass", as it is described in [FAH 85], is a rough analytical estimation of the integrals involved in (3.91), which has the advantages of clearly describing the physical phenomenon and giving an idea of the effect of the fluid without the use of a computer.

4. *Comments on the numerical difficulties*

The integrals $\mathcal{A}(U_m, U_q^*)$ and $\int_\Sigma (1+\varepsilon)U_m U_q$, which appear in the case of the *in vacuo* plate and the fluid-loaded plate as well, are easy to evaluate: a Simpson's algorithm, for example, is well adapted. The main difficulty is to estimate the double integral defined by (3.84) $\beta_{\omega_n}(U_m, U_q^*)$ with a good accuracy. Indeed, the integration mesh must be adapted to the oscillations of the basis functions U_m, but also to those of the Green's kernel $g(|x_1 - x_1'|, |x_2 - x_2'|)$. Furthermore, this kernel has a singularity at $(x_1 = -x_1', x_2 = x_2')$. A powerful method consists of combining a Simpson's algorithm with an analytical approximation around the singularity.

5. *Numerical results*

The results presented here correspond to a plate with a constant thickness equal to $h_0 = 0.005$ m. The plate dimensions are $2L_1 = 0.350$ m and $2L_2 = 0.500$m. The material (steel) has a density $\mu = 7800\text{kg m}^{-3}$, a Young's modulus $E = 2 \times 10^{11}$ Pa and a Poisson's ratio $\nu = 0.3$.

The fluid considered is either air with $\mu_0 = 1.3\,\text{kg m}^{-3}$ and $c_0 = 340\,\text{m s}^{-1}$; or water with $\mu_0 = 1000\,\text{kg m}^{-3}$ and $c_0 = 1500\,\text{m s}^{-1}$. The computations account for 20 basis functions which have up to 4 nodal lines parallel to the x_1-direction and up to 3 nodal lines parallel to the x_2-axis. Tables 3.3 and 3.4 show the convergence of the iterative procedure. In Table 3.5 the Warburton's approximation of the first 12 resonance frequencies and the corresponding light-fluid approximation are compared to the "exact" values (obtained by iterations on the exact equations) for a plate in air. The norm of the difference between the exact resonance mode W_n and its Warburton's approximation W_n^W is also given. Table 3.6 presents the same results for the plate in water (the light-fluid approximation, which is of course meaningless in that case, is not mentioned).

The first comment is that the method proposed – the iterative procedure – is quite efficient. The second comment is that, in both cases, the Warburton's approximation provides reasonably accurate results for the resonance frequencies and for the resonance modes as well. For the plate in air, the light-fluid approximation is an efficient method, which has the advantage of being less time-consuming.

Figures 3.17 and 3.18 compare the *in vacuo* modes 1-4 and 3-2 to the corresponding modes for the plate loaded by water. In addition, the difference between the *in vacuo* and water-loaded modes is shown. It appears that, though this difference is not negligible, the shape of a water-loaded mode is quite similar to the shape of the corresponding air-loaded (or *in vacuo*) one.

The last calculations concern a plate with variable thickness. Figure 3.19(a) shows the variation of the plate described by the following expression:

$$h = h_0 \left[1 + \frac{1}{14} \left(\frac{x_1}{L_1} + 1 \right)^2 \cos(\pi x_1 / L_1) \left(\frac{x_2}{L_2} + 1 \right)^2 \cos(\pi x_2 / L_2) \right]$$

The thickness h varies from h_0 to $1.5h_0$. Figures 3.19(b) and (c) present the mode 1-4 for the plate in air and in water.

in vacuo plate	$f^{(0)} = 1443.6$ Hz
air-loaded plate	$f^{(i)}$ (in Hz)
1st iteration	$1441.1(1 - \imath 0.000\ 761)$
2nd iteration	$1441.1(1 - \imath 0.000\ 755)$
3rd iteration	$1441.1(1 - \imath 0.000\ 755)$
4th iteration	$1441.1(1 - \imath 0.000\ 755)$
5th iteration	$1441.1(1 - \imath 0.000\ 755)$
$\|V^{(5)}\|/\|V^{(0)}\|$	$0.565\ 10^{-6}$

Table 3.4. *Resonance frequency of mode 1-3 in air: convergence of the iterative procedure*

The results are different: the mode shape of the water-loaded plate is strongly different from the mode shape of the air-loaded plate. This is not surprising at all. Indeed, the sensitivity of the plate to the fluid loading increases as its thickness decreases. Thus, the regions of the plate where the thickness is maximum are less affected than the regions of minimum thickness.

in vacuo plate	$f^{(0)} = 1443.6$ Hz
water-loaded plate	$f^{(i)}$ (in Hz)
1st iteration	$855.3(1 - \imath 0.000\,003)$
2nd iteration	$891.3(1 - \imath 0.000\,606)$
3rd iteration	$891.2(1 - \imath 0.000\,612)$
4th iteration	$891.2(1 - \imath 0.000\,613)$
5th iteration	$891.2(1 - \imath 0.000\,612)$
$\\|V^{(5)}\\|/\\|V^{(0)}\\|$	$0.967\ 10^{-7}$

Table 3.5. *Resonance frequency of mode 1-3 in water: convergence of the iterative procedure*

mode index	*vacuum*	air (exact)	air (Warburton)	$\\|W_n - W_n^W\\|$	Light fluid approximation
0 0	275.2	274.1× $(1 - \imath 0.00204)$	274.9× $(1 - \imath 0.00200)$	0.00031	274.1× $(1 - \imath 0.00208)$
0 1	440.0	438.7× $(1 - \imath 0.00077)$	440.4× $(1 - \imath 0.00075)$	0.00027	438.7× $(1 - \imath 0.00079)$
1 0	663.8	662.2× $(1 - \imath 0.00104)$	664.6× $(1 - \imath 0.00101)$	0.00031	662.2× $(1 - \imath 0.00105)$
0 2	715.2	713.7× $(1 - \imath 0.00047)$	716.5× $(1 - \imath 0.00048)$	0.00131	713.6× $(1 - \imath 0.00048)$
1 1	817.2	815.4× $(1 - \imath 0.00077)$	818.8× $(1 - \imath 0.00074)$	0.00033	815.4× $(1 - \imath 0.00079)$
1 2	1076.8	1074.7× $(1 - \imath 0.00068)$	1078.2× $(1 - \imath 0.00066)$	0.00065	1074.6× $(1 - \imath 0.00069)$
0 3	1093.6	1091.7× $(1 - \imath 0.00044)$	1095.7× $(1 - \imath 0.00045)$	0.00150	1091.7× $(1 - \imath 0.00045)$
2 0	1252.7	1250.7× $(1 - \imath 0.00088)$	1254.3× $(1 - \imath 0.00088)$	0.00081	1250.6× $(1 - \imath 0.00089)$
2 1	1403.3	1401.1× $(1 - \imath 0.00090)$	1405.3× $(1 - \imath 0.00089)$	0.00096	1401.0× $(1 - \imath 0.00091)$
1 3	1443.6	1441.1× $(1 - \imath 0.00075)$	1444.8× $(1 - \imath 0.00075)$	0.00082	1441.1× $(1 - \imath 0.00078)$
0 4	1571.2	1568.6× $(1 - \imath 0.00066)$	1573.9× $(1 - \imath 0.00066)$	0.00268	1568.5× $(1 - \imath 0.00068)$
2 2	1656.0	1653.5× $(1 - \imath 0.00099)$	1654.7× $(1 - \imath 0.00098)$	0.00252	1653.5× $(1 - \imath 0.00101)$

Table 3.6. *Resonance frequencies (in Hz) for the plate in air: iterative procedure applied to exact equations and to the Warburton's approximated equations; light-fluid approximation*

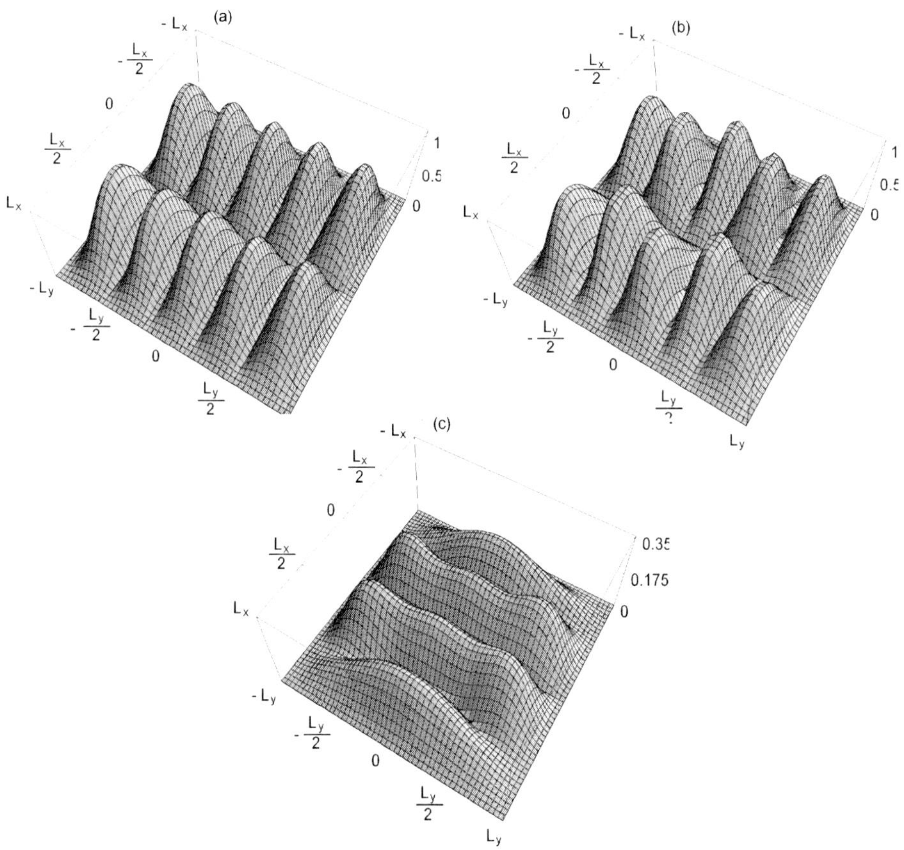

Figure 3.17. *Mode 1-4: (a) in vacuo mode; (b) water-loaded mode; (c) difference between the in vacuo and the water-loaded modes*

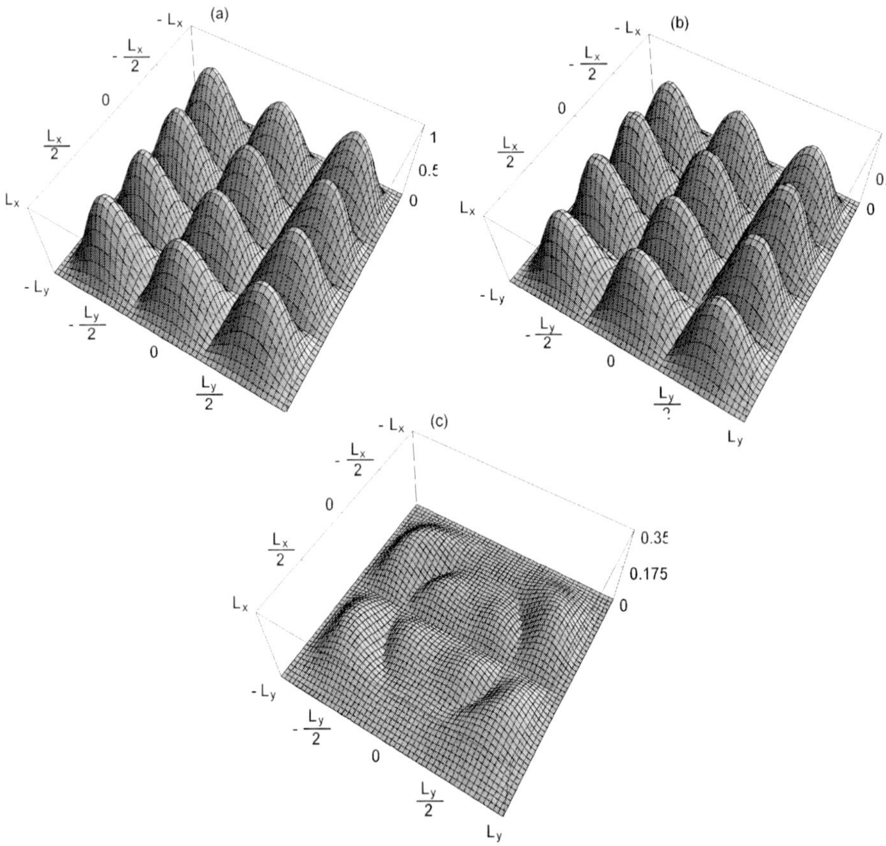

Figure 3.18. *Mode 3-2: (a) in vacuo mode; (b) water-loaded mode; (c) difference between the in vacuo and the water-loaded modes*

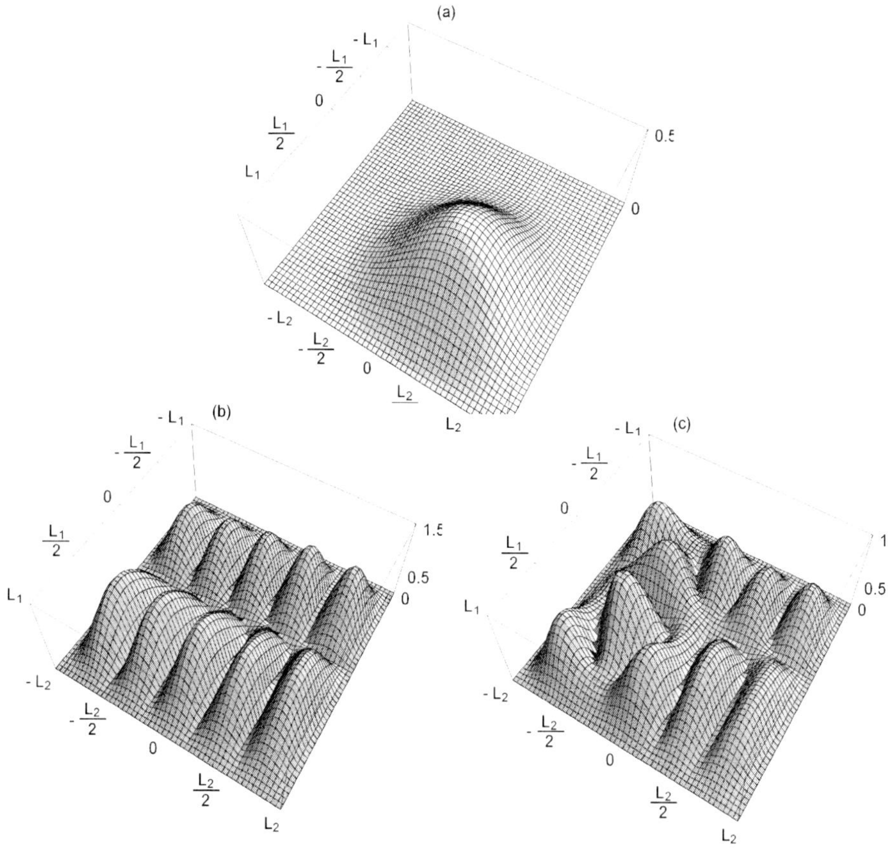

Figure 3.19. *Mode 1-4 of a plate with variable thickness: (a) variation of the plate thickness; (b) air-loaded mode; (c) water-loaded mode*

mode index	vacuum	water (exact)	water (Warburton)	$\|W_n - W_n^{(W)}\|$
0 0	275.2	94.6 (1-$\imath$ 0.01681)	95.4 (1-$\imath$ 0.01627)	0.02912
0 1	440.0	202.4 (1-$\imath$ 0.00011)	203.3 (1-$\imath$ 0.00017)	0.02442
1 0	663.8	329.5 (1-$\imath$ 0.00034)	203.1 (1-$\imath$ 0.00042)	0.02930
0 2	715.2	371.2 (1-$\imath$ 0.00839)	358.3 (1-$\imath$ 0.02375)	0.19230
1 1	817.2	441.5 (1-$\imath$ 0.00027)	442.3 (1-$\imath$ 0.00006)	0.02673
1 2	1076.8	624.4 (1-$\imath$ 0.00001)	620.3 (1-$\imath$ 0.00064)	0.07404
0 3	1093.6	628.6 (1-$\imath$ 0.00158)	616.0 (1-$\imath$ 0.00294)	0.15805
2 0	1252.7	709.6 (1-$\imath$ 0.01395)	671.4 (1-$\imath$ 0.05011)	0.23586
2 1	1403.3	841.6 (1-$\imath$ 0.00094)	824.8 (1-$\imath$ 0.00346)	0.13886
1 3	1443.6	891.2 (1-$\imath$ 0.00061)	881.2 (1-$\imath$ 0.00000)	0.07679
0 4	1571.2	968.3 (1-$\imath$ 0.00651)	946.4 (1-$\imath$ 0.02934)	0.24222
2 2	1656.0	1042.1 (1-$\imath$ 0.00057)	1023.1 (1-$\imath$ 0.01405)	0.19341

Table 3.7. *Resonance frequencies (in Hz) for the plate in water: iterative procedure applied to exact equations and to the Warburton's approximated equations*

3.6.3. *Application: Transient Response of a Plate Struck by a Hammer*

As an illustration of the preceding sections, we present a comparison between the numerical prediction and the experimental response of a plate excited by an impact hammer (a very classical tool for experimental studies of vibrating structures) which provides an impact point force of very short duration. The scheme of the experimental device is shown in Figure 3.20.

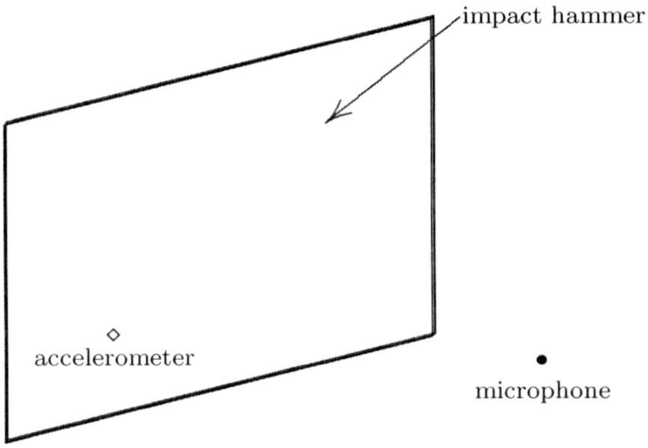

Figure 3.20. *Scheme of the experimental device: the plate is excited by an impact hammer, and the transient response is recorded by an accelerometer on the plate and a microphone off the plate*

The impact force is, of course, proportional to the acceleration of the hammer head. It is modeled by the following function

$$\begin{aligned}\tilde{\psi}(t) &= \frac{A}{2}\left(1 - \cos\frac{2\pi\, t}{T_1}\right) \text{ for } 0 < t < \frac{T_1}{2} \\ &= \frac{A}{2}\left(1 + \cos\frac{2\pi(t - T_1)}{T_2}\right) \text{ for } \frac{T_1}{2} < t < \frac{T_1 + T_2}{2}\end{aligned}$$

where $(T_1 + T_2)/2$ is the duration of the impact. The parameters A, T_1 and T_2 are adjusted so that the theoretical curve is as close as possible to the experimental acceleration of the hammer head. Thus, the excitation force is known up to a multiplicative factor. Figure 3.21 shows an example of the modeled force.

Figures 3.22 show a comparison between the measured and computed acceleration (a), and a comparison between the measured and computed acoustic pressures at two different points ((b) and (c)).

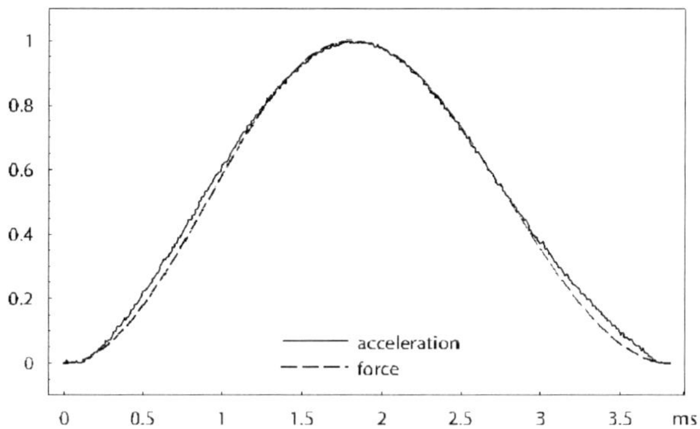

Figure 3.21. *Hammer head – measured acceleration and modeled force: the amplitude scale is arbitrary and the time origin has no meaning*

A few comments must be made. The experimental plate does not have a very constant thickness and is not perfectly homogenous and isotropic. Thus, the mechanical data must be fitted by experiments. A simple method is to choose the plate rigidity, which includes both the plate thickness and the Young's modulus so that the real parts of the first ten computed resonance frequencies are as close as possible to the experimental ones. The computation of the transient regime requires, in theory, an infinity of resonance modes, but, in practice, the number of modes used in the calculations is finite. The number of modes which are necessary for an accurate prediction of the plate response is, in some way, determined by the spectrum of the excitation force. Indeed, let ω_{max} be the maximum angular frequency with a non-negligible amplitude contained in the excitation force spectrum. To correctly describe the response of the plate, it is necessary to include at least all the modes with an angular resonance frequency less than or equal to ω_{max}. In practice, it is useful to include modes up to a slightly higher resonance frequency.

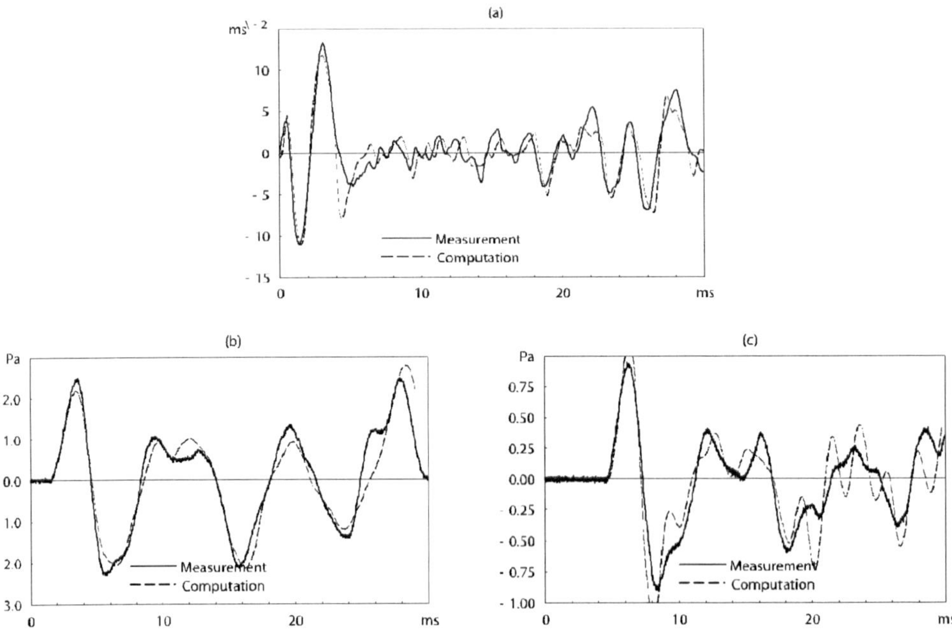

Figure 3.22. *Response of a rectangular clamped plate excited by an impact hammer with a rubber head: acceleration (a); acoustic pressure recorded at two different points (b) and (c)*

3.7. General Method for the Harmonic Regime: Classical Variational Formulation and Green's Representation of the Plate Displacement

Up to now we have shown how to compute the resonance modes and the response of a rectangular plate. This technique can easily be transposed to the case of a circular plate for which the plate displacement can be expressed in separate variables. For more complex shapes, other techniques must be used. In particular, methods based on the representation of the plate displacement by boundary integrals are frequently used.

We consider a baffled plate, in contact with a fluid on one side or on both sides, and excited by a harmonic force. The plate displacement W and the radiated acoustic pressure P satisfy the following system of equations:

$$\begin{gathered}
[\Delta^2 - \lambda^4]W(M) = \frac{1}{D}\big(F(M) - \bar{P}(M)\big) \quad \text{for} \quad M \in \Sigma \\
[\Delta + k_0^2]P(Q) = 0 \quad Q \in \Omega \\
\partial_z P(M) = \omega^2 \mu_0 W(M) \quad \text{for} \quad M \in \Sigma, \quad = 0 \quad \text{for} \quad M \in \overline{\Sigma}
\end{gathered} \tag{3.93}$$

where $\bar{P}(M)$ is the step of the acoustic pressure across the plane $z = 0$ defined by $\bar{P}(M) = \lim_{z>0, z\to 0}\big(P(x, y, z) - P(x, y, -z)\big)$, $\lambda^4 = \mu_s h\omega^2/D$ and $k_0 = \omega/c_0$. This set of equations is completed by a Sommerfeld condition for the acoustic pressure and two boundary conditions for the plate displacement. For example, if the plate is clamped along its boundary, we have:

$$\operatorname{Tr} W(M) = 0 \quad , \quad \operatorname{Tr} \partial_n W(M) = 0 \quad \text{for} \quad M \in \partial\Sigma$$

1. *Classical variational formulation*

The problem can be stated in a variational form. Let $w(M)$ be any function of the functional space $\mathbf{H}(\Sigma)$ which $W(M)$ belongs to (in particular, $w(M)$ satisfies the same boundary conditions as $W(M)$), and $\Psi(M)$ be any function of $\mathbf{L}^2(\Sigma)$, the space of square integrable functions on Σ. The functions $W(M)$ and $\bar{P}(M')$ satisfy the following system of equation:

$$\begin{gathered}
\mathcal{A}(W, w) - \lambda^4 \int_\Sigma W w^* + \frac{1}{D}\int_\Sigma \bar{P} w^* = \frac{1}{D}\int_\Sigma F w^* \\
\int_\Sigma \left[\bar{P} + \varepsilon\omega^2 \mu_0 \int_\Sigma \frac{\mathrm{e}^{\imath k_0 r}}{2\pi r} W\right]\Psi^* = 0
\end{gathered} \tag{3.94}$$

where $\varepsilon = 1$ if the fluid is present on one side of the plate, and $\varepsilon = 2$ if the fluid is present on both sides. System (3.94) can be solved by any classical finite element method.

2. *Green's representation of the plate displacement and Boundary Integrals formulation*

The Green's representation of the plate displacement, as given in section 2.3.7, is written:

$$W(M) = W_0(M) - \int_\Sigma \bar{P}(M')\Gamma(M, M')\,\mathrm{d}\Sigma(M')$$

$$+ D \int_{\partial\Sigma} \Big[\big(\operatorname{Tr} \Delta W(M') - (1 - \nu) \operatorname{Tr} \partial_{s^2} W(M')\big) \operatorname{Tr} \partial_n \Gamma(M, M')$$

$$- \left(\mathrm{Tr}\, \partial_n \Delta W(M') + (1-\nu)\partial_{\bar{s}}\, \mathrm{Tr}\, \partial_n \partial_s W(M') \right) \mathrm{Tr}\, \Gamma(M, M') \Big] \mathrm{d}\bar{s}(M')$$

$$\text{with} \quad W_0(M) = \int_{\Sigma} F(M')\Gamma(M, M')\, \mathrm{d}\Sigma(M') \tag{3.95}$$

The pressure step $\bar{P}(M)$ is related to the plate displacement by:

$$\bar{P}(M) + \varepsilon\omega^2\mu_0 \int_{\Sigma} \frac{\mathrm{e}^{\imath k_0 r(M,M')}}{2\pi r(M, M')} W(M')\, \mathrm{d}\Sigma(M') = 0 \tag{3.96}$$

which is a Boundary Integral Equation for the acoustic pressure step. These two equations are completed by the boundary conditions on $W(M)$ along $\partial\Sigma$, which provide two Boundary Integral Equations. We are left with four unknown functions:

$$W(M) \quad \text{and} \quad \bar{P}(M) \quad \text{for} \quad M \in \Sigma$$

$$\varpi_2(M) = \mathrm{Tr}\, \Delta W(M) - (1-\nu)\, \mathrm{Tr}\, \partial_{s^2} W(M) \quad \text{for} \quad M \in \partial\Sigma$$

$$\varpi_3(M) = \mathrm{Tr}\, \partial_n \Delta W(M) + (1-\nu)\partial_{\bar{s}}\, \mathrm{Tr}\, \partial_n \partial_s W(M) \quad \text{for} \quad M \in \partial\Sigma$$

System (3.95,3.96), together with the Boundary Integral Equations deduced from the boundary conditions, can be solved by a classical collocation method (including finite elements and boundary elements), the unknown functions being approximated, for example, by piecewise constant functions.

From a practical point of view, both formulations of the problem – variational and Boundary Integrals – are interesting. Starting with a computer program which solves the problem for an *in vacuo* plate, it is only necessary to add the coupling terms to obtain a program for the fluid-loaded case. Thus, the choice between one formulation or the other depends on the already existing program for the *in vacuo* plate, if any does exist. The choice is only a matter of taste.

3. *Example: response of a plate excited by a turbulent boundary layer*

To illustrate the use of the Boundary Element formulation presented here above, we consider the example of a baffled plate excited by the wall pressure due to a turbulent boundary layer. This example was proposed in reference [FIL 01].

A rectangular baffled plate, with dimensions L_1 and L_2, located in the plane $x_3 = 0$, is in contact with water on its side $x_3 = 0^+$, and a *vacuum* on the other. It is assumed that, in the half-space $x_3 > 0$, there exists a turbulent flow, with speed at infinity U_∞ parallel to the x_1-axis. The turbulent pressure exerted by the flow on the plate is a space-time random process, characterized by a cross-power density spectrum [CRI 92] $\Phi_p(x_1, x_2; x_1', x_2'; \omega)$ where ω is the angular frequency. The aim of the problem is to relate the response of the plate – cross power spectra of the plate velocity and of the radiated acoustic pressure – to the characteristics of the flow.

Among the many models of turbulent wall pressure (see, for example, [BLA 86]) the one proposed by Corcos is adopted here:

$$\Phi_p(x_1, x_2; x_1', x_2'; \omega) = \Phi_p^0(\omega) e^{-\alpha_{x_1}\omega|x_1-x_1'|/U_c} e^{-\alpha_{x_2}\omega|x_2-x_2'|/U_c} e^{\imath\omega(x_2-x_2')/U_c} \quad (3.97)$$

with $\alpha_{x_2} = 0.1$, $\alpha_{x_1} = 7\alpha_{x_2}$ and $U_c = 0.7U_\infty$. Following reference [DUR 99-1], the spectrum $\Phi_p^0(\omega)$ is defined by:

$$\log_{10}\Phi_e(\omega) = -5.1 - 0.9\log_{10} f_e - 0.34(\log_{10} f_e)^2 - 0.04(\log_{10} f_e)^3$$

$$\text{with} \quad f_e = \frac{\omega\delta}{2\pi U_\infty}, \quad q_0 = \frac{1}{2\mu_0 U_\infty^2}, \quad \Phi_p^0(\omega) = \frac{q_0\delta}{U_\infty}\Phi_e(\omega)$$

where $\delta = 0.10$ m is the turbulent boundary layer thickness.

As shown in many papers (see, for example, [FIL 97] – the proof is elementary), the cross-spectrum density $S_V(M, M'; \omega)$ of the plate velocity is given by:

$$S_V(M, M'; \omega) = -\omega^2 \int_\Sigma \int_\Sigma W(Q, M; \omega)\Phi_p(Q, Q'; \omega)W^*(Q', M'; \omega)\,dQ\,dQ' \quad (3.98)$$

where $W(Q, M; \omega)$ is the plate displacement of the fluid-loaded plate due to a harmonic unit point force δ_Q and $W^\star$ is the complex conjugate of W.

The data for the plate are: $E = 2.0\ 10^{11}$Pa, $\nu = 0.3$, $\mu_s = 7800$ kg m^{-3}, $L_1 = 1$ m, $L_2 = 0.7$ m, $h = 0.005$ m. The data for the fluid are: $\mu_0 = 1000$ kg m^{-3}, $c_0 = 1500$ m s^{-2}. The fluid velocity is small compared to the sound speed and does not influence the sound propagation. The density variations due to the turbulent flow can also be neglected. Thus, the acoustic pressure is almost the same as if the fluid was homogenous and at rest. This is an important simplification for the theoretical formulation of the problem and for numerical applications. Based on these assumptions, the results, obtained by the preceding method, are presented in Figure 3.23: we compare the response of the fluid-loaded plate with the response of the *in vacuo* plate to the same random wall pressure described by (3.22). The plate acts as a filter

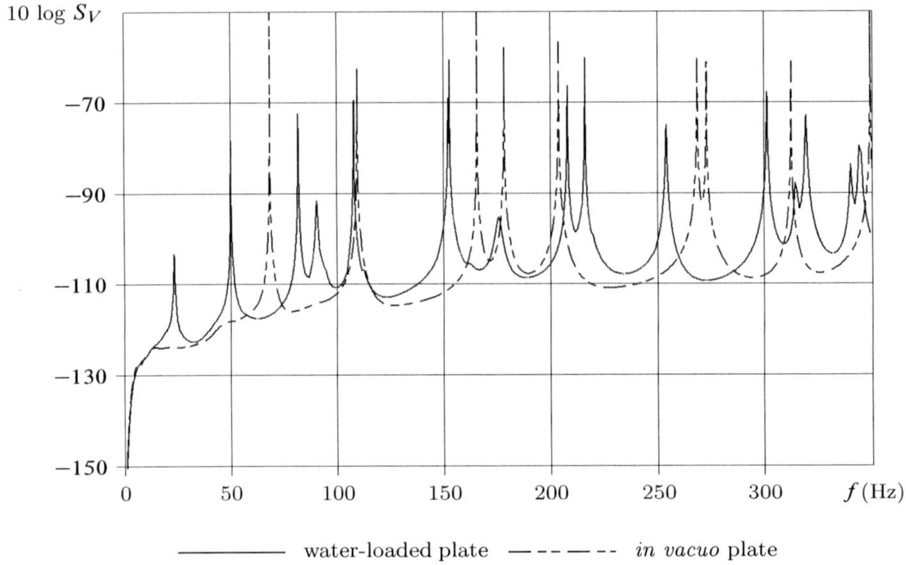

Figure 3.23. *Power spectral density of the velocity of the plate at* $x = 0.9L_x$, $y = 0.9L_y$

on the random excitation: indeed, the velocity cross-spectrum density presents peaks which correspond to the real part of the resonance frequencies of the *in vacuo* or fluid-loaded plate. The comparison between the two curves clearly points out the influence of the fluid on the plate response. The fluid-loading has two effects: the added mass effect, that shifts the peaks toward the low frequencies domain, and the damping effect, that reduces their amplitude.

3.8. Baffled Plate Closing a Cavity

Let us consider a plate which occupies a domain Σ in the plane $z = 0$, extended by an infinite perfectly rigid baffle $\overline{\Sigma}$ and excited by a harmonic force F. The half-space $\Omega^e = (z > 0)$ is occupied by a first fluid with density μ_0^e and sound velocity c_0^e. The plate closes a cavity Ω^i, with boundary $\partial\Omega^i = \Sigma \cup \partial\Omega_1^i$, which lies in the region $z < 0$ and which contains a second fluid with density μ_0^i and sound velocity c_0^i. The unit vector, normal to $\partial\Omega^i$ and pointing out to the exterior of Ω^i, is denoted by $\vec{n}$. The mechanical properties of the plate and of the fluids are as in the preceding section. For simplicity, the plate is assumed to have a constant thickness.

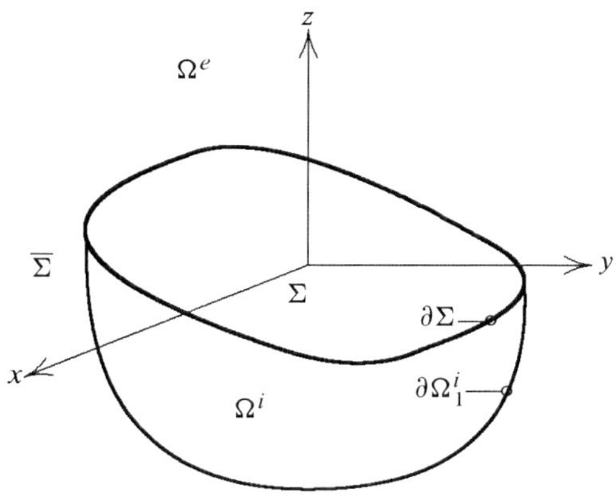

Figure 3.24. *Sketch of the baffled fluid-loaded plate closing a cavity*

3.8.1. *Equations Governing the Harmonic Motion of the Plate-Cavity-External Fluid System*

The plate displacement is denoted by W. The acoustic pressure in Ω^e is denoted by P^e, while in Ω^i it is denoted by P^i. The basic equations are:

$$
\begin{aligned}
&\left(\Delta + k_0^{e\,2}\right) P^e(Q) = 0 \quad \text{for } Q \in \Omega^e \\
&\left(\Delta + k_0^{i\,2}\right) P^i(Q) = 0 \quad \text{for } Q \in \Omega^i \\
&\left(\Delta^2 - \lambda^4\right) W(M) - \frac{\bar{P}(M)}{D} = \frac{F(M)}{D} \quad \text{for} \quad M \in \Sigma \\
&\operatorname{Tr} \partial_z P^e(M) = \omega^2 \mu_0^e W(M) \text{ for } M \in \Sigma\ ,\ = 0 \text{ for } M \in \overline{\Sigma} \\
&\operatorname{Tr} \partial_z P^i(M) = \omega^2 \mu_0^i W(M) \quad \text{for} \quad M \in \Sigma
\end{aligned}
\tag{3.99}
$$

with:

$$
\bar{P}(M) = \operatorname{Tr} P^i(M) - \operatorname{Tr} P^e(M)\ ,
$$

$$
k_0^{e\,2} = \frac{\omega^2}{c_0^{e\,2}}\ ,\ k_0^{i\,2} = \frac{\omega^2}{c_0^{i\,2}}\ ,\ \lambda^4 = \frac{\mu_s h_0 \omega^2}{D}
$$

The acoustic pressure P^e satisfies a Sommerfeld condition at infinity, while P^i is assumed to satisfy a homogenous Robin boundary condition (also called "impedance boundary condition") along the boundary $\partial\Omega_1^i$

$$
\operatorname{Tr} \partial_n P^i + \alpha \operatorname{Tr} P^i = 0 \quad \text{on} \quad \partial\Omega_1^i
$$

where the parameter α can be real or complex. The plate displacement W satisfies two boundary conditions along $\partial\Sigma$. Because the system is dissipative (energy is lost at infinity and, possibly, through $\partial\Omega_1^i$), the solution exists and is unique for any real angular frequency ω.

As done in section 3.6, let $\mathcal{G}_\omega^e(Q, Q')$ be the Green's function of the Helmholtz equation defined in Ω^e, which satisfies the homogenous Neumann condition on $\Sigma \cup \overline{\Sigma}$, that is:

$$\mathcal{G}_\omega^e(Q, Q') = -\frac{e^{ik_e r(Q,Q')}}{4\pi r(Q, Q')} - \frac{e^{ik_e r(Q,Q'')}}{4\pi r(Q, Q'')}$$

where Q'' is the point symmetric to Q' with respect to the plane $z = 0$. The acoustic pressure P^e is given by

$$P^e(Q) = \omega^2 \mu_0^e \int_\Sigma W(M')\mathcal{G}_\omega^e(Q, M')\,\mathrm{d}M'$$

The system of partial differential equations (3.99) is replaced by the system of integro-differential equations:

$$\begin{aligned}
&\left(\Delta + \frac{\omega^2}{c_0^{i\,2}}\right) P^i(Q) = 0 \quad \text{for} \quad Q \in \Omega^i \\
&\Delta^2 W(M) - \frac{\mu_s h_0 \omega^2}{D}\left[W(M) - \frac{\mu_0^e}{\mu_s h_0}\int_\Sigma W(M')\mathcal{G}_\omega^e(M, M')\,\mathrm{d}M'\right] \\
&\qquad\qquad - \frac{\mathrm{Tr}\, P^i(M)}{D} = \frac{F(M)}{D} \quad \text{for} \quad M \in \Sigma \\
&\mathrm{Tr}\,\partial_z P^i(M) = \omega^2 \mu_0^i W(M) \quad \text{for} \quad M \in \Sigma
\end{aligned} \tag{3.100}$$

It is also useful to write the variational form of these equations. Let us recall that the acoustic pressure P^i belongs to the space $\mathcal{H}(\Omega^i)$ of functions having the following properties: a/ they are square integrable on Ω^i together with their first order derivatives, b/ they satisfy the homogenous Neumann condition on Σ and c/ they satisfy the boundary condition $\mathrm{Tr}\,\partial_n \psi + \alpha\, \mathrm{Tr}\,\psi = 0$ on $\partial\Omega_1^i$. The space of functions having the properties a/ and b/ and satisfying the conjugate boundary condition $\mathrm{Tr}\,\partial_n \psi + \alpha^*\, \mathrm{Tr}\,\psi = 0$ on $\partial\Omega_1^i$ is denoted by $\widetilde{\mathcal{H}}(\Omega^i)$. Let us denote again the $\mathbf{L}^2$-scalar product on Σ by $\langle W, U\rangle$, and define the $\mathbf{L}^2$-scalar products in Ω^i by:

$$\langle\langle \varphi, \psi \rangle\rangle = \int_{\Omega^i} \varphi(Q)\psi^*(Q)\,\mathrm{d}Q \quad , \quad \langle\langle \vec{u}, \vec{v} \rangle\rangle = \int_{\Omega^i} \vec{u}(Q).\vec{v}^*(Q)\,\mathrm{d}Q$$

Introducing the form $\beta^e_\omega(W,U)$ defined by:

$$\beta^e_\omega(W,U) = \iint\limits_\Sigma W(M)\mathcal{G}^e_\omega(M,M')U^*(M')\,\mathrm{d}M\,\mathrm{d}M'$$

the variational expression of equations (3.100) writes:

$$\begin{aligned} &-\langle\langle \overrightarrow{\nabla} P^i, \overrightarrow{\nabla}\psi\rangle\rangle + \frac{\omega^2}{c_0^{i\,2}}\langle\langle P^i,\psi\rangle\rangle + \omega^2\mu_0^i\langle W,\ \mathrm{Tr}\,\psi\rangle = 0 \\ &\mathcal{A}(W,U) - \mu_s h_0\omega^2\Big[\langle W,U\rangle - \frac{\mu_0^e}{\mu_s h_0}\beta^e_\omega(W,U)\Big] - \langle\,\mathrm{Tr}\,P^i, U\rangle = \langle F,U\rangle \end{aligned} \tag{3.101}$$

where U is any function in the space $\mathbf{H}(\Sigma)$ and ψ is any function in the space $\widetilde{\mathcal{H}}(\Omega^i)$.

As done for the baffled plate, we can define resonance modes and resonance frequencies for the system cavity-plate-external fluid.

DEFINITION 3.5 (RESONANCE MODES OF THE SYSTEM PLATE-CAVITY).– *The* resonance modes (W_n, P^i_n) *of the fluid-loaded plate closing a cavity are the non-zero solutions of the homogenous variational system of equations*

$$\begin{aligned} &-\langle\langle \overrightarrow{\nabla} P^i_n, \overrightarrow{\nabla}\psi\rangle\rangle + \frac{\omega_n^2}{c_0^{i\,2}}\langle\langle P^i_n,\psi\rangle\rangle + \omega_n^2\mu_0^i\langle W_n,\ \mathrm{Tr}\,\psi\rangle = 0 \\ &\mathcal{A}(W_n,U) - \mu_s h_0\omega_n^2\Big[\langle W_n,U\rangle - \frac{\mu_0^e}{\mu_s h_0}\beta^e_{\omega_n}(W_n,U)\Big] - \langle\,\mathrm{Tr}\,P^i_n, U\rangle = 0 \\ &\qquad n = 0,1,2,\ldots,\infty \end{aligned}$$

The corresponding angular frequencies ω_n are called resonance angular frequencies *of the system cavity-plate-fluid.*

It can be shown that the resonance modes and resonance frequencies have the same properties as those of the baffled fluid-loaded plate. In particular:

- ◇ the imaginary part of any resonance frequency is negative;
- ◇ if $\omega_n = \Omega_n - \imath\tau_n$ is a resonance frequency, then $-\omega_n^* = -\Omega_n - \imath\tau_n$ is also a resonance frequency;
- ◇ the solution of equations (3.101) can be expanded into a series of the resonance modes.

3.8.2. *Integro-differential Equation for the Plate Displacement and Matched Asymptotic Expansions*

In a first step, we are not interested in discussing a general method for solving equations (3.100) or (3.101). We will focus on the case of a "light" fluid occupying the cavity domain. We have shown that the baffled fluid-loaded plate problem can be reduced to an equation for the plate displacement only. For numerical applications, this has the advantage of having a single unknown function to compute. It seems interesting to develop a similar method for the problem of a fluid-loaded plate closing a cavity.

Let $\mathcal{G}_\omega^{\Omega^i}(Q, Q')$ be the Green's function of the Helmholtz equation in Ω^i, which satisfies the boundary conditions

$$\operatorname{Tr}\partial_n \mathcal{G}_\omega^{\Omega^i}(M, Q') + \alpha \operatorname{Tr}\mathcal{G}_\omega^{\Omega^i}(M, Q') = 0 \quad \text{for} \quad M \in \partial\Omega_1^i$$

$$\operatorname{Tr}\partial_z \mathcal{G}_\omega^{\Omega^i}(M, Q') = 0 \quad \text{for} \quad M \in \Sigma$$

This Green's function is defined for any ω different from the resonance frequencies ω_k of the interior problem. The interior pressure writes:

$$P^i(Q) = -\mu_0^i\omega^2 \int_\Sigma W(M')\mathcal{G}_\omega^{\Omega^i}(Q, M')\,\mathrm{d}M'$$

Introducing this function into equation (3.100), we obtain an integro-differential equation for W only:

$$\Delta^2 W(M) - \frac{\mu_s h_0 \omega^2}{D}\Big[W(M) - \frac{\mu_0^e}{\mu_s h_0}\int_\Sigma W(M')\mathcal{G}_\omega^e(M, M')\,\mathrm{d}M' - \frac{\mu_0^i}{\mu_s h_0}\int_\Sigma W(M')\mathcal{G}_\omega^{\Omega^i}(M, M')\,\mathrm{d}M'\Big] = \frac{F(M)}{D}$$

$$\text{for} \quad M \in \Sigma \quad \text{and} \quad \omega \neq \omega_k \qquad (3.100')$$

Similarly, we introduce the integral form

$$\beta_\omega^i(W, U) = \iint_\Sigma W(M)\mathcal{G}_\omega^{\Omega^i}(M, M')U^*(M')\,\mathrm{d}M\,\mathrm{d}M'$$

to obtain the variational form of equation (3.100'):

$$\mathcal{A}(W, U) - \mu_s h_0 \omega^2\Big[\langle W, U\rangle - \frac{\mu_0^e}{\mu_s h_0}\beta_\omega^e(W, U) - \frac{\mu_0^i}{\mu_s h_0}\beta_\omega^i(W, U)\Big] = \langle F, U\rangle \quad \text{for} \quad \omega \neq \omega_k \quad (3.101')$$

If the part $\partial\Omega_1^i$ of the cavity boundary is absorbing (that is, if $\Im(\alpha) \neq 0$), the resonance frequencies of the cavity have a non-zero imaginary part and the Green's function $\mathcal{G}_\omega^{\Omega^i}(M, M')$ is defined for any real frequency. Thus, equations (3.100') and (3.101') are always valid and can be solved by any numerical method. They fail to be valid for any frequency if the resonance frequencies of the cavity are real (that is, if $\Im(\alpha) = 0$).

Nevertheless, the response of the system, which exists and is unique for any real angular frequency ω, is a continuous function, even in the vicinity of a real resonance frequency ω_k.

Let us now assume that the cavity has real resonance frequencies. If the interior fluid is a gas, we can assume that the parameter $\varepsilon = \mu_0^i/\mu_s h_0$ is "small". The meaning of "small" will become obvious in the following. With this assumption, the technique of "matched asymptotic expansions", as it is developed in [NAY 73], can be used to obtain an approximate expression of the solution W which is continuous in the vicinity of a resonance frequency ω_k. We must define three different approximations: a/ the "outer expansion", which is valid for ω different from any resonance frequency ω_k, and which is based on the hypothesis that the parameter ε is small; b/ the "inner expansion", which is valid in the vicinity of and at ω_k and which is obtained by assuming that the angular frequency has the form $\omega = \omega_k(1+\chi)$, with $\chi \ll 1$ and $\chi = \zeta\varepsilon$; c/ a linear combination of these two expansions which is continuous around ω_k.

1. *The outer expansion (valid away from ω_k)*

The plate displacement is sought as a formal series of the successive powers of ε, that is:

$$W(\omega) = W^{(0)}(\omega) + \varepsilon W^{(1)}(\omega) + \varepsilon^2 W^{(2)}(\omega) + \cdots$$

This series is introduced into equation (3.101'). We obtain a sequence of equations corresponding to the cancellation of the successive powers of ε:

$$\mathcal{A}(W^{(0)}, U) - \mu_s h_0 \omega^2 \Big[\langle W^{(0)}, U\rangle - \frac{\mu_0^e}{\mu_s h_0} \beta_\omega^e(W^{(0)}, U)\Big] = \langle F, U\rangle \qquad (3.102)$$

$$\mathcal{A}(W^{(1)}, U) - \mu_s h_0 \omega^2 \Big[\langle W^{(1)}, U\rangle - \frac{\mu_0^e}{\mu_s h_0} \beta_\omega^e(W^{(1)}, U)\Big] =$$
$$= -\mu_s h_0 \omega^2 \beta_\omega^i(W^{(0)}, U) \quad \text{for} \quad \omega \neq \omega_k \qquad (3.102')$$
$$\cdots\cdots\cdots$$

The displacement $W^{(0)}$, given by equation (3.102), is the response of the plate, loaded by the external fluid only, to the force F; we denote it by:

$$W^{(0)}(\omega) = \mathcal{C}_\omega^{-1}(F)$$

where $\mathcal{C}_\omega^{-1}$ is the inverse of the operator $\mathcal{C}_\omega$ governing the displacement of the plate, loaded by the external fluid only, and defined by:

$$\mathcal{C}_\omega(W) = \Delta^2 W - \frac{\mu_s h_0 \omega^2}{D}\Big[W - \frac{\mu_0^e}{\mu_s h_0}\int_\Sigma W \mathcal{G}_\omega^e\Big]$$

The second term, $W^{(1)}$, is calculated as a function of the first one, and so on for the other terms.

The acoustic pressure inside the cavity is deduced from the plate displacement by using its Green's representation, which is valid if $\omega \neq \omega_k$ for any value of k. We can express it as a formal series:

$$P^i(\omega) = P^{i\,(0)}(\omega) + \varepsilon P^{i\,(1)}(\omega) + \varepsilon^2 P^{i\,(2)}(\omega) + \cdots$$

$$\text{with}\quad P^{i\,(0)}(\omega) = 0 \quad , \quad P^{i\,(1)}(\omega) = -\mu_s h_0 \omega^2 \int_\Sigma W^{(0)} \mathcal{G}_\omega^{\Omega^i} \quad , \quad \cdots$$

2. *The inner expansion (valid around ω_k)*

Let us first recall the expansion of the Green's function $\mathcal{G}_\omega^{\Omega^i}(Q, Q')$ as a series of the resonance modes Φ_q of the cavity:

$$\mathcal{G}_\omega^{\Omega^i}(Q, Q') = c_0^{i\,2} \sum_{q=1, q\neq k}^{\infty} \frac{\Phi_q(Q)\Phi_q(Q')}{\omega^2 - \omega_q^2} + c_0^{i\,2} \frac{\Phi_k(Q)\Phi_k(Q')}{(\omega - \omega_k)(\omega + \omega_k)}$$

In this expression, the resonance modes are assumed to have a unit norm. We now assume that $\omega = \omega_k(1 + \chi)$ is close to the resonance frequency ω_k of the cavity and we write χ as $\chi = \varepsilon\zeta$. We look again for a series expansion of W in terms of the successive powers of ε:

$$W\big(\omega_k(1+\varepsilon\zeta)\big) = \overline{W}^{(0)}\big(\omega_k(1+\varepsilon\zeta)\big) + \varepsilon \overline{W}^{(1)}\big(\omega_k(1+\varepsilon\zeta)\big) + \varepsilon^2 \overline{W}^{(2)}\big(\omega_k(1+\varepsilon\zeta)\big) + \cdots$$

Equation (3.101'), written for $\omega = \omega_k$, leads to a sequence of equations for the functions $\overline{W}^{(n)}$, the first one being:

$$\begin{aligned}\mathcal{A}(\overline{W}^{(0)}, U) - \mu_s h_0 \omega_k^2 \Big[\langle \overline{W}^{(0)}, U\rangle - \frac{\mu_0^e}{\mu_s h_0} \beta_{\omega_k}^e(\overline{W}^{(0)}, U)\Big] \\ + \mu_s h_0 \frac{c_0^{i\,2}}{2\zeta} \int_\Sigma \overline{W}^{(0)}(M) \,\mathrm{Tr}\, \Phi_k(M)\, \mathrm{d}M \times \\ \times \int_\Sigma \mathrm{Tr}\, \Phi_k(M') U^*(M')\, \mathrm{d}M' = \langle F, U\rangle \quad (3.103)\end{aligned}$$

The solution of equation (3.103) is given by:

$$\overline{W}^{(0)}\big(\omega_k(1+\varepsilon\zeta)\big) = \mathcal{C}_{\omega_k}^{-1}(F) - \mu_s h_0 \frac{{c_0^i}^2}{2\zeta}\langle \overline{W}^{(0)}, \operatorname{Tr}\Phi_k^*\rangle \mathcal{C}_{\omega_k}^{-1}(\operatorname{Tr}\Phi_k) \qquad (3.104)$$

where the coefficient $\langle \overline{W}^{(0)}, \operatorname{Tr}\Phi_k^*\rangle$ is determined by taking the scalar product of expression (3.104) by $\operatorname{Tr}\Phi_k^*$:

$$\langle \overline{W}^{(0)}, \operatorname{Tr}\Phi_k^*\rangle = \langle \mathcal{C}_{\omega_k}^{-1}(F), \operatorname{Tr}\Phi_k^*\rangle - \mu_s h_0 \frac{{c_0^i}^2}{2\zeta}\langle \overline{W}^{(0)}, \operatorname{Tr}\Phi_k^*\rangle\langle \mathcal{C}_{\omega_k}^{-1}(\operatorname{Tr}\Phi_k), \operatorname{Tr}\Phi_k^*\rangle$$

$$\Longrightarrow \langle \overline{W}^{(0)}, \operatorname{Tr}\Phi_k^*\rangle = \frac{2\zeta\langle \mathcal{C}_{\omega_k}^{-1}(F), \operatorname{Tr}\Phi_k^*\rangle}{2\zeta + \mu_s h_0 {c_0^i}^2 \langle \mathcal{C}_{\omega_k}^{-1}(\operatorname{Tr}\Phi_k), \operatorname{Tr}\Phi_k^*\rangle}$$

We finally arrive at:

$$\begin{aligned}\overline{W}^{(0)}\big(\omega_k(1+\varepsilon\zeta)\big) &= \mathcal{C}_{\omega_k}^{-1}(F)\\ &\quad - \mu_s h_0 {c_0^i}^2 \frac{\langle \mathcal{C}_{\omega_k}^{-1}(F), \operatorname{Tr}\Phi_k^*\rangle}{2\zeta + \mu_s h_0 {c_0^i}^2 \langle \mathcal{C}_{\omega_k}^{-1}(\operatorname{Tr}\Phi_k), \operatorname{Tr}\Phi_k^*\rangle}\mathcal{C}_{\omega_k}^{-1}(\operatorname{Tr}\Phi_k) \qquad (3.105)\end{aligned}$$

This expression is defined for $\zeta = 0$, that is, for $\omega = \omega_k$.

With this approximation of W, the general expression of the interior pressure leads to:

$$\begin{aligned}P^i \simeq -\mu_s h_0 {c_0^i}^2 \varepsilon\omega_k^2(1+\zeta\varepsilon)^2\Big[&\frac{\Phi_k}{2\zeta\varepsilon\omega_k^2}\langle W^{(0)}, \operatorname{Tr}\Phi_k^*\rangle\\ &+ \sum_{q=1,q\neq k}^{\infty} \frac{\Phi_q}{\omega^2-\omega_q^2}\langle W^{(0)}, \operatorname{Tr}\Phi_q^*\rangle\Big]\end{aligned}$$

which gives the following zero-order approximation:

$$\begin{aligned}\overline{P^i}^{(0)}\big(\omega_k(1+\varepsilon\zeta)\big) &= -\frac{\mu_s h_0 {c_0^i}^2}{2\zeta}\langle \overline{W}^{(0)}, \operatorname{Tr}\Phi_k^*\rangle\Phi_k\\ &= -\mu_s h_0 {c_0^i}^2 \frac{\langle \mathcal{C}_{\omega_k}^{-1}(F), \operatorname{Tr}\Phi_k^*\rangle}{2\zeta + \mu_s h_0 {c_0^i}^2 \langle \mathcal{C}_{\omega_k}^{-1}(\operatorname{Tr}\Phi_k), \operatorname{Tr}\Phi_k^*\rangle}\Phi_k \qquad (3.106)\end{aligned}$$

REMARK 1.– While the outer approximation of the acoustic pressure P^i is of order ε, the inner approximation is of order 1.

REMARK 2.– It is known that the response of the cavity has very sharp peaks (possibly infinite) around each of its own resonance frequencies. Thus, the complete system cavity-plate-external fluid has two kinds of resonance frequencies: a sequence governed by the *in vacuo* plate resonance frequencies and a sequence governed by the resonance frequencies of the cavity.

3. *The continuous expansion*

We now combine the outer and inner expansions to obtain an expansion which remains continuous in the vicinity of the resonance frequency ω_k. The outer approximation can be considered as a function of the parameter χ through the angular frequency ω. Similarly, the inner approximation is a function of ε through the parameter $\chi = \varepsilon\zeta$. The lowest order term of the continuous approximation of $\widehat{W}(\omega)$ in the vicinity of $\omega = \omega_k(1+\chi) = \omega_k(1+\varepsilon\zeta)$ is given by:

$$\begin{aligned}\widehat{W}\big(\omega_k(1+\chi)\big) &= W^{(0)}\big(\omega_k(1+\chi)\big) + \overline{W}^{(0)}\big(\omega_k(1+\varepsilon\zeta)\big) - \lim_{\chi\to 0} W^{(0)}\big(\omega_k(1+\chi)\big)\\ &= W^{(0)}\big(\omega_k(1+\chi)\big) + \overline{W}^{(0)}\big(\omega_k(1+\varepsilon\zeta)\big) - \lim_{\zeta\to\infty} \overline{W}^{(0)}\big(\omega_k(1+\varepsilon\zeta)\big)\end{aligned}$$

Indeed we have:

$$\begin{aligned}\widehat{W}\big(\omega_k(1+\chi)\big) &= \mathcal{C}^{-1}_{\omega_k(1+\chi)}(F)\\ &\quad - \varepsilon\mu_s h_0 {c_0^i}^2 \frac{\langle \mathcal{C}^{-1}_{\omega_k}(F),\ \mathrm{Tr}\,\Phi_k^*\rangle}{2\chi + \varepsilon\mu_s h_0 {c_0^i}^2 \langle \mathcal{C}^{-1}_{\omega_k}(\mathrm{Tr}\,\Phi_k),\ \mathrm{Tr}\,\Phi_k^*\rangle}\mathcal{C}^{-1}_{\omega_k}(\mathrm{Tr}\,\Phi_k)\end{aligned}$$

This function is continuous around $\chi = 0$ or equivalently around $\omega = \omega_k$. For $\chi = \mathcal{O}(1)$, the leading term is the first one, while, for $\chi \to 0$, both terms must be accounted for.

Similarly, the continuous approximation for the interior sound pressure is:

$$\begin{aligned}\widehat{P^i}\big(\omega_k(1+\chi)\big) &= {P^i}^{(0)}\big(\omega_k(1+\chi)\big) + \overline{P^i}^{(0)}\big(\omega_k(1+\varepsilon\zeta)\big) - \lim_{\chi\to 0} {P^i}^{(0)}\big(\omega_k(1+\chi)\big)\\ &= {P^i}^{(0)}\big(\omega_k(1+\chi)\big) + \overline{P^i}^{(0)}\big(\omega_k(1+\varepsilon\zeta)\big) - \lim_{\zeta\to\infty} \overline{P^i}^{(0)}\big(\omega_k(1+\varepsilon\zeta)\big)\\ &= -\varepsilon\mu_s h_0 {c_0^i}^2 \frac{\langle \mathcal{C}^{-1}_{\omega_k}(F),\ \mathrm{Tr}\,\Phi_k^*\rangle}{2\chi + \varepsilon\mu_s h_0 {c_0^i}^2 \langle \mathcal{C}^{-1}_{\omega_k}(\mathrm{Tr}\,\Phi_k),\ \mathrm{Tr}\,\Phi_k^*\rangle}\Phi_k\end{aligned}$$

The interior sound pressure is $\mathcal{O}(\varepsilon)$ for $\chi = \mathcal{O}(1)$ and $\mathcal{O}(1)$ for $\chi \to 0$.

4. *A few remarks on the numerical calculation of the system response*

Equations (3.100') and (3.101'), which are very similar to the equations governing a baffled plate loaded by fluids extending up to infinity, can be solved by exactly the same numerical methods. The computer programs are quite similar: the difference lies only in the Green's functions used. However, around the cavity resonance frequencies, instabilities can appear and the equations are no more valid at any real resonance frequency of the cavity. Thus, the matched asymptotic expansions method provides an excellent numerical approximation of the system response, which remains valid at each real resonance frequency of the cavity. Nevertheless, the frequency interval within which numerical instabilities appear is often very small. In such a case, a good approximation of the system response at a cavity resonance frequency ω_k can be simply obtained by a linear interpolation between its response at $\omega_k(1-\varepsilon)$ and at $\omega_k(1+\varepsilon)$, with $\varepsilon = \mathcal{O}(10^{-2})$ or so.

3.8.3. *Boundary Integral Representation of the Interior Acoustic Pressure*

A much more general method to evaluate the interior pressure is to use its Green's representation. Let $\mathcal{G}^i_\omega(Q, Q')$ be the Green's function of the Helmholtz equation defined in the half-space $z < 0$ and which satisfies the Neumann condition on $z = 0$, that is

$$\mathcal{G}^i_\omega(Q, Q') = -\frac{e^{\imath k_i r(Q,QM')}}{4\pi r(Q, Q')} - \frac{e^{\imath k_i r(Q,Q'')}}{4\pi r(Q, Q'')}$$

where Q'' is the point symmetric to Q' with respect to the plane $z = 0$. The Green's representation of the pressure P^i is given by:

$$\begin{aligned} P^i(Q) = -\omega^2\mu_0^i \int_\Sigma W(M')\,\mathrm{Tr}\,\mathcal{G}^i_\omega(Q, M')\,d\sigma(M') + \int_{\partial\Omega_1^i} \mathrm{Tr}\,P^i(M')\times \\ \times\Big[\mathrm{Tr}\,\partial_{n'}\mathcal{G}^i_\omega(Q, M') + \alpha\,\mathrm{Tr}\,\mathcal{G}^i_\omega(Q, M')\Big]\,d\sigma(M') \quad \text{for} \quad Q \in \Omega^i \qquad (3.107) \end{aligned}$$

In this expression, the boundary conditions satisfied by P^i on Σ and on $\partial\Omega_1^i$ have been introduced. We have two unknown functions to determine: $W(M)$ on Σ, and $\mathrm{Tr}\,P^i(M)$ on $\partial\Omega_1^i$. To do so, we take the value of $P^i(Q)$, as given by (3.107), along $\partial\Omega_1^i$ and obtain

$$\begin{aligned} &-\omega^2\mu_0^i \int_\Sigma W(M')\,\mathrm{Tr}\,\mathcal{G}^i_\omega(M, M')\,d\sigma(M') \\ &\qquad -\frac{\mathrm{Tr}\,P^i(M)}{2} + \int_{\partial\Omega_1^i} \mathrm{Tr}\,P^i(M')\Big[\mathrm{Tr}\,\partial_{n'}\mathcal{G}^i_\omega(M, M') \\ &\qquad + \alpha\,\mathrm{Tr}\,\mathcal{G}^i_\omega(M, M')\Big]\,d\sigma(M') = 0 \qquad \text{for} \qquad M \in \partial\Omega_1^i \quad (3.108) \end{aligned}$$

in which the discontinuity of the double layer potential has been accounted for. The second equation is obtained by introducing expression (3.107) into the second equation (3.100):

$$D\Delta^2 W(M) - \mu_s h_0 \omega^2 \Big[W(M) - \int_{\Sigma} W(M') \Big(\frac{\mu_0^e}{\mu_s h_0} \mathcal{G}_\omega^e(M,M') + \frac{\mu_0^i}{\mu_s h_0} \mathcal{G}_\omega^i(M,M') \Big) \, d\sigma(M') \Big] - \int_{\partial\Omega_1^i} \operatorname{Tr} P^i(M') \Big[\operatorname{Tr} \partial_{n'} \mathcal{G}_\omega^i(M,M') + \alpha \operatorname{Tr} \mathcal{G}_\omega^i(M,M') \Big] \, d\sigma(M') = F(M) \quad \text{for} \quad M \in \Sigma \qquad (3.108')$$

Two different types of approximations are necessary to obtain a numerical approximation of the solution $\big(W(M),\ \operatorname{Tr} P^i(M)\big)$:

- ◇ As done in previous examples, if the plate geometry is simple, the plate displacement can be approximated by a truncated series of *in vacuo* resonance modes of the plate. Finite elements can also be chosen, with the advantage of being convenient for complex geometries.
- ◇ The acoustic pressure along $\partial\Omega_1^i$ is approximated by either a truncated series of basis functions or – and this is the more commonly used method – by boundary elements.

A system of linear algebraic equations is thus obtained, which provides an approximation of the solution. If the excitation is due to a sound source instead of a force applied on the plate, the changes to be done are obvious.

The values of ω for which the determinant of this system of equations is zero are approximations of the resonance frequencies of the system cavity–plate–exterior fluid. As for the baffled plate loaded by a fluid, this is not a classical eigenvalue problem. Any numerical method which can be used requires to have an idea of the location of the eigenvalues. The physics of the problem and experimental observations suggest that two types of resonance frequencies are expected: resonance frequencies which are close to the plate resonance frequencies; and resonance frequencies which are close to the cavity resonance frequencies. The plate and the cavity resonance frequencies can be used as initial values in a numerical algorithm.

3.8.4. *Comparison between Numerical Predictions and Experiments*

To illustrate this problem, the following example has been considered. The cavity is a parallelepiped with dimensions $L_x = 0.420$ m, $L_y = 0.610$ m and $L_z = 0.490$ m. The cavity wall in the $z = 0$ plane is made of a thin plate with dimensions $L_x = 0.420$ m, $L_y = 0.610$ m and $h = 0.0025$ m, which is clamped along its boundary. The cavity lies in the $z < 0$ half-space. Both fluids are air. A point source is located outside the cavity. Figure 3.25 presents a scheme of the geometry of the problem (the notations are as before). The computed sound field inside the cavity is compared with experimental results.

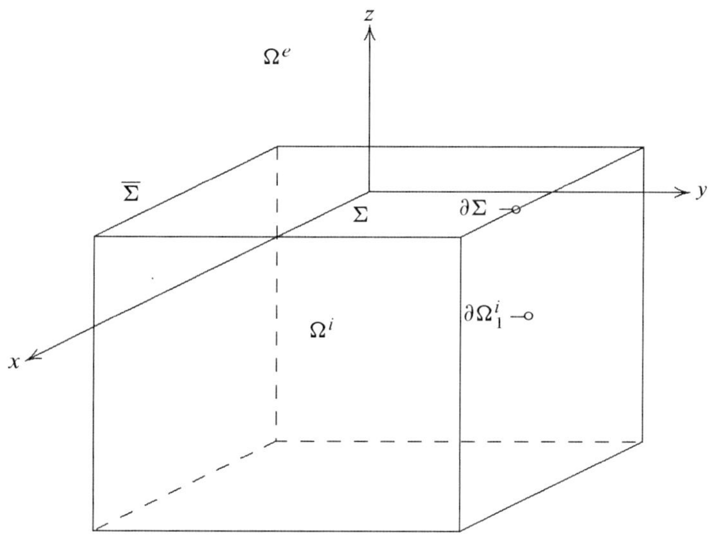

Figure 3.25. *Sketch of the baffled plate closing a parallelepipedic cavity*

In the experiment, there is no baffle extending the plate. Nevertheless, for simplicity, the calculations are conducted for a baffled plate. Because the aim is to look at the sound transmission from the exterior domain Ω^e to the interior of the cavity Ω^i, the effect of the baffle is quite small and, thus, this validates the simplified experimental assembly which has been adopted.

The walls of the cavity are assumed to be slightly absorbing (absorption coefficient less than 10%). The plate is made of an aluminum alloy. It has a rigidity $D = 94.6$ Pa m^3. The surface density is $\mu_s h = 6.62$ kg/m^2. The point source (a small loudspeaker) is located on the axis of the plate at $x = 0, y = 0, z = 2$ m .

This example is due to [GUI 08] whose thesis work concerns noise annoyance transmitted from external sources to the interior of train wagons.

The computer program used solves equation (3.100′). The cavity Green's function $\mathcal{G}^i_\omega$ is approximated by a classical image method and the plate displacement is expanded into a series of *in vacuo* modes.

The first resonance frequencies of the plate have been computed for three situations: *in vacuo*; plate with baffle and air-loading on both sides by a fluid extending up to infinity; plate with baffle, air-loading on one side by a fluid extending up to infinity and air filled cavity on the other side. The first four values are presented in Table 3.8: for the fluid-loaded plate and the cavity-plate-fluid system, a second line shows the relative shift of the resonance frequencies with respect to the *in vacuo* frequencies. It appears that the first resonance frequency is lowered when the plate is loaded by fluids extending up to infinity, while it increases slightly when the cavity is present.

The physical interpretation is simple. In the first case, the fluid induces an added mass effect, at least under the critical frequency of the plate. Conversely, in the second case, at the first resonance mode, the plate moves almost like a piston on which the cavity acts like a spring: this is equivalent to a stiffness increase of the plate. For higher order modes, the rigidity increase is almost zero because the modes shape present space oscillations, the number of which increases with the mode order (this is a well-known result, which is particularly clearly explained in [FAH 85]).

	mode 0 0	mode 0 1	mode 1 0	mode 0 2
in vacuo	93.49	147.67	226.68	238.52
air-loaded	90.48	145.06	223.33	235.62
	-3.2%	*-1.8%*	*-1.5%*	*-1.2%*
cavity-loaded	96.50	145.39	223.36	235.33
	+3.2%	*-1.5%*	*-1.5%*	*-1.3%*

Table 3.8. *Computed resonance frequencies (in Hz) of the plate*

In Figure 3.26, experimental spectra of the pressure in the cavity, recorded at three different points with a frequency step of 1 Hz, are compared with calculated spectra. Figure 3.27 presents a third octave analysis of the same results. It can be observed that these computed spectra agree rather well with measurements, the difference is slightly smoothed by the third octave analysis.

It appears that the resonance frequencies are not predicted accurately enough. This discrepancy is due to the difficulty of modeling the acoustic properties of the cavity walls with a fair correct accuracy. Furthermore, it is not at all easy to obtain a perfect clamping of the plate. So, it is impossible to account precisely for the few energy which is lost through the boundary material of the cavity and through the plate boundaries. As a consequence of this difference between the exact acoustic and vibratory properties of the

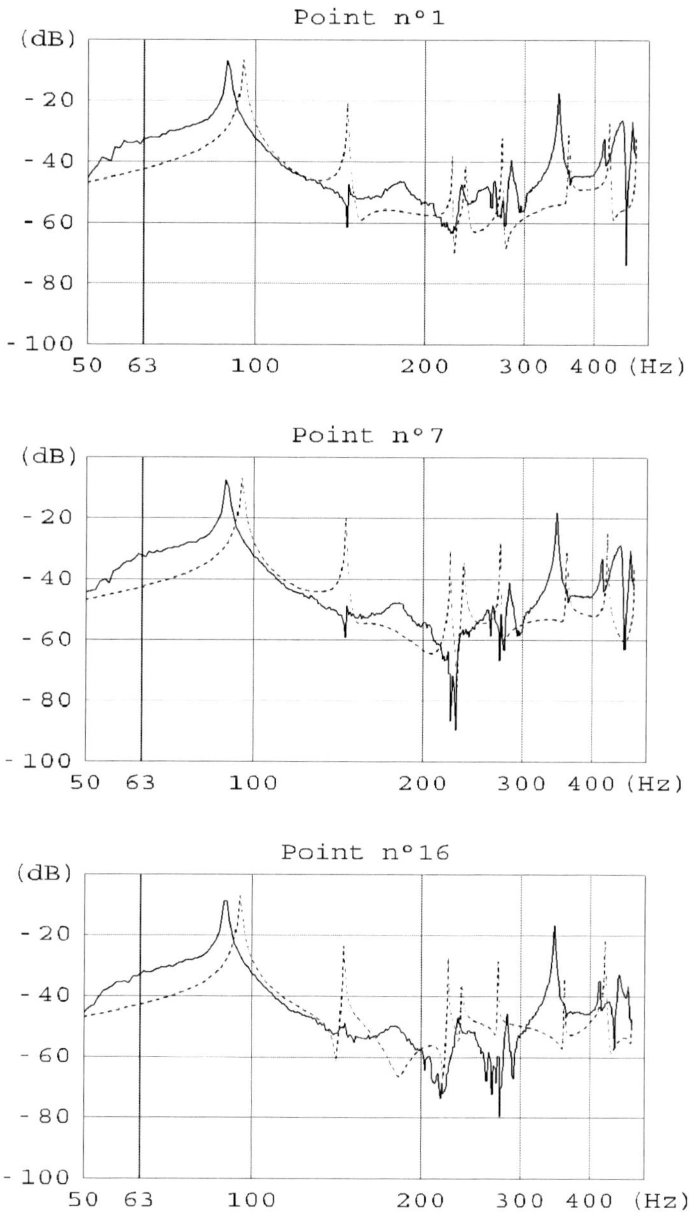

Figure 3.26. *Comparison between the pressure spectrum computed by solving (3.100′) (dashed lines) and the experimental spectrum (solid lines): the points are located along the line* $x = 0.12$ m, $z = -0.13$ m *and successively at* $y = -0.305$ m, $y = -0.185$ m *and* $y = 0.095$ m

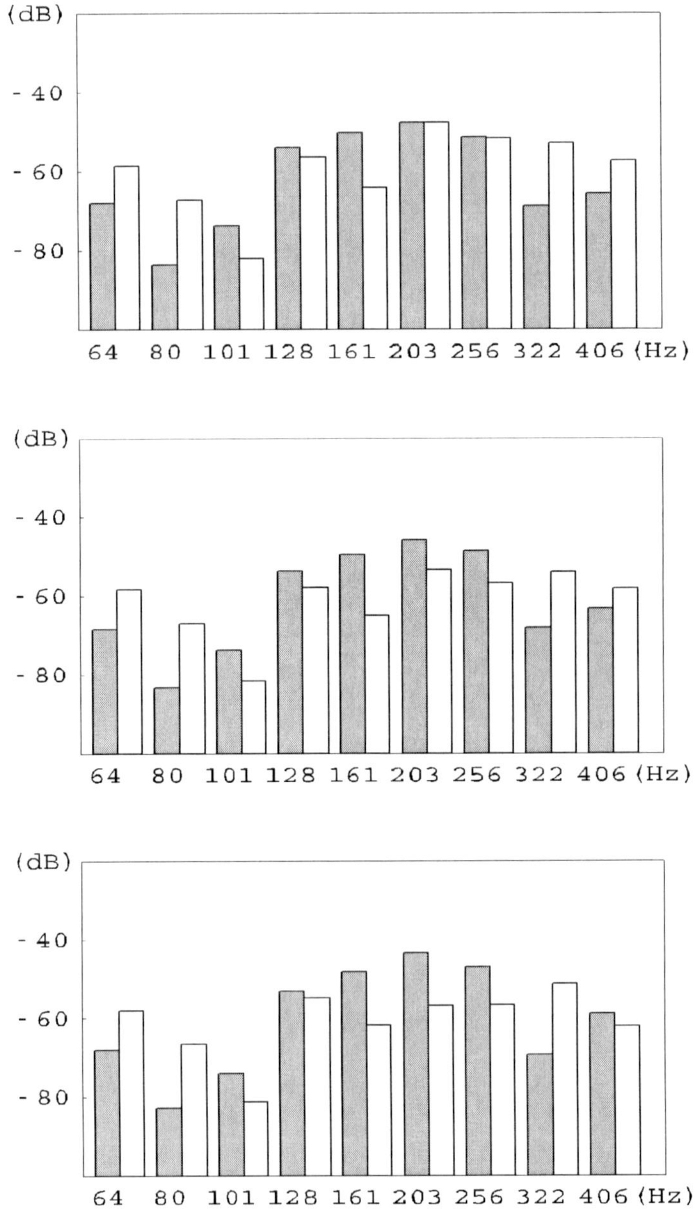

Figure 3.27. *Comparison between the third octave analysis of the pressure spectrum computed by solving (3.100′) (gray bars) and the experimental spectrum (white bars): the points are located along the line* $x = 0.12$ m, $z = -0.13$ m *and successively at* $y = -0.305$ m, $y = -0.185$ m *and* $y = 0.095$ m

experimental assembly and the modeled ones, the experimental resonance frequencies are a little lower than the theoretical ones.

However, the author of this study had in mind the annoyance due to noises transmitted through a train wagon window from the exterior to the interior. Even if the experimental data are not very accurately modeled, the variations of the transmitted noise characteristics due to variations of the mechanical plate properties will be well described. For such a reason, numerical simulations, which are always much cheaper than experiments, are a good investigation tool. They can be improved by a parameter matching technique which enables us to refine the numerical model.

3.9. Cylindrical Finite Length Baffled Shell Excited by a Turbulent Internal Flow

The excitation of a vibrating structure by a turbulent flow, inducing noise radiation, is a phenomenon which occurs in many real life situations. In particular, it appears inside cars, high speed trains, planes, etc. It is important to be able to predict noise annoyance and to evaluate numerically the efficiency of the various techniques of noise reduction before developing scale or full size experiments. When academic examples are studied, it is easy to account for all the physical parameters and to evaluate their respective importance. Then, it is not too difficult to extrapolate the conclusions to realistic situations, using, in general, a trial and experiment method. Here is the main interest of this example.

Let us consider a thin, elastic, cylindrical shell of finite length occupying the domain $\Sigma \equiv \{\rho = R, 0 \leq \varphi < 2\pi, -L/2 < z < +L/2\}$. It is extended up to infinity by two perfectly rigid cylindrical surfaces Σ^- and Σ^+, with same radius (see Figure 3.28). Its thickness is constant and denoted by h. The material is characterized by a density μ_s, a Young's modulus E and a Poisson's ratio ν.

The interior domain $\Omega^i \equiv \{0 \leq \rho < R, 0 \leq \varphi < 2\pi, -\infty < z < +\infty\}$ and the exterior domain $\Omega^e \equiv \{R < \rho, 0 \leq \varphi < 2\pi, -\infty < z < +\infty\}$ contain possibly different fluids characterized by a density μ_0^i (respectively μ_0^e) and a sound speed c_0^i (respectively c_0^e).

The shell is excited by a turbulent boundary layer induced by a fluid flow in the interior domain. It is assumed that the flow speed is small enough so that the acoustic waves in Ω^i propagate as if the fluid were homogenous and at rest.

As in section 3.5.2, the components of the shell displacement are denoted by (u, v, w). The interior and exterior acoustic pressures are denoted by p^i

and p^e, respectively. The boundary layer pressure exerted by the fluid on the shell is a random process, characterized by a cross power density spectrum $S_F(Q_0, Q_0'; \omega)$ (for details concerning the excitation of structures by turbulent flows, the reader can refer to classical books, for example, [BLA 86]; the following papers [BUL 96, CHA 87, COR 63] present fundamental studies). Let $\big(u(Q, Q_0; \omega), v(Q, Q_0; \omega), w(Q, Q_0; \omega)\big)$ be the response of the fluid-loaded

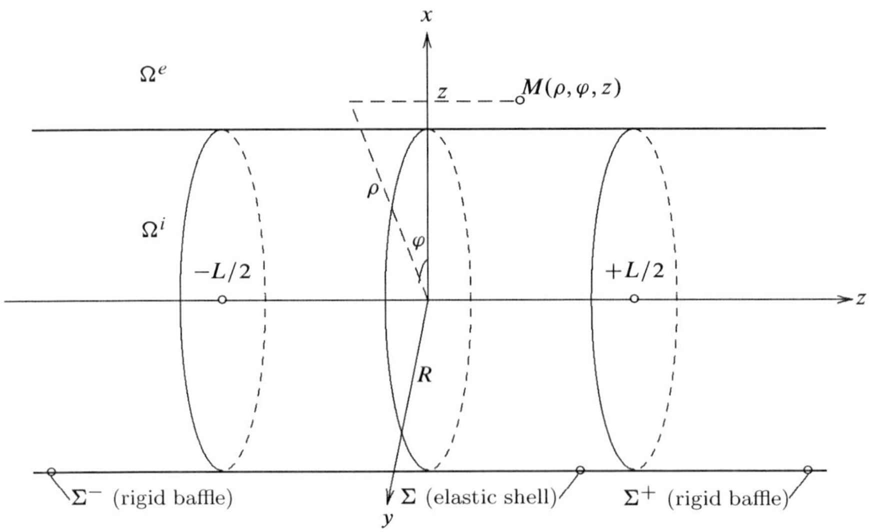

Figure 3.28. *Geometry of the baffled cylindrical shell and the coordinate system*

shell at a point Q to a harmonic unit point force, normal to its surface at a point Q_0, and $p^i(M, Q_0; \omega)$ and $p^e(M, Q_0; \omega)$ the induced acoustic pressures. It is easily established that the cross power spectral density $S_u(Q, Q'; \omega)$ due to the random excitation, which characterizes the first component of the shell displacement, is given by (see [BLA 86]):

$$S_u(Q, Q'; \omega) = \iint_{\Sigma\ \Sigma} u(Q, Q_0; \omega) S_F(Q_0, Q_0'; \omega) u^*(Q', Q_0'; \omega)\, \mathrm{d}Q_0\, \mathrm{d}Q_0'$$

Similar expressions are established for $S_v(Q, Q'; \omega)$, $S_w(Q, Q'; \omega)$, $S_{p^i}(M, M'; \omega)$ and $S_{p^e}(M, M'; \omega)$. The problem is then reduced to computing the system response to a harmonic unit point force, normal to the shell surface, for a large couple of points (Q, Q') and a wide range of angular frequencies ω.

3.9.1. *Basic Equations and Green's Representations of the Exterior and Interior Acoustic Pressures for a Normal Point Force*

Let us consider a harmonic excitation of the shell by a point unit force δ_Q acting at a point Q, and normal to the shell surface. Using the same notations as in section 3.5.2, the acoustic pressures and the shell displacement satisfy the following system of equations:

$$
\begin{gathered}
\left(\Delta + k_0^{e,i\,2}\right) p^{e,i} = 0 \quad \text{in} \quad \Omega^{e,i} \quad , \quad \text{with} \quad k_0^{e,i} = \frac{\omega}{c_0^{e,i}} \\
\frac{Eh}{1-\nu^2}\mathcal{M}^c \begin{pmatrix} u \\ v \\ w \end{pmatrix} - \omega^2 \mu_s h \begin{pmatrix} u \\ v \\ w \end{pmatrix} = \begin{pmatrix} 0 \\ 0 \\ \delta_Q \end{pmatrix} + \begin{pmatrix} 0 \\ 0 \\ \operatorname{Tr} p^i \end{pmatrix} - \begin{pmatrix} 0 \\ 0 \\ \operatorname{Tr} p^e \end{pmatrix} \quad \text{on} \quad \Sigma \\
\operatorname{Tr} \frac{\partial_\rho p^e}{\mu_0^e} = \operatorname{Tr} \frac{\partial_\rho p^i}{\mu_0^i} = \omega^2 w \quad \text{on} \quad \Sigma \quad , \quad = 0 \quad \text{on} \quad \Sigma^- \cup \Sigma^+
\end{gathered}
\tag{3.109}
$$

The shell displacement components satisfy boundary conditions along the two circles $z = -L/2$ and $z = +L/2$; in the numerical example which will be given, the shell is clamped along its boundaries. Both acoustic pressures satisfy an "outgoing wave" condition which ensures the uniqueness of the solution.

This problem can be transformed into a boundary value problem for the shell displacement and the acoustic pressure inside the domain $\Omega_0^i = \{0 \le \rho < R, 0 \le \varphi < 2\pi, -L/2 < z + L/2\}$. We do not detail this transformation which can be found in [MAT 99]. The resulting system of equations has a unique solution for any angular real frequency, but it leads to rather heavy numerical calculations. We propose to develop a much simpler method.

As done in the example of a baffled plate closing a cavity, we express the acoustic pressures by their Green's representations:

$$
p^e(M;\omega) = \omega^2 \mu^e \int_\Sigma w(M';\omega)\mathcal{G}_\omega^e(M,M')dM'
$$

$$
p^i(M;\omega) = -\omega^2 \mu^i \int_\Sigma w(M';\omega)\mathcal{G}_\omega^i(M,M')dM'
$$

As in section 3.5.2, the Green's function $\mathcal{G}_\omega^e(M,M')$ of the exterior problem can be expressed as the inverse Fourier transform of the function

$$
\begin{aligned}
\widehat{\mathcal{G}}^e(\rho,\rho';\varphi,\varphi';\zeta,z') = -\frac{\imath}{4}\Bigg[& H_0(Kd) \\
& - \sum_{n=-\infty}^{+\infty} \frac{J_n'(KR)}{H_n'(KR)} H_n(K\rho) H_n(K\rho') \mathrm{e}^{\imath n(\varphi-\varphi')}\Bigg] \mathrm{e}^{-2\imath\pi\zeta z'}
\end{aligned}
$$

$$\text{with} \quad K^2 = \frac{\omega^2}{c_0^{e\,2}} - 4\pi^2\varsigma^2 \quad \text{and} \quad \Im(K) < 0$$

It is defined for any real angular frequency. The Green's function $\mathcal{G}^i_\omega(M, M')$ of the interior problem is given by the series (see, for example, [MOR 53]):

$$\mathcal{G}^i_\omega(\rho, \varphi, z; \rho', \varphi', z') = \sum_{n\ m} \frac{\psi_{nm}(\rho, \varphi)\psi^*_{nm}(\rho', \varphi')\mathrm{e}^{\imath\chi_{nm}|z-z'|}}{2\imath\chi_{nm}} \tag{3.110}$$

where $\psi_{nm}(\rho, \varphi)$ is the normal mode of order $(n\ m)$ of the cylindrical waveguide Ω^i and χ_{nm} the corresponding eigenwavenumber. We have:

$$\psi_{nm}(\rho, \varphi) = A_{nm}J_n(\kappa_{nm}\rho)\mathrm{e}^{\imath n\varphi} \quad \text{with } \kappa_{nm} \text{ defined by } J'_n(\kappa_{nm}R) = 0$$

$$\chi^2_{nm} = k^{i\,2} - \kappa^2_{nm}$$

where A_{nm} is an arbitrary coefficient which is generally chosen such that $\|\psi_{nm}\| = 1$. The parameters χ_{nm} depend on the angular frequency ω. Because the wavenumbers κ_{nm} are real, there exists a sequence of angular frequencies $\omega_{nm} = c^i_0\kappa_{nm}$ for which $\chi_{nm} = 0$. For these frequencies, called the "cut-off frequencies", the Green's function $\mathcal{G}^i_\omega$ is not defined. Thus, for any $\omega \neq \omega_{nm}$, the initial boundary value problem (3.109) can be replaced by a matrix equation (system of 3 equations) for the shell displacement vector only:

$$\frac{Eh}{1-\nu^2}\mathcal{M}^c\begin{pmatrix} u \\ v \\ w \end{pmatrix} - \omega^2\mu_s h\left[\begin{pmatrix} u \\ v \\ w \end{pmatrix} - \frac{\mu^i}{\mu_s h}\begin{pmatrix} 0 \\ 0 \\ \int_\Sigma \mathcal{G}^i_\omega w \end{pmatrix} - \frac{\mu^e}{\mu_s h}\begin{pmatrix} 0 \\ 0 \\ \int_\Sigma \mathcal{G}^e_\omega w \end{pmatrix}\right] = \begin{pmatrix} 0 \\ 0 \\ \delta_Q \end{pmatrix} \quad \text{on } \Sigma \tag{3.111}$$

This equation fails to be valid for the cut-off frequencies of the waveguide. Nevertheless, if the interior fluid is a gas, that is, if we can assume that the parameter $\varepsilon = \mu^i_0/\mu_s h$ is small, it is possible to use the method of matched asymptotic expansions and thus define a solution which remains continuous in the vicinity of the cut-off frequencies.

3.9.2. *Numerical Methods for Solving Equations (3.111)*

In a first step, the shell displacement is sought as a Fourier series of the angular variable φ, that is:

$$\begin{pmatrix} u \\ v \\ w \end{pmatrix} = \sum_{n=-\infty}^{+\infty}\begin{pmatrix} u_n \\ v_n \\ w_n \end{pmatrix}\mathrm{e}^{\imath n\varphi}$$

Denoting by (z_Q, φ_Q) the coordinates of the excitation point, system (3.111) is thus replaced by the infinite sequence of systems:

$$\frac{Eh}{1-\nu^2}\mathcal{M}_n^c\begin{pmatrix}u_n\\v_n\\w_n\end{pmatrix} - \omega^2\mu_s h\left[\begin{pmatrix}u_n\\v_n\\w_n\end{pmatrix} - \frac{\mu^i}{\mu_s h}\begin{pmatrix}0\\0\\2\pi R\int_{-L/2}^{+L/2}\mathcal{G}_{\omega n}^i w_n\end{pmatrix}\right.$$
$$\left. - \frac{\mu^e}{\mu_s h}\begin{pmatrix}0\\0\\2\pi R\int_{-L/2}^{+L/2}\mathcal{G}_{\omega n}^e w_n\end{pmatrix}\right] = \begin{pmatrix}0\\0\\\delta_{z-z_Q}\ \mathrm{e}^{-\imath n\varphi_Q}\end{pmatrix}$$
$$\text{for } -L/2<z<+L/2 \quad (3.111')$$

where $\mathcal{G}_{\omega n}^i$ and $\mathcal{G}_{\omega n}^e$ are the n-th angular components of $\mathcal{G}_\omega^i$ and $\mathcal{G}_\omega^e$. The factor $2\pi R$ comes from the integration with respect to φ.

A numerical approximation of (u_n, v_n, w_n) can be obtained using any classical finite element method: the variational form of equations (3.111′) is then required. Let (U, V, W) be any vector function which has the same regularity properties and which satisfies the same boundary conditions as (u_n, v_n, w_n) does. This variational form writes:

$$\mathcal{A}_n^C(u_n, v_n, w_n; U, W, W) - 2\pi R\omega^2\mu_s h\int_{-L/2}^{+L/2}\left[\begin{pmatrix}u_n\\v_n\\w_n\end{pmatrix}\cdot\begin{pmatrix}U\\V\\W\end{pmatrix}\right.$$
$$\left. - \frac{\mu^i}{\mu_s h}\begin{pmatrix}0\\0\\2\pi R\int_{-L/2}^{+L/2}\mathcal{G}_{\omega n}^i w_n\end{pmatrix}\cdot\begin{pmatrix}U\\V\\W\end{pmatrix} - \frac{\mu^e}{\mu_s h}\begin{pmatrix}0\\0\\2\pi R\int_{-L/2}^{+L/2}\mathcal{G}_{\omega n}^e w_n\end{pmatrix}\cdot\begin{pmatrix}U\\V\\W\end{pmatrix}\right]$$
$$= 2\pi R\int_{-L/2}^{+L/2}\begin{pmatrix}0\\0\\\delta_{z-z_Q}\ \mathrm{e}^{-\imath n\varphi_Q}\end{pmatrix}\cdot\begin{pmatrix}U\\V\\W\end{pmatrix} \quad (3.111'')$$

The bilinear form $\mathcal{A}_n^C(u_n, v_n, w_n; U, W, W)$ is deduced from (1.34′) by replacing (u, v, w) by $(u_n, v_n, w_n)\mathrm{e}^{\imath n\varphi}$, $(\hat{u}, \hat{v}, \hat{w})$ by $(U, V, W)\mathrm{e}^{\imath n\varphi}$ and integrating with respect to φ, that is:

$$\mathcal{A}_n^C(u_n, v_n, w_n; U, W, W) = 2\pi R \int_{-L/2}^{L/2} \mathrm{d}z$$

$$\frac{Eh}{1-\nu^2}\Bigg[\nu\Big(u_{n,z} + \frac{\imath n v_n + w}{R}\Big)\Big(U_{,z}^* + \frac{-\imath n V^* + W^*}{R}\Big)$$

$$+ (1-\nu)\Big(u_{n,z}\,U_{,z}^* + \frac{\imath n v_n + w}{R}\frac{-\imath n V^* + W^*}{R}\Big)$$

$$+ \frac{1-\nu}{2}\Big(\frac{\imath n u_n}{R} + v_{n,z}\Big)\Big(\frac{-\imath n U^*}{R} + V_{,z}^*\Big)$$

$$+ \frac{h^2}{12}\Big[\nu\Big(w_{n,zz} - \frac{n^2 w_n}{R^2}\Big)\Big(W_{,zz}^* - \frac{n^2 W^*}{R^2}\Big)$$

$$+ (1-\nu)\Big(w_{n,zz}\,W_{,zz}^* + \frac{n^2 w_{n,z}\,W_{,z}^*}{R^2} + \frac{n^2 w_{n,z}\,W_{,z}^*}{R^2} + \frac{n^4 w_n W^*}{R^4}\Big)\Big]\Bigg]$$

A more interesting approximation consists of expanding the components of the displacement (u_n, v_n, w_n) into a series of the components (U_{nj}, V_{nj}, W_{nj}) of the *in vacuo* resonance modes of the shell $[(U_{nj}, V_{nj}, W_{nj})\exp(\imath n\varphi), j = 1, \cdots, \infty]$. Indeed, this set of functions is a basis of the functional space which the vector (u_n, v_n, w_n) belongs to.

When the fluids are gases, their influence on the shell displacement is rather small and a perturbation method, restricted to the first order term only, will generally provide a good approximation of the result.

3.9.3. *Comparison Between Numerical Results and Experimental Data*

It is first necessary to check the accuracy of the numerical method used to compute the response of the fluid-loaded shell. This is achieved by comparing the computed resonance frequencies to the experimental ones.

Then, the response of the shell to an interior turbulent flow can be investigated: the numerical prediction of the cross power spectrum density of the shell acceleration is compared to the experimental one.

1. *Resonance frequencies of the fluid-loaded shell*

The resonance angular frequencies of the fluid-loaded shell are the values of ω for which equations (3.111′) or (3.111″) with a zero second member have a non-zero solution. Numerically, this leads us to look for the values of the angular frequency which minimize the determinant of the linear system which approximates the exact equations. Because both fluids are gases (air), the fluid-loaded resonance frequencies are close to the *in vacuo* ones. Thus, a Newton procedure which starts from the successive *in vacuo* resonance frequencies converges very rapidly to the fluid-loaded ones.

f_{nm}	$m = 1$	$m = 2$	$m = 3$	$m = 4$
	$(1 - \imath 2.4\ 10^{-4})$			
$n = 1$	$1975\times$			
	$(1 - \imath 4.0\ 10^{-4})$			
$n = 2$	$941\times$	$2157\times$		
	$(1 - \imath 3.2\ 10^{-4})$	$(1 - \imath 4.6\ 10^{-4})$		
Exp.	**942**			
$n = 3$	$565\times$	$1260\times$	$2170\times$	$3198\times$
	$(1 - \imath 3.5\ 10^{-4})$	$(1 - \imath 2.4\ 10^{-4})$	$(1 - \imath 3.7\ 10^{-4})$	$(1 - \imath 4.7\ 10^{-4})$
Exp.	**565**	**1301**		
$n = 4$	$573\times$	$922\times$	$1499\times$	$2208\times$
	$(1 - \imath 1.7\ 10^{-4})$	$(1 - \imath 0.2)$	$(1 - \imath 0.4)$	$(1 - \imath 2.3\ 10^{-4})$
Exp.	**570**	**937**	**1544**	
$n = 5$	$794\times$	$944\times$	$1267\times$	$1735\times$
	$(1 - \imath 2.5\ 10^{-4})$	$(1 - \imath 2.1\ 10^{-4})$	$(1 - \imath 2.4\ 10^{-4})$	$(-\imath 2.3\ 10^{-4})$
Exp.	**794**	**959**	**1311**	
$n = 6$	$1120\times$	$1189\times$	$1355\times$	$1636\times$
	$(1 - \imath 2.7\ 10^{-4})$	$(1 - \imath 2.5\ 10^{-4})$	$(1 - \imath 2.2\ 10^{-4})$	$(1 - \imath 2.4\ 10^{-4})$
Exp.	**1131**	**1207**	**1390**	
$n = 7$	$1558\times$	$1650\times$	$1814\times$	$2057\times$
	$(1 - \imath 1.9\ 10^{-4})$	$(1 - \imath 2.4\ 10^{-4})$	$(1 - \imath 2.8\ 10^{-4})$	$(1 - \imath 2.4\ 10^{-4})$
Exp.	**1564**			

Table 3.9. *Resonance frequencies of the air-loaded shell: comparison between the computed (1st and 2nd lines) and the experimental ones (3rd line,* **Exp.***)*

The vibrating shell is made of stainless steel; it is 0.46 m long, its diameter is 0.125 m and its thickness is $5\ 10^{-4}$ m. The material parameters are: $\mu_s = 7850$ kg m^{-3}, $E = 19.5\ 10^7$ Pa, $\nu = 0.33$. Table 3.9 shows a few results which are compared with experimental data, whenever available.

2. *Response of the shell to the excitation by a turbulent boundary layer*

This example is issued from [MAT 99] for the numerical prediction and from [DUR 99-1, DUR 99-2] for the experiment. The vibrating shell is inserted between two long and rigid pipes in which an air flow is established. This experimental assembly is isolated from any possible vibrations and located in an anechoïc room in order to eliminate noise – other than the noise radiated by the shell.

As in the case of the plate (section 3.7-paragraph 3), the turbulent wall pressure exerted by the flow on the shell is described by a Corcos's model:

$$S_F(Q, Q'; \omega) = S(\omega) e^{-|z|/\ell_z(\omega)} e^{-|R\varphi|/\ell_\varphi(\omega)} e^{\imath \vartheta(\xi,\omega)}$$

The various parameters, which depend on the angular frequency ω and the

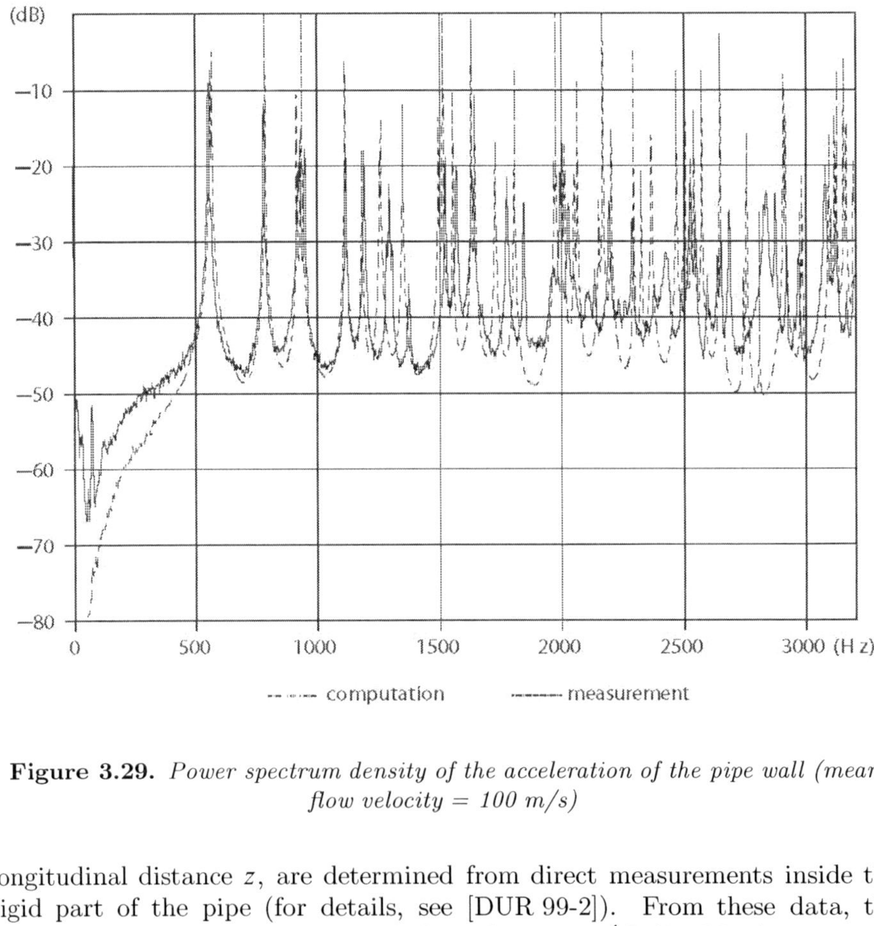

Figure 3.29. *Power spectrum density of the acceleration of the pipe wall (mean flow velocity = 100 m/s)*

longitudinal distance z, are determined from direct measurements inside the rigid part of the pipe (for details, see [DUR 99-2]). From these data, the power density spectrum of the radial acceleration $\omega^4 S_u(M, M; \omega)$ at a point M has been calculated and compared with the experimental one. Figure 3.29 shows that the numerical predictions are in good agreement with the measured spectrum.

It is clear that this problem requires an important amount of calculation, but this is no longer the main difficulty. The computation power and the accuracy of modern computers is continuously increasing, so, such results can be obtained within a computation time which is already short and will be shorter in the future. Nevertheless, it must be kept in mind that the rounding errors can increase with the amount of calculations and, thus, the choice of the numerical method is very important.

3.10. Radiation by a Finite Length Cylindrical Shell Excited by an Internal Acoustic Source

In this section, we consider a thin elastic shell, closed by two perfectly rigid discs, immersed in water and filled with air. The system is driven by an internal acoustic source. The shell has a length L, is clamped at one end and is freely supported along the other end. This study was developed in [SEI 97, SEI 99]. The aim of the authors was mainly to predict the far-field acoustic pressure radiated by the shell.

The radiated acoustic pressure is modeled by a hybrid layer potential that is a function of the form

$$\int_{\Sigma} \mu(Q)\Big[\partial_{n(Q)}G^e(M,Q) + \alpha G^e(M,Q)\Big]\,\mathrm{d}\sigma$$

where Σ is the shell surface, $\mu(Q)$ is the layer density, $G^e(M,Q)$ is the free-field Green's function of the Helmholtz equation, and α is a constant with a non-zero imaginary part.

The shell displacement is expressed as the sum of the field generated by the acoustic pressures (internal and external) and that due to boundary sources.

As done for the case of a plate closing a cavity, use is made of the Green's function of the interior Neumann problem to express the acoustic pressure inside the shell in terms of the acoustic source and of the shell normal displacement. Of course, the same difficulty appears: this representation fails to be valid for each frequency equal to one of the resonance frequencies of the shell interior. However, because the inner fluid is air, a *light fluid* approximation can be adopted and, around each resonance frequency, matched asymptotic expansions are developed.

3.10.1. *Statement of the Problem*

A thin elastic shell occupies the cylindrical surface Σ, having the z-axis as symmetry axis. Its radius is R, and it extends from $-L/2$ to $+L/2$. It is bounded by two circles C^+ and C^- and closed by perfectly rigid bulkheads corresponding to the disks Σ^+ and Σ^- (see Figure 3.30). Let Ω^i be the interior domain which contains a gas (air). The exterior domain Ω^e is occupied by a liquid (water).

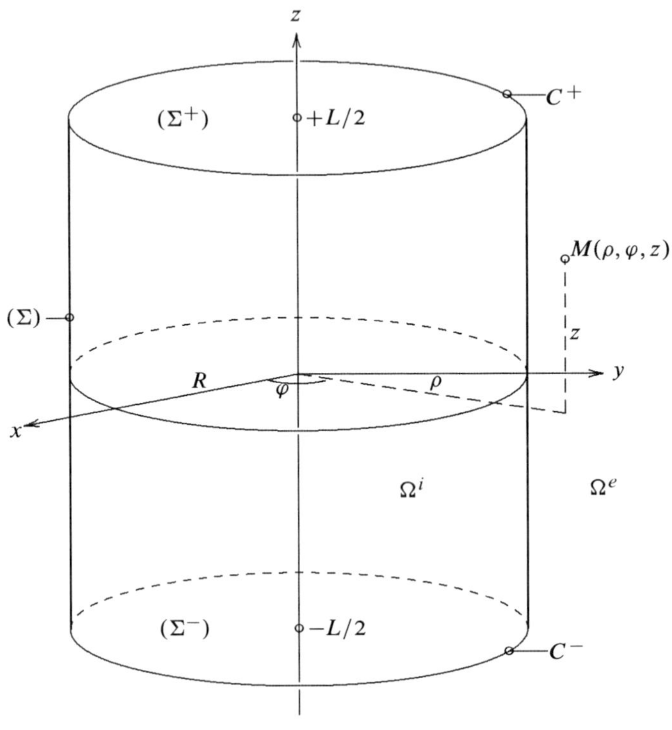

Figure 3.30. *Geometry of the structure*

The shell has a thickness h. It is made of a material with density μ_s, Young's modulus E and Poisson's ratio ν. The interior and external fluids are characterized by densities μ_0^i and μ_0^e, and sound velocities c_0^i and c_0^e. The shell is clamped along C^+ and freely supported along C^-.

The components of the shell displacement vector W are: u, parallel to the z-axis; v, orthogonal to the z-axis and tangent to the surface Σ; w, normal to the surface Σ and positive if pointing out to Ω^e.

The acoustic pressures are denoted by p^i in Ω^i, and p^e in Ω^e. Cylindrical coordinates (ρ, φ, z) are adopted.

Using again the same notations as in section 3.5.2, the system motion is governed by the following system of partial differential equations:

$$\left\{\frac{Eh}{1-\nu^2}\mathcal{M}^c - \mu_s h\omega^2 \mathbf{I}\right\}\begin{pmatrix} u \\ v \\ w \end{pmatrix} + \begin{pmatrix} 0 \\ 0 \\ \mathrm{Tr}\, p^e - \mathrm{Tr}\, p^i \end{pmatrix} = 0\,, \quad \text{on } \Sigma \tag{3.112}$$

$$\left(\Delta + k^{e2}\right) p^e = 0\,, \quad \text{in } \Omega^e \tag{3.112'}$$

$$\left(\Delta + k^{i2}\right) p^i = f\,, \quad \text{in } \Omega^i \tag{3.112''}$$

In these equations $k^{i,e} = \omega / c_0^{i,e}$; $\mathbf{I}$ is the identity matrix.

To simplify the expressions of the boundary conditions, we adopt the notation $f,_x$ for the derivative $\partial_x f$ of f with respect to variable x. The boundary conditions write:

◇ for the acoustic pressures

$$\frac{\mathrm{Tr}\, p^i,_\rho}{\mu_0^i} = \frac{\mathrm{Tr}\, p^e,_\rho}{\mu_0^e} = \omega^2 w\,, \quad \text{on } \Sigma \tag{3.113}$$

$$\mathrm{Tr}\, p^i,_z = \mathrm{Tr}\, p^e,_z = 0\,, \quad \text{on } \Sigma^- \cup \Sigma^+ \tag{3.113'}$$

◇ for the shell displacement

$$\mathrm{Tr}\, u = \mathrm{Tr}\, v = \mathrm{Tr}\, w = \mathrm{Tr}\, w,_z = 0 \quad \text{on} \quad C^- \tag{3.114}$$

$$\left.\begin{aligned} &\mathrm{Tr}\, u,_z + \frac{\nu}{R}\,\mathrm{Tr}\left[v,_\varphi + w\right] = 0 \\ &\frac{\mathrm{Tr}\, u,_\varphi}{R} + \mathrm{Tr}\, v,_z = 0 \\ &\mathrm{Tr}\, w,_{zz} + \frac{\nu}{R^2}\,\mathrm{Tr}\, w,_{\varphi\varphi} = 0 \\ &\mathrm{Tr}\, w,_{zzz} + \frac{\mathrm{Tr}\, w,_{z\varphi\varphi}}{R^2} + \frac{1-\nu}{R^2}(\mathrm{Tr}\, w,_{z\varphi}),_\varphi = 0 \end{aligned}\right\} \quad \text{on } C^+ \tag{3.114'}$$

As usual, the symbol " Tr " means that the value of the function is taken along the boundary of the domain. The uniqueness of the solution is ensured by a Sommerfeld condition on p^e.

3.10.2. *Boundary Integral Representations of the Radiated Pressure and of the Shell Displacement*

1. *Boundary integral representation of the radiated pressure*

Let $\mathcal{G}_\omega^e(M, M')$ be the Green's kernel of Helmholtz equation (3.112′) which satisfies the Sommerfeld condition at infinity. The acoustic pressure p^e can be represented by a hybrid layer potential [FIL 99]:

$$p^e(M) = \int_{\Sigma\cup\Sigma^+\cup\Sigma^-} \varpi(P)\big(G^e(M,P) - \imath\partial_{\vec{n}(P)}G^e(M,P)\big)\,\mathrm{d}\sigma(P)$$
$$\text{with}\quad G^e(M,P) = -\frac{\mathrm{e}^{\imath k^e r(M,P)}}{4\pi r(M,P)} \tag{3.115}$$

where $\vec{n}(P)$ is the unit vector normal to the boundary of Ω^i and pointing out to Ω^e. The values of $p^e(M)$ and of its normal derivative must be calculated on the surfaces Σ, Σ^+ and Σ^-. It must be recalled that the value of $p^e(M)$ involves a Cauchy principal value which does not present any numerical difficulty. The value on Σ of the normal derivative $\partial_n p^e(M)$ of $p^e(M)$ requires to calculate a Hadamard finite part (see section 3.4).

2. *Boundary Integral representation of the shell displacement*

Using the *in vacuo* shell operator Green's tensor, the shell displacement can be represented in terms of the pressure step and of boundary sources by a Green's formula. By using the expressions given in section 2.4.2, the Green's tensor $\Gamma^c(M,M')$ of the shell operator is the matrix which satisfies:

$$\left\{\frac{Eh}{1-\nu^2}\mathcal{M}^c - \mu_s h\omega^2\mathbf{I}\right\}\Gamma^c(M,M') = \begin{pmatrix}\delta_{M'}(M) & 0 & 0\\ 0 & \delta_{M'}(M) & 0\\ 0 & 0 & \delta_{M'}(M)\end{pmatrix}$$

with $\delta_{M'}(M)$ is the Dirac measure located at the point M' (the uniqueness of $\Gamma^c(M,M')$ is ensured by adding an out-going wave condition). Denoting by (z,φ) and (z',φ') the respective coordinates of M and M' on Σ, its Fourier series expansion with respect to the angular variable φ is defined by

$$\Gamma^c(M,M') = \sum_{n=-\infty}^{+\infty} \Gamma^{cn}(z,z')\mathrm{e}^{\imath(\varphi-\varphi')}$$

and the components $\Gamma^{cn}(z,z')$ are given by (2.81).

Let us introduce the following eight matrix operators:

$$
\begin{aligned}
&\ell_1 = \big[\,\mathrm{Tr}\,,0,0\big] \quad , \quad \ell_2 = \big[0,\ \mathrm{Tr}\,,0\big] \\
&\ell_3 = \big[0,0,\ \mathrm{Tr}\,\big] \quad , \quad \ell_4 = \big[0,0,\ \mathrm{Tr}\,\partial_z\big] \\
&\mathcal{L}_1 = \left[\,\mathrm{Tr}\,\partial_z\,, \frac{\nu}{R}\,\mathrm{Tr}\,\partial_\varphi, \frac{\nu}{R}\,\mathrm{Tr}\,\right] \quad , \quad \mathcal{L}_2 = \frac{1-\nu}{2}\left[\frac{\mathrm{Tr}\,\partial_\varphi}{R},\ \mathrm{Tr}\,\partial_z\,, 0\right] \\
&\mathcal{L}_3 = \left[0,0, -\,\mathrm{Tr}\,\partial_{zzz} - \frac{\mathrm{Tr}\,\partial_{z\varphi\varphi}}{R^2} - \frac{1-\nu}{R^2}\partial_\varphi\big(\,\mathrm{Tr}\,\partial_{z\varphi}\big)\right] \\
&\mathcal{L}_4 = \left[0,0,\ \mathrm{Tr}\,\partial_{zz} + \frac{\nu}{R^2}\,\mathrm{Tr}\,\partial_{\varphi\varphi}\right]
\end{aligned}
\tag{3.116}
$$

which, when applied to any vector, are defined as matrix products:

$$
\ell_1 \begin{pmatrix} u \\ v \\ w \end{pmatrix} = \big[\,\mathrm{Tr}\,,0,0\big] \begin{pmatrix} u \\ v \\ w \end{pmatrix} = \begin{pmatrix} \mathrm{Tr}\,u \\ 0 \\ 0 \end{pmatrix} \quad \ldots
$$

These operators are the boundary operators, which appear in the variational form of the shell equation. Let us denote by ${}^t Z$ the transposed of any matrix Z. Accounting for boundary conditions (3.114) and (3.114′), the Green's representation of the shell displacement writes:

$$
\begin{aligned}
U(M) = &- \int_\Sigma \Gamma^c(M, M') \begin{pmatrix} 0 \\ 0 \\ \mathrm{Tr}\,p^e - \,\mathrm{Tr}\,p^i \end{pmatrix} (M')\,\mathrm{d}\sigma(M') \\
&- \frac{Eh}{1-\nu^2} \int_{C-} \Big\{ {}^t\big[\ell_1 \Gamma^c(M,Q)\big]\mathcal{L}_1 U(Q) + {}^t\big[\ell_2 \Gamma^c(M,Q)\big]\mathcal{L}_2 U(Q) \\
&+ \frac{h^2}{12}\,{}^t\big[\ell_3 \Gamma^c(M,Q)\big]\mathcal{L}_3 U(Q) + \frac{h^2}{12}\,{}^t\big[\ell_4 \Gamma^c(M,Q)\big]\mathcal{L}_4 U(Q)\Big\}\,\mathrm{d}s(Q) \\
&+ \frac{Eh}{1-\nu^2} \int_{C+} \Big\{ {}^t\big[\mathcal{L}_1 \Gamma^c(M,Q)\big]\ell_1 U(Q) + {}^t\big[\mathcal{L}_2 \Gamma^c(M,Q)\big]\ell_2 U(Q) \\
&+ \frac{h^2}{12}\,{}^t\big[\mathcal{L}_3 \Gamma^c(M,Q)\big]\ell_3 U(Q) + \frac{h^2}{12}\,{}^t\big[\mathcal{L}_4 \Gamma^c(M,Q)\big]\ell_4 U(Q)\Big\}\,\mathrm{d}s(Q)
\end{aligned}
\tag{3.117}
$$

It depends on the function ϖ related to $\mathrm{Tr}\,p^e$ by (3.115), on the pressure p^i and on eight boundary sources.

3.10.3. *Green's Representation of the Interior Acoustic Pressure and Matched Asymptotic Expansions*

1. *Green's representation of the interior acoustic pressure*

Because of the cylindrical geometry, the Green's function of the Neumann problem in Ω^i is known (see, for example, [MOR 53]) and can be written as:

$$\mathcal{G}^i(M, M') = \sum_{m,n,p} \frac{\psi_{mnp}(M)\psi^*_{mnp}(M')}{{k^i}^2 - k^2_{mnp}} \tag{3.118}$$

where $\psi_{mnp}(M)$ are the normalized eigenmodes of the Neumann problem and k_{mnp} the corresponding eigenwavenumbers. Then, if $k^i \neq k_{mnp}$, the pressure p^i is written:

$$p^i(M) = \int_{\Omega^i} \mathcal{G}^i(M, M') f(M')\, \mathrm{d}\Omega^i(M') + \mu_0^i \omega^2 \int_{\Sigma \cup \Sigma^- \cup \Sigma^+} \mathcal{G}^i(M, Q) w(Q)\, \mathrm{d}\sigma(Q) \tag{3.119}$$

This expression, together with the integral representations of the external pressure and of the shell displacement provides a system of Boundary Integral Equations which is equivalent, for almost any frequency, to the initial system of partial differential equations.

For each eigenwavenumber k_{mnp}, this representation is no longer valid, and in its vicinity any numerical procedure becomes unstable. For these two reasons, it is necessary to look for a representation which is valid in the whole frequency range.

2. *The light fluid approximation and the matched asymptotic expansions method*

The fact that the inner fluid is a gas allows us to develop approximate formulae which tend asymptotically to the exact solution for $\varepsilon = \mu_0^i/\mu_s h \to 0$, as done in [NOR 84], for example; it is hoped that these expressions remain a good approximation, for $\varepsilon \ll 1$ but finite. The solution is sought as a formal series of ε^a and it can be shown that $a = 1$.

The problem can be split into two coupled problems: the vibro-acoustic response of the shell coupled to the external fluid, which is uniquely determined for any real frequency because of the absorption effect of the fluid; a Neumann problem for the Helmholtz equation inside Ω^i, which has real resonance frequencies.

Let $\mathcal{G}^e_\omega(M, M')$ be the Green's function of the Neumann problem in Ω^e: it exists and it will be considered as known (although a numerical procedure is necessary to get an approximation). The governing equations thus write:

$$\left\{\frac{Eh}{1-\nu^2}\mathcal{M}^c - \mu_s h\omega^2 \mathbf{I}\right\}\begin{pmatrix} u \\ v \\ w \end{pmatrix} - \mu_0^e\omega^2 \begin{pmatrix} 0 \\ 0 \\ \mathcal{K}^e w \end{pmatrix} - \begin{pmatrix} 0 \\ 0 \\ \operatorname{Tr} p^i \end{pmatrix} = 0 \quad \text{on } \Sigma \qquad (3.120)$$

$$\text{with} \quad \mathcal{K}^e w(M) = \operatorname{Tr} \int_\Sigma w(M')\mathcal{G}^e(M, M')\,\mathrm{d}\sigma(M') \qquad (3.120')$$

$$\left(\Delta + {k^i}^2\right)p^i = \varepsilon \tilde{f}\,, \quad \text{in } \Omega^i \qquad (3.121)$$

$$\operatorname{Tr} {p^i}_{,\rho} = -\varepsilon\mu_s h\omega^2 w\,, \quad \text{on } \Sigma \qquad (3.122)$$

where $\tilde{f}$ is defined by $f = \varepsilon \tilde{f}$. The boundary conditions on W and p^i complete this system.

The outer expansion.– Here, it is assumed that the driving frequency is different from any resonance frequency of the interior problem. Let us look for a formal series expansion of the solution:

$$\begin{aligned} W &= W^0 + \varepsilon W^1 + \dots \\ p^i &= {p^i}^0 + \varepsilon {p^i}^1 + \dots \end{aligned} \qquad (3.123)$$

This expansion is introduced into equations (3.120, 3.121, 3.122). The successive terms in ε^n are made equal to 0. This leads to the following boundary value problems:

◇ for order 0:

$$\left\{\frac{Eh}{1-\nu^2}\mathcal{M}^c - \mu_s h\omega^2 \mathbf{I}\right\}\begin{pmatrix} u^0 \\ v^0 \\ w^0 \end{pmatrix}$$

$$- \mu_0^e\omega^2 \begin{pmatrix} 0 \\ 0 \\ \mathcal{K}^e w^0 \end{pmatrix} - \begin{pmatrix} 0 \\ 0 \\ \operatorname{Tr} {p^i}^0 \end{pmatrix} = 0\,, \quad \text{on } \Sigma \qquad (3.124)$$

$$\left(\Delta + {k^i}^2\right){p^i}^0 = 0\,, \text{ in } \Omega^i$$

$$\operatorname{Tr} {p^i}^0_{,\rho} = 0\,, \text{ on } \Sigma$$

◇ for order 1:

$$\left\{\frac{Eh}{1-\nu^2}\mathcal{M}^c - \mu_s h\omega^2\mathbf{I}\right\}\begin{pmatrix}u^1\\ v^1\\ w^1\end{pmatrix}$$

$$- \mu_0^e\omega^2\begin{pmatrix}0\\ 0\\ \mathcal{K}^e w^1\end{pmatrix} - \begin{pmatrix}0\\ 0\\ \mathrm{Tr}\, {p^i}^1\end{pmatrix} = 0\,, \quad \text{on } \Sigma \qquad (3.125)$$

$$(\Delta + {k^i}^2){p^i}^1 = \tilde{f}\,, \text{ in } \Omega^i$$

$$\mathrm{Tr}\, {p^i}^1_{,\rho} = -\mu_s h\omega^2 w^0\,, \text{ on } \Sigma$$

Obviously, the zero-order term $(W^0, {p^i}^0)$ is identically zero, while the first order term $(W^1, {p^i}^1)$ is uniquely determined by solving first the Helmholtz equation and then the fluid-loaded shell equation. It must be observed that the solution of the problem is $\mathcal{O}(\varepsilon)$.

The inner expansion.– If k^i is equal to one eigenwavenumber, say k_{rst}, the former approximation is not defined ($\mathcal{G}^i_\omega$ is not defined for these wavenumbers and neither are ${p^i}^0$ and ${p^i}^1$). Thus, another expansion must be used. Let us define a second parameter α by writing $k^i = k_{rst}(1+\varepsilon\alpha)$. This is introduced into equations (3.120, 3.121, 3.122) and, as in the preceding section, the solution is looked at as a formal series of ε

$$W = \widetilde{W}^0 + \varepsilon\widetilde{W}^1 + \ldots$$
$$p^i = {\tilde{p}^i}{}^0 + \varepsilon {\tilde{p}^i}{}^1 + \ldots \qquad (3.126)$$

The following zero and first order equations are obtained as:

◇ order 0:

$$\left\{\frac{Eh}{1-\nu^2}\mathcal{M}^c - \mu_s h {c_0^i}^2 k_{rst}^2\mathbf{I}\right\}\begin{pmatrix}\tilde{u}^0\\ \tilde{v}^0\\ \tilde{w}^0\end{pmatrix}$$

$$- \mu_0^e {c_0^i}^2 k_{rst}^2\begin{pmatrix}0\\ 0\\ \mathcal{K}^e \tilde{w}^0\end{pmatrix} - \begin{pmatrix}0\\ 0\\ \mathrm{Tr}\, {\tilde{p}^i}{}^0\end{pmatrix} = 0, \text{ on } \Sigma \qquad (3.127)$$

$$(\Delta + k_{rst}^2){\tilde{p}^i}{}^0 = 0\,, \text{ in } \Omega^i$$

$$\mathrm{Tr}\, {\tilde{p}^i}{}^0_{,\rho} = 0\,, \text{ on } \Sigma$$

◇ order 1:

$$\left\{\frac{Eh}{1-\nu^2}\,\mathcal{M}^c - \mu_s h {c_0^i}^2 k_{rst}^2 \mathbf{I}\right\} \begin{pmatrix} \tilde{u}^1 \\ \tilde{v}^1 \\ \tilde{w}^1 \end{pmatrix}$$

$$- \mu_0^e {c_0^i}^2 k_{rst}^2 \begin{pmatrix} 0 \\ 0 \\ \mathcal{K}^e \tilde{w}^1 \end{pmatrix} - \begin{pmatrix} 0 \\ 0 \\ \operatorname{Tr} \tilde{p}^{i^1} \end{pmatrix} = 0, \text{ on } \Sigma \quad (3.128)$$

$$(\Delta + k_{rst}^2)\tilde{p}^{i^1} = \tilde{f} - 2\alpha k_{rst}^2 \tilde{p}^{i^0}, \text{ in } \Omega^i$$

$$\operatorname{Tr} \tilde{p}^{i^1}_{,\rho} = -\mu_s h {c_0^i}^2 k_{rst}^2 \tilde{w}^0, \text{ on } \Sigma$$

Obviously, equations (3.127) lead to $\tilde{p}^{i^0} = A\psi_{rst}$, where the coefficient A is undetermined. Thus, $\widetilde{W}^0$ is uniquely determined up to the same factor A. As a consequence, w^0 can be written in the form $A\mathcal{C}(\psi_{rst})$, where $\mathcal{C}(\psi_{rst})$ is the response of the shell loaded by the external fluid, to an excitation with components $(0, 0, \psi_{rst})$. In order to determine the coefficient A, let us write the variational form of the Helmholtz equation satisfied by p^{i^1}. We have:

$$\int_{\Omega^i} \left(-\nabla p^{i^1} \nabla\psi^* + k_{rst}^2 p^{i^1} \psi^*\right) - \mu_s h {c_0^i}^2 k_{rst}^2 A \int_\Sigma \mathcal{C}(\psi_{rst}) \operatorname{Tr} \psi^* \, d\sigma$$

$$= \int_{\Omega^i} (\tilde{f} - 2\alpha k_{rst}^2 A \psi_{rst}) \psi^* \, d\Omega^i \quad (3.129)$$

where W^0 has been replaced by its expression in terms of ψ_{rst}. This relationship must be satisfied for any function ψ belonging to the convenient functional space. In particular, it remains valid for $\psi = \psi_{rst}$. Accounting for the equation satisfied by the resonance mode ψ_{rst}, the value of A is given by:

$$A = \frac{\int_{\Omega^i} \tilde{f} \psi_{rst}^* \, d\Omega^i}{k_{rst}^2 \left[2\alpha \|\psi_{rst}\|^2 - \mu_s h {c_0^i}^2 \int_\Sigma \mathcal{C}(\psi_{rst}) \operatorname{Tr} \psi_{rst}^* \, d\sigma\right]} \quad (3.130)$$

This shows that, in the neighborhood of a resonance frequency, the zero-order approximation of the shell normal displacement is $\mathcal{O}(1)$. As done for the plate closing a cavity, a continuous expansion is obtained by combining the inner and outer expansions.

It can be observed that, in practical applications, the zero order expansions are generally sufficient. A simpler approximation of the system response at a resonance frequency ω_{rst} of the interior domain is obtained by an interpolation between the responses at frequencies a little bit lower and a little bit higher than ω_{rst}. This approximation, which depends on the interval on which the interpolation is evaluated, is, of course, less accurate than the approximation provided by the matched asymptotic expansions.

3.10.4. *Directivity Pattern of the Radiated Acoustic Pressure*

Acousticians are often interested in the directivity pattern of sound sources, that is, in the far-field acoustic level as a function of the angular direction. In the last paragraph of section 3.5.1, we have established the expression of the far-field sound pressure radiated by an infinite plate, that is, the asymptotic behavior of the sound pressure as a function of the angular variables. We will do the same for the more complex case of the finite length shell.

Let (R, θ, φ) be the spherical coordinates of a point M and $p(R, \theta, \varphi)$ be the acoustic pressure radiated by a source located at finite distance. The directivity pattern of the source is the following limit:

$$\psi(\theta, \varphi) = 4\pi \lim_{R\to\infty} \big(R \times p(R, \theta, \varphi)\big) \tag{3.131}$$

The acoustic level directivity pattern $N(\theta, \varphi) = 20 \log_{10}\left|\psi(\theta, \varphi)\right|$ is more commonly used. This is a way to compare the acoustic radiation of a source to that of a point unit isotropic source located at the coordinates origin. Let us establish the analytical expression of such a directivity pattern in the case of the vibrating structure considered in this section.

In expression (3.115), the acoustic pressure is given as a boundary integral which involves the Green's function for the Helmholtz equation in the infinite three-dimensional space and its normal derivative:

$$G^e(M, P) = -\frac{e^{\imath k^e r(M,P)}}{4\pi r(M, P)} \quad , \quad \partial_{\vec{n}(P)} G^e(M, P) = -\vec{n}(P) \cdot \overrightarrow{\nabla}_P \frac{e^{\imath k^e r(M,P)}}{4\pi r(M, P)}$$

We denote by (R, θ, φ) the coordinates of the point M and by (R', θ', φ') those of the point P located on the shell surface $\Sigma \cup \Sigma^+ \cup \Sigma^-$ (the z-axis corresponds to $\theta = 0$). Let us recall the expansion of the function $G^e(M, P)$ into a series of spherical harmonics used in equation (3.52):

$$G^e(M, P) = -\frac{\imath k^e}{4\pi} \sum_{n=0}^{\infty} (2n+1) \sum_{m=0}^{n} \varepsilon_m \frac{(n-m)!}{(n+m)!} \cos\, m(\varphi - \varphi') P_n^m(\cos\theta)$$
$$\times P_n^m(\cos\theta') \begin{vmatrix} j_n(k^e R') h_n(k^e R) & \text{if } R > R' \\ j_n(k^e R) h_n(k^e R') & \text{if } R < R' \end{vmatrix} \tag{3.132}$$

Because we are interested in the case $R \to \infty$, we use the series valid for $R > R'$. The spherical Hankel functions are expressed by the following sum:

$$h_n\left(k^e R\right) = \frac{\mathrm{e}^{\imath k R}}{\imath^{n+1} k^e R} \sum_{p=0}^{n} \frac{(n+p)!}{p!(n-p)!} \left(\frac{-1}{2\imath k^e R}\right)^p \tag{3.133}$$

Keeping the first term only, we have:

$$G^e(M,P) = -\frac{\mathrm{e}^{\imath k^e R}}{4\pi R} \sum_{n=0}^{\infty} (2n+1) \sum_{m=0}^{n} \varepsilon_m \frac{(n-m)!}{(n+m)!} \cos\, m(\varphi - \varphi') P_n^m(\cos\theta) \times P_n^m(\cos\theta') j_n\left(k^e R'\right) + \mathcal{O}\left(R^{-2}\right) \tag{3.134}$$

Using the relationship

$$\exp\left[-\imath k^e R'\left(\sin\theta \sin\theta' \cos(\varphi - \varphi') + \cos\theta\cos\theta'\right)\right] = \sum_{n=0}^{\infty} (2n+1) \sum_{m=0}^{n} \varepsilon_m \frac{(n-m)!}{(n+m)!} \cos\, m(\varphi-\varphi') P_n^m(\cos\theta) P_n^m(\cos\theta') j_n\left(k^e R'\right)$$

we finally obtain:

$$G^e(M,P) = -\frac{\mathrm{e}^{\imath k^e R}}{4\pi R} \exp\left[-\imath k^e R'\left(\sin\theta\sin\theta'\cos(\varphi-\varphi') + \cos\theta\cos\theta'\right)\right] + \mathcal{O}\left(R^{-2}\right) \tag{3.135}$$

The asymptotic expression of the gradient, with respect to the coordinates of P, of the Green's function is also needed. It is obtained by taking the gradient of expression (3.135). In spherical coordinates, its components are:

$$\frac{\partial G^e(M,P)}{\partial R'} \simeq -\imath k^e \left(\sin\theta\sin\theta'\cos(\varphi-\varphi') + \cos\theta\cos\theta'\right) \overline{G}^e(M,P)$$

$$\frac{1}{R'} \frac{\partial G^e(M,P)}{\partial \theta'} \simeq -\imath k^e \left(\sin\theta\cos\theta'\cos(\varphi-\varphi') - \cos\theta\sin\theta'\right) \overline{G}^e(M,P)$$

$$\frac{1}{R'\sin\theta'} \frac{\partial \mathcal{G}^e(M,P)}{\partial \varphi'} \simeq -\imath k^e \sin\theta \sin(\varphi-\varphi') \overline{G}^e(M,P)$$

with $$\overline{G}^e(M,P) = -\frac{\mathrm{e}^{\imath k^e R}}{4\pi R} \exp\left[-\imath k^e R'\left(\sin\theta\sin\theta'\cos(\varphi-\varphi') + \cos\theta\cos\theta'\right)\right] \tag{3.135'}$$

These asymptotic expressions are introduced into expression (3.115) and, after integration, the far-field approximation of the radiated pressure is obtained as a function of the form $-\mathrm{e}^{\imath k^e R}/4\pi R \times \Phi(\theta, \varphi)$. Of course, in the present case, the integrals cannot be performed analytically, a numerical evaluation is necessary.

3.10.5. *Numerical Method, Results and Concluding Remarks*

1. *Numerical method*

A first simplification of the equations comes from the geometry of the physical system which has an axis of symmetry. It is possible to expand the data and the solution into Fourier series with respect to the angular variable φ. Let us denote by ζ the longitudinal variable on the shell surface. Thus, each unknown function $f(\varphi, \zeta)$ along $\Sigma^- \cup \Sigma \cup \Sigma^+$ is replaced by a sequence of unknown functions $f_n(\zeta)$, which depend on one variable only. Each unknown function $g(\varphi)$ along C^- and C^+ is replaced by a sequence of constants g_n.

Gathering the integral representations of p^e, W and p^i together with the continuity and boundary conditions, a sequence of systems of integral and algebraic equations is obtained, which determines the Fourier components of the unknown functions. In [SEI 97, SEI 99], the functions were approximated by piecewise constant functions and a standard collocation method was adopted.

2. *Numerical results and remarks*

The shell and the fluids have the following characteristics:

- ◇ shell dimensions: $L = 5.8$ m; $h = 0.01$ m; $R = 0.266\ 5$ m;
- ◇ shell mechanical properties: $\mu_s = 2800$ kg/m^3; $E = 0.72\ 10^{11}$Pa; $\nu = 0.32$;
- ◇ external fluid mechanical properties: $\mu_0^e = 1000$ kg/m^3; $c_0^e = 1460$ m/s;
- ◇ internal fluid mechanical properties: $\mu_0^i = 1.3$ kg/m^3; $c_0^i = 340$ m/s.

The acoustic source is located at the center of the shell interior domain Ω^i and the excitation frequency is equal to, successively, 235 Hz, 783 Hz and 1000 Hz.

The directivity pattern of the radiated pressure has been computed as a function of the angular spherical coordinate θ. Although the shell displacement does not satisfy identical boundary conditions at both ends, the directivity pattern is not significantly sensitive to this lack of symmetry. Thus, Figure 3.31 presents the results for an angular variation $[0, \pi/2]$ only. Furthermore, other calculations have pointed out that the position of the acoustic source along the shell axis does not have a significant influence on the pressure directivity pattern.

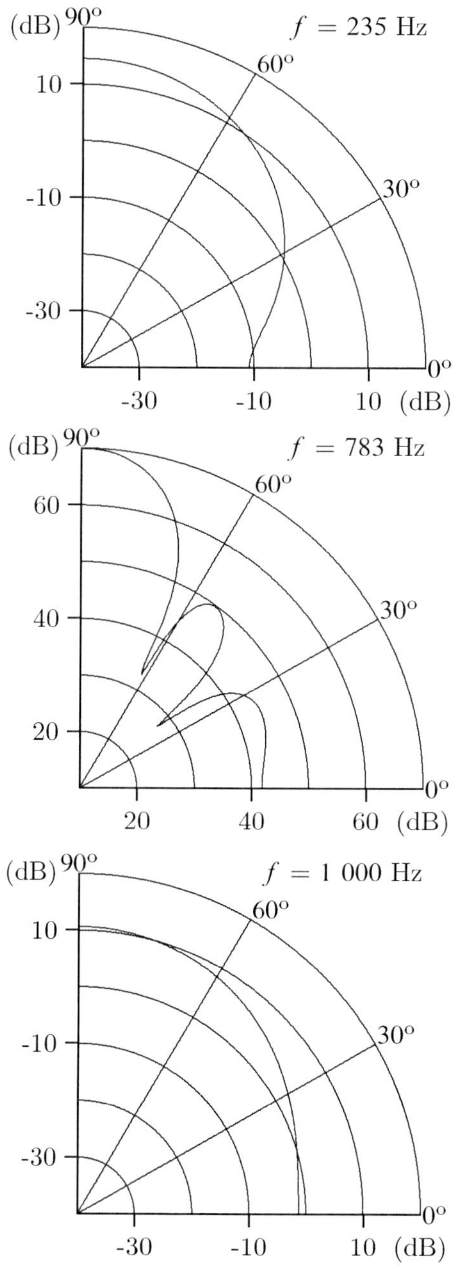

Figure 3.31. *Directivity pattern of the shell excited by a point isotropic source at its center* $L = 5.8$ m; $h = 0.01$ m; $R = 0.2665$ m; $\mu_s = 2800$ kg/m^3; $E = 0.72\ 10^{11}$Pa; $\nu = 0.32$; $\mu_0^e = 1000$ kg/m^3; $c_0^e = 1460$ m/s; $\mu_0^i = 1.3$ kg/m^3; $c_0^i = 340$ m/s; *frequencies* = 235 Hz, 783 Hz *and* 1000 Hz

This example shows again that, for a structure containing a gas and embedded in a fluid of higher density, matched asymptotic expansions, based on the Green's representation of the interior acoustic pressure, are a powerful tool which reduces the computation effort by a significant amount. There are, of course, not so many geometries for which a Green's function is known analytically. However, such analytical representations can be replaced by any other approximation – analytical or purely numerical – and some computation effort will still be saved.

This example shows once more that the use of the Green's representation of the interior acoustic pressure leads to powerful numerical methods, though this representation fails to be valid at each of the resonance frequencies of the cavity. When the interior fluid is a gas, it is simple to overcome such a difficulty by using matched asymptotic expansions. For a heavy fluid, similar expansions can be developed.

This example also shows that the transformation of the initial boundary value problem into a system of Boundary Integral Equations provides efficient numerical methods which have two main advantages:

1. all the unknown functions – the three components of the shell displacement and the two acoustic pressures – are approximated by analytic functions in terms of boundary elements;

2. more information – as, for example, the directivity pattern of the shell – can be deduced analytically.

The last advantage is that the effort made for the analytical transformation of the boundary value problem into Boundary Integral Equations is paid by a reduction of the amount of numerical effort.

3.11. Diffraction of a Transient Acoustic Wave by a *Line 2'* Shell

This final example deals with the response of a finite length shell, immersed into a heavy fluid (water), to a transient incident acoustic wave.

The shell *Line 2'* surface Σ is composed of three elementary surfaces Σ_1, Σ_2 and Σ_3. The elements Σ_1 and Σ_3 are two identical hemi-spherical end-caps, which close the extremities of a cylindrical element Σ_2 with length $2L$. They all have the same mean radius R. The shell has a constant thickness h which is assumed to be a few percent of R. It is made of a homogenous material, characterized by a density μ_s, a Young's modulus E and a Poisson's ratio ν.

The structure is immersed into a fluid – water – characterized by a density μ_0 and a sound velocity c_0 and which occupies Ω, the infinite domain exterior to the shell. It is excited by a transient incident wave.

A coordinate system (spherical or cylindrical) is associated with each element (see Figure 3.38a). It is also convenient to define a unique coordinate system on the shell surface. The location of a point M is defined by the angular variable φ and a curvilinear abscissa s which runs from $(-\pi R/2 - L)$ to $(+\pi R/2 + L)$; the unit vector, normal to the shell surface and pointing out to the exterior is denoted by $\vec{n}$ (see Figure 3.38b).

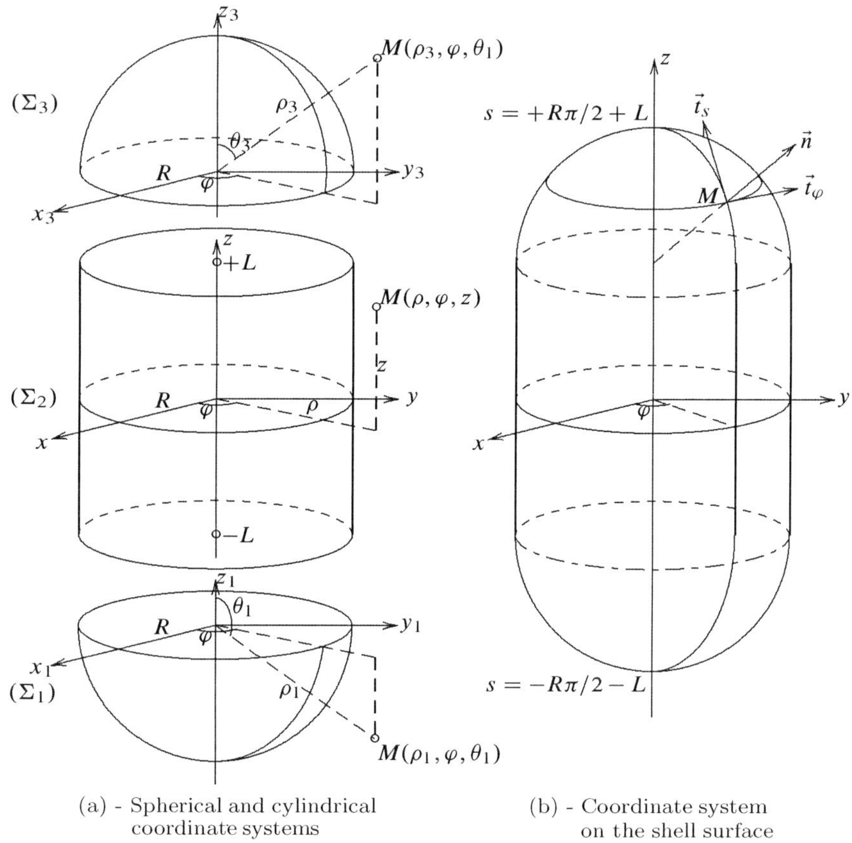

Figure 3.32. *Geometry of Line 2': (a) the three space coordinate systems; (b) the unique coordinate system on the shell surface*

As will be seen in the following, the response of the system is sought as a series of the fluid-loaded resonance modes of the shell and of the corresponding radiated acoustic pressure components. The details concerning this example are presented in [MAU 99, FIL 01, MAU 01].

The first step is thus to determine the resonance frequencies and modes of the fluid-loaded shell. For that purpose, the acoustic pressure is expressed as the radiation of a source supported by the shell surface. The equation of continuity between the shell normal velocity and the normal fluid particle velocity provides a first equation which relates the source density and the normal component of the shell displacement. The equation governing the shell displacement provides a system of three equations relating the acoustic surface source density and the three components of the shell displacement. This system is completed by boundary conditions at the apexes of the shell (points $s = -\pi R/2 - L$ and $s = +\pi R/2 + L$). Because the shell has a circular cylindrical symmetry around the z-axis, every unknown function can be expanded into a Fourier series of the angular variable φ. Thus, we are left with a sequence of systems of equations which depend on the variable s only. This is an important simplification which saves a large amount of computing time.

3.11.1. *Statement of the Problem*

Use is made of the variational form of the equations as they have been established in the first chapter. In the surface coordinate system, the displacement of M is the vector $\tilde{u}\vec{t}_s + \tilde{v}\vec{t}_\varphi + \tilde{w}\vec{n}$ where: $\vec{t}_s$ is the unit vector tangent to the shell and in the plane defined by M and the z-axis; $\vec{t}_\varphi$ is the unit vector tangent to the shell and orthogonal to the z-axis; and $\vec{n}$ is the unit vector normal to the shell surface and pointing out to its exterior; these vectors form a direct trihedral.

The shell-fluid system is assumed to be excited by an incident acoustic field $\tilde{p}^i$, which is zero for $t < 0$ and is a square integrable function on any finite space domain and any finite time interval: this corresponds to an incident field of finite power. The diffracted field is denoted by $\tilde{p}$. The values, on the shell surface, of the normal derivatives of $\tilde{p}^i$ and $\tilde{p}$ are, as usual, denoted by $\operatorname{Tr}\partial_n \tilde{p}^i$ and $\operatorname{Tr}\partial_n \tilde{p}$.

The diffracted acoustic pressure $\tilde{p}(M,t)$ satisfies the wave equation and the boundary condition which follow:

$$\begin{aligned} &\left[\Delta - \frac{1}{c_0^2}\frac{\partial^2}{\partial t^2}\right]\tilde{p}(M,t) = 0 \quad \text{for} \quad M \in \Omega \\ &\operatorname{Tr}\partial_n \tilde{p}(M,t) = -\mu_0 \ddot{\tilde{w}} - \operatorname{Tr}\partial_n \tilde{p}_i(M,t) \quad \text{for} \quad M \in \Sigma \end{aligned} \tag{3.136}$$

Furthermore, $\tilde{p}(M,t)$ and $\partial_t \tilde{p}(M,t)$ are zero for $t < 0$; $\tilde{p}(M,t)$ satisfies an outgoing wave condition.

The variational form of the equations for the shell-fluid system is written as:

$$\frac{Eh}{1-\nu^2}\int_0^{+\infty}\Big[\mathcal{A}(\tilde{u},\tilde{v},\tilde{w};\delta u,\delta v,\delta w)(t)$$
$$+\mu_s h\int_\Sigma [\ddot{\tilde{u}}\delta\tilde{u}^* + \ddot{\tilde{v}}\delta\tilde{v}^* + \ddot{\tilde{w}}\delta\tilde{w}^*](M,t)\,\mathrm{d}M\Big]\,\mathrm{d}t$$
$$+\int_0^{+\infty}\int_\Sigma [\,\mathrm{Tr}\,\tilde{p}\delta\tilde{w}^*](M,t)\,\mathrm{d}M\,\mathrm{d}t = -\int_0^{+\infty}\int_\Sigma [\,\mathrm{Tr}\,\tilde{p}^i\delta\tilde{w}^*](M,t)\,\mathrm{d}M\,\mathrm{d}t \quad (3.137)$$
$$\mu_0\int_0^{+\infty}\int_\Sigma [\ddot{\tilde{w}}\delta\tilde{\psi}^*](M,t)\,\mathrm{d}M\,\mathrm{d}t + \int_0^{+\infty}\int_\Sigma [\,\mathrm{Tr}\,\partial_n\tilde{p}\delta\tilde{\psi}^*](M,t)\,\mathrm{d}M\,\mathrm{d}t$$
$$= -\int_0^{+\infty}\int_\Sigma [\,\mathrm{Tr}\,\partial_n\tilde{p}^i\delta\tilde{\psi}^*](M,t)\,\mathrm{d}M\,\mathrm{d}t\ ,\quad \forall\delta\tilde{u},\delta\tilde{v},\delta\tilde{w},\delta\tilde{\psi}$$

The first equality is the energy balance of the shell. To obtain the second equality, the energy balance in the fluid – as expressed in equation (3.25) and integrated over the variable t – is written for a domain Ω_R between the shell surface and a sphere of large radius R. The boundary condition on Σ is introduced. Then, an integration by parts over Ω_R is performed, which leaves boundary integrals along Σ and the sphere of radius R. This last integral disappears by taking the limit for $R \to \infty$ and using the outgoing wave condition.

The functional $\mathcal{A}$ is the bilinear form associated with the potential energy of the shell It has different expressions on the various shell elements. Using the formulae obtained in sections 1.4 and 1.5, we obtain the expressions of $\mathcal{A}$ for the cylindrical and the spherical elements of the shell:

◇ on the cylindrical element Σ_2, $\mathcal{A}$ is given by:

$$\mathcal{A}_2(\tilde{u},\tilde{v},\tilde{w};\delta\tilde{u},\delta\tilde{v},\delta\tilde{w}) =$$
$$\int_0^{2\pi} R\,\mathrm{d}\varphi\int_{-L}^{L}\mathrm{d}s\,\Big\{\tilde{u}_{,s}\,\delta\tilde{u}^*_{,s} + \frac{1}{R^2}\left(\tilde{v}_{,\varphi}+\tilde{w}\right)\left(\delta\tilde{v}^*_{,\varphi}+\delta\tilde{w}^*\right) + \frac{\nu}{R}\left(\tilde{v}_{,\varphi}+\tilde{w}\right)\delta\tilde{u}^*_{,s}$$
$$+\frac{\nu}{R}\tilde{u}_{,s}\left(\delta\tilde{v}^*_{,\varphi}+\delta\tilde{w}^*\right) + \frac{1-\nu}{2}\left(\frac{\tilde{u}_{,\varphi}}{R}+\tilde{v}_{,s}\right)\left(\frac{\delta\tilde{u}^*_{,\varphi}}{R}+\delta\tilde{v}^*_{,s}\right)$$
$$+\frac{h^2}{12}\Big[\tilde{w}_{,ss}\,\delta\tilde{w}^*_{,ss} + \frac{1}{R^4}\tilde{w}_{,\varphi\varphi}\,\delta\tilde{w}^*_{,\varphi\varphi} + \frac{\nu}{R^2}\tilde{w}_{,\varphi\varphi}\,\delta\tilde{w}^*_{,ss}$$

$$+\frac{\nu}{R^2}\tilde{w}_{,ss}\,\delta\tilde{w}^*_{,\varphi\varphi}+\frac{1-\nu}{R^2}\tilde{w}_{,s\varphi}\,\delta\tilde{w}^*_{,s\varphi}+\frac{1-\nu}{R^2}\tilde{w}_{,\varphi s}\,\delta\tilde{w}^*_{,\varphi s}\Big]\Big\} \qquad (3.138)$$

◇ on the spherical elements $\Sigma_i\,(i=1,3)$, $\mathcal{A}$ is given by:

$$\begin{aligned}
\mathcal{A}_i(\tilde{u},\tilde{v},\tilde{w};\delta\tilde{u},\delta\tilde{v},\delta\tilde{w}) &= \int_0^{2\pi} R\,\mathrm{d}\varphi\int_{a_i}^{b_i}\sin\theta_i\,\mathrm{d}s\Big\{\Big(\tilde{u}_{,s}+\frac{\tilde{w}}{R}\Big)\Big(\delta\tilde{u}^*_{,s}+\frac{\delta\tilde{w}^*}{R}\Big)\\
&+\frac{1}{R^2}\Big(-\tilde{u}\,\mathrm{cotg}\,\theta_i+\frac{\tilde{v}_{,\varphi}}{\sin\theta_i}+\tilde{w}\Big)\Big(-\delta\tilde{u}^*\,\mathrm{cotg}\,\theta_i+\frac{\delta\tilde{v}^*_{,\varphi}}{\sin\theta_i}+\delta\tilde{w}^*\Big)\\
&+\frac{\nu}{R}\Big(-\tilde{u}\,\mathrm{cotg}\,\theta_i+\frac{\tilde{v}_{,\varphi}}{\sin\theta_i}+\tilde{w}\Big)\Big(\delta\tilde{u}^*_{,s}+\frac{\delta\tilde{w}^*}{R}\Big)\\
&+\frac{\nu}{R}\Big(\tilde{u}_{,s}+\frac{\tilde{w}}{R}\Big)\Big(-\delta\tilde{u}^*\,\mathrm{cotg}\,\theta_i+\frac{\delta\tilde{v}^*_{,\varphi}}{\sin\theta_i}+\delta\tilde{w}^*\Big)\\
&+\frac{1-\nu}{2}\Big(\frac{\tilde{u}_{,\varphi}}{R\sin\theta_i}+\frac{\tilde{v}\,\mathrm{cotg}\,\theta_i}{R}+\tilde{v}_{,s}\Big)\Big(\frac{\delta\tilde{u}^*_{,\varphi}}{R\sin\theta_i}+\frac{\delta v^*\,\mathrm{cotg}\,\theta_i}{R}+\delta v^*_{,s}\Big)\\
&+\frac{h^2}{12}\Big[\tilde{w}_{,ss}\,\delta\tilde{w}^*_{,ss}+\frac{1}{R^4}\Big(\frac{\tilde{w}_{,\varphi\varphi}}{\sin^2\theta_i}-R\tilde{w}_{,s}\,\mathrm{cotg}\,\theta_i\Big)\Big(\frac{\delta\tilde{w}^*_{,\varphi\varphi}}{\sin^2\theta_i}-R\delta\tilde{w}^*_{,s}\,\mathrm{cotg}\,\theta_i\Big)\\
&+\frac{\nu}{R^2}\Big(-R\tilde{w}_{,s}\,\mathrm{cotg}\,\theta_i+\frac{\tilde{w}_{,\varphi\varphi}}{\sin^2\theta_i}\Big)\delta\tilde{w}^*_{,ss}\\
&+\frac{\nu}{R^2}\tilde{w}_{,ss}\Big(-R\delta\tilde{w}^*_{,s}\,\mathrm{cotg}\,\theta_i+\frac{\delta\tilde{w}^*_{,\varphi\varphi}}{\sin^2\theta_i}\Big)\\
&+\frac{1-\nu}{R^4\sin^2\theta_i}\Big(R\tilde{w}_{,s\varphi}+\tilde{w}_{,\varphi}\,\mathrm{cotg}\,\theta_i\Big)\Big(R\delta\tilde{w}^*_{,s\varphi}+\delta\tilde{w}^*_{,\varphi}\,\mathrm{cotg}\,\theta_i\Big)\\
&+\frac{1-\nu}{R^4\sin^2\theta_i}\Big(R\tilde{w}_{,\varphi s}+\tilde{w}_{,\varphi}\,\mathrm{cotg}\,\theta_i\Big)\Big(R\delta\tilde{w}^*_{,\varphi s}+\delta\tilde{w}^*_{,}\varphi\,\mathrm{cotg}\,\theta_i\Big)\Big]\Big\} \qquad (3.139)
\end{aligned}$$

where the integration bounds are:

$$a_1=-\pi R/2-L \quad , \quad b_1=-L \quad ; \quad a_3=+L \quad , \quad b_3=\pi R/2+L$$

and the angles θ_i are related to s by

$$\pi\geq\theta_1=\pi/2-(s+L)/R\geq\pi/2 \quad ; \quad \pi/2\geq\theta_3=\pi/2-(s-L)/R\geq 0$$

Let us remark that because of the choice adopted for the unit vector $\vec{t}_s$, which is opposite to the classical unit vector $\vec{e}_\theta$, the sign of u is the opposite of the sign used in the first chapter.

In the previous equations, the unknown functions $\tilde{u}$, $\tilde{v}$, $\tilde{w}$ and $\mathrm{Tr}\,\tilde{p}$ (together with the test functions $\delta\tilde{u}$, $\delta\tilde{v}$, $\delta\tilde{w}$ and $\delta\tilde{\psi}$) belong to convenient functional spaces corresponding to the following properties:

◇ $\tilde{u}$, $\tilde{v}$, $\tilde{w}$ and $\mathrm{Tr}\,\tilde{p}$ are square integrable on Σ and on any finite time interval;

◇ the first order derivatives of $\tilde{u}$ and $\tilde{v}$, and the derivatives of order 1 and 2 of $\tilde{w}$ with respect to the space variables s and φ are square integrable on Σ and on any finite time interval;

◇ as shown in section 1.5.2, $\tilde{u}$, $\tilde{v}$, and $\tilde{w}$ must satisfy regularity conditions at the apexes. We propose expressions slightly different from, but equivalent to (1.32) and (1.32′), by choosing the shell displacement of the form:

$$\begin{aligned}
\tilde{u}(s,\varphi) &= \tilde{u}_1(s)\mathrm{e}^{\imath\varphi} + \tilde{u}_{-1}(s)\mathrm{e}^{-\imath\varphi} \\
&\qquad + (s - L - R\pi/2)(s + L + R\pi/2)\widetilde{U}(s,\varphi) \\
\tilde{v}(s,\varphi) &= \tilde{v}_0(s) + \tilde{v}_1(s)\mathrm{e}^{\imath\varphi} + \tilde{v}_{-1}(s)\mathrm{e}^{-\imath\varphi} \\
&\qquad + (s - L - R\pi/2)(s + L + R\pi/2)\widetilde{V}(s,\varphi) \\
\tilde{w}(s,\varphi) &= \tilde{w}_0(s) + (s - L - R\pi/2)(s + L + R\pi/2)\widetilde{W}(s,\varphi)
\end{aligned} \tag{3.140}$$

$$\text{with} \int_0^{2\pi} \widetilde{U}(s,\varphi)\mathrm{e}^{\pm\imath\varphi}\,\mathrm{d}\varphi = \int_0^{2\pi} \widetilde{V}(s,\varphi)\,\mathrm{d}\varphi = \int_0^{2\pi} \widetilde{V}(s,\varphi)\mathrm{e}^{\pm\imath\varphi}\,\mathrm{d}\varphi = \int_0^{2\pi} \widetilde{W}(s,\varphi)\,\mathrm{d}\varphi = 0$$

thus, the regularity conditions at the apexes are:

$$\begin{gathered}
-\tilde{u}_1(s) + \imath\,\mathrm{sgn}(s)\,\tilde{v}_1(s) = 0 \quad , \quad \tilde{u}_1{}_{,s}(s) = 0 \\
-\tilde{u}_{-1}(s) - \imath\,\mathrm{sgn}(s)\,\tilde{v}_{-1}(s) = 0 \quad , \quad \tilde{u}_{-1}{}_{,s}(s) = 0 \\
v_0(s) = 0 \\
\widetilde{U}(s,\varphi) = \widetilde{V}_{,\varphi}(s,\varphi) = 0 \\
w_0{}_{,s}(s) = w_0{}_{,ss}(s) = \widetilde{W}(s,\varphi) = \widetilde{W}_{,s}(s,\varphi) = 0 \\
\text{at } s = \pm(L + \pi R/2)
\end{gathered} \tag{3.140′}$$

(the time variable has been omitted);

◇ along the circles $s = -L$ and $s = +L$, the shell displacement components and the efforts densities must be continuous. This is expressed by eight continuity conditions along each line, which are valid for any φ (this variable will be omitted):

1. continuity of the displacement components and of the derivative of

the radial displacement:

$$\begin{aligned}
&\lim_{\varepsilon\to 0}\{\tilde{u}(\pm(L-\varepsilon))-\tilde{u}(\pm(L+\varepsilon))\}=0\\
&\lim_{\varepsilon\to 0}\{\tilde{v}(\pm(L-\varepsilon))-\tilde{v}(\pm(L+\varepsilon))\}=0\\
&\lim_{\varepsilon\to 0}\{\tilde{w}(\pm(L-\varepsilon))-\tilde{w}(\pm(L+\varepsilon))\}=0\\
&\lim_{\varepsilon\to 0}\{\partial_s\tilde{w}(\pm(L-\varepsilon))-\partial_s\tilde{w}(\pm(L+\varepsilon))\}=0
\end{aligned}\tag{3.141}$$

2. continuity of tangential forces and of the bending momentum:

$$\begin{aligned}
&\lim_{\varepsilon\to 0}\left[\tilde{u},_s+\frac{\nu}{R}\left(\tilde{v},_\varphi+\tilde{w}\right)\right](\pm(L-\varepsilon))\\
&\qquad=\lim_{\varepsilon\to 0}\left[\tilde{u},_s+\frac{\nu}{R}\left(\tilde{v},_\varphi+w\right)+\frac{\tilde{w}}{R}\right](\pm(L+\varepsilon))\\
&\lim_{\varepsilon\to 0}\left(\frac{\tilde{u},_\varphi}{R}+\tilde{v},_s\right)(\pm(L-\varepsilon))\\
&\qquad=\lim_{\varepsilon\to 0}\left(\frac{\tilde{u},_\varphi}{R}+\tilde{v},_s\right)(\pm(L+\varepsilon))\\
&\lim_{\varepsilon\to 0}\left(\tilde{w},_{ss}+\frac{\nu}{R^2}\tilde{w},_{\varphi\varphi}\right)(\pm(L-\varepsilon))\\
&\qquad=\lim_{\varepsilon\to 0}\left(\tilde{w},_{ss}+\frac{\nu}{R^2}\tilde{w},_{\varphi\varphi}\right)(\pm(L+\varepsilon))
\end{aligned}\tag{3.141'}$$

3. continuity of the sum of the shearing force and the derivative of the twisting momentum:

$$\begin{aligned}
&\lim_{\varepsilon\to 0}\left\{\left(\tilde{w},_{sss}+\frac{\tilde{w},_{\varphi\varphi s}}{R^2}\right)+\frac{1-\nu}{R^2}\tilde{w},_{\varphi\varphi s}\right\}(\pm(L-\varepsilon))=\\
&\lim_{\varepsilon\to 0}\left\{\left(\tilde{w},_{sss}+\frac{\tilde{w},_{\varphi\varphi s}}{R^2}+\nu\frac{\tilde{w},_s}{R^2}\right)+\frac{1-\nu}{R^2}\tilde{w},_{\varphi\varphi s}\right\}(\pm(L+\varepsilon))
\end{aligned}\tag{3.141''}$$

Because the propagation domain of the acoustic pressure is infinite, it is convenient to express the diffracted pressure as the integral over the shell surface of an unknown source density $\widetilde{\varpi}(M,t)$. Thus, all the unknown functions are defined on a bounded space domain. For this purpose, as done in section 3.10, we use a hybrid layer potential representation of $p(M,t)$ [FIL 99]:

$$\tilde{p}(M,t)=\int_0^{+\infty}dt'\int_\Sigma\widetilde{\varpi}(M',t')\big(\widetilde{G}(M,M';t-t')+\epsilon\,\partial_{n'}\widetilde{G}(M,M';t-t')\big)dM'$$

$$\text{with}\quad\widetilde{G}(M,M';t-t')=-\frac{\delta\big(t-t'-r(M,M')/c_0\big)}{4\pi r(M,M')}$$

where $\widetilde{G}(M, M'; t-t')$ is the free-field Green's function of the wave equation satisfying the outgoing wave condition; $r(M, M')$ is the distance between M and M'; $\widetilde{\varpi}(M', t')$ is an unknown square-integrable layer density; ϵ is an arbitrary constant. The Green's function involving a Dirac distribution, the above integrals must be understood as integrals of generalized functions – see [CRI 92], Chapter 2 – also known as duality products [SCH 61]. Let us introduce the following boundary operators:

$$\tilde{\kappa}_1(\widetilde{\varpi}) = \mathrm{Tr}\left[\int_0^{+\infty} \mathrm{d}t' \int_\Sigma \widetilde{\varpi}(M', t')\big(\widetilde{G}(M, M'; t-t') + \epsilon\, \partial_{n'}\widetilde{G}(M, M'; t-t')\big) dM'\right]$$

$$\tilde{\kappa}_2(\widetilde{\varpi}) = \mathrm{Tr}\,\partial_n\left[\int_0^{+\infty} \mathrm{d}t' \int_\Sigma \widetilde{\varpi}(M', t')\big(\widetilde{G}(M, M'; t-t') + \epsilon\, \partial_{n'}\widetilde{G}(M, M'; t-t')\big)\, \mathrm{d}M'\right]$$

The variational equations (3.137) thus become:

$$\frac{Eh}{1-\nu^2}\int_0^{+\infty}\Big[\mathcal{A}(\tilde{u}, \tilde{v}, \tilde{w}; \delta\tilde{u}, \delta\tilde{v}, \delta\tilde{w})(t) + \mu_s h\int_\Sigma [\ddot{\tilde{u}}\delta\tilde{u}^* + \ddot{\tilde{v}}\delta\tilde{v}^* + \ddot{\tilde{w}}\delta\tilde{w}^*](M, t)\,\mathrm{d}M\Big]\,\mathrm{d}t + \int_0^{+\infty}\int_\Sigma [\tilde{\kappa}_1(\widetilde{\varpi})\delta\tilde{w}^*](M, t)\,\mathrm{d}M\,\mathrm{d}t$$
$$= -\int_0^{+\infty}\int_\Sigma [\,\mathrm{Tr}\,\tilde{p}^i\delta\tilde{w}^*](M, t)\,\mathrm{d}M\,\mathrm{d}t$$

$$\mu_0\int_0^{+\infty}\int_\Sigma [\ddot{\tilde{w}}\delta\tilde{\psi}^*](M, t)\,\mathrm{d}M\,\mathrm{d}t + \int_0^{+\infty}\int_\Sigma [\tilde{\kappa}_2(\widetilde{\varpi})\delta\tilde{\psi}^*](M, t)\,\mathrm{d}M\,\mathrm{d}t$$
$$= -\int_0^{+\infty}\int_\Sigma [\,\mathrm{Tr}\,\partial_n\tilde{p}^i\delta\tilde{\psi}^*](M, t)\,\mathrm{d}M\,\mathrm{d}t \qquad \forall\delta\tilde{u}, \delta\tilde{v}, \delta\tilde{w}, \delta\tilde{\psi} \quad (3.142)$$

Finally, because all the functions involved in (3.142) are 2π-periodic with respect to the variable φ, this equation can be replaced by a sequence of variational equations for the Fourier components of the unknown functions. Let $f(s, \varphi)$ be any function defined on Σ; its Fourier series is written as:

$$f(s, \varphi) = \sum_{n=-\infty}^{+\infty} f_n(s)\mathrm{e}^{in\varphi}$$

The Fourier components of the operators $\mathcal{A}$ are denoted by $\mathcal{A}^n$: they are obtained by replacing the derivation operator ∂_φ by $\imath n$. The Fourier components of $\tilde{\kappa}_1$ and $\tilde{\kappa}_2$ are denoted by $\tilde{\kappa}_1^n$ and $\tilde{\kappa}_2^n$: they are deduced from the Fourier components of the free-field Green's function for the wave equation, which are known in spherical and in cylindrical coordinates as inverse time Fourier transforms of the components of the free-field Green's kernel for the Helmholtz equation (see, for example, [MOR 53]). The variational equations are replaced by the following set of one-dimensional variational equations:

$$\frac{Eh}{1-\nu^2}\int_0^{+\infty}\Big[\mathcal{A}^n(\tilde{u}_n,\tilde{v}_n,\tilde{w}_n;\delta\tilde{u}_n,\delta\tilde{v}_n,\delta\tilde{w}_n)(t)$$
$$+\mu_s h\int_{\mathcal{L}}[\ddot{\tilde{u}}_n\delta\tilde{u}_n^*+\ddot{\tilde{v}}_n\delta\tilde{v}_n^*+\ddot{\tilde{w}}_n\delta\tilde{w}_n^*](s,t)\,\mathrm{d}s\Big]\,\mathrm{d}t+\int_0^{+\infty}\int_{\mathcal{L}}[\tilde{\kappa}_1^n(\tilde{\varpi}_n)\delta\tilde{w}_n^*](s,t)\,\mathrm{d}s\,\mathrm{d}t$$
$$=-\int_0^{+\infty}\int_{\mathcal{L}}[\operatorname{Tr}\tilde{p}_n^i\delta\tilde{w}_n^*](s,t)\,\mathrm{d}s\,\mathrm{d}t$$

$$\mu_0\int_0^{+\infty}\int_{\mathcal{L}}[\ddot{\tilde{w}}_n\delta\tilde{\psi}_n^*](s,t)\,\mathrm{d}s\,\mathrm{d}t+\int_0^{+\infty}\int_{\mathcal{L}}[\tilde{\kappa}_2^n(\tilde{\varpi}_n)\delta\tilde{\psi}_n^*](s,t)\,\mathrm{d}s\,\mathrm{d}t$$
$$=-\int_0^{+\infty}\int_{\mathcal{L}}[\operatorname{Tr}\partial_n\tilde{p}_n^i\delta\tilde{\psi}_n^*](s,t)\,\mathrm{d}s\,\mathrm{d}t$$
$$\forall\delta\tilde{u}_n,\delta\tilde{v}_n,\delta\tilde{w}_n,\delta\tilde{\psi}_n\quad\text{and}\quad-\infty<n<+\infty\qquad(3.143)$$

where $\mathcal{L}$ is the interval $[-L-R\pi/2,+L+R\pi/2]$ of variation of s. The functions $\tilde{u}_n$, $\tilde{v}_n$, w_n, $\delta\tilde{u}_n$, $\delta\tilde{v}_n$ and δw_n satisfy the continuity conditions and the regularity conditions at the apexes, deduced from (3.140), (3.140′), (3.141), (3.141′) and (3.141″).

3.11.2. *Resonance Modes and Response of the System to an Incident Transient Acoustic Wave*

1. *Equations governing the resonance modes of the system*

Let us consider now a harmonic time dependence ($\mathrm{e}^{-\imath\omega t}$). Then, all functions are of the form $\tilde{f}=f\mathrm{e}^{-\imath\omega t}$, where, as already used, the symbol without the over-sign "$\sim$" denotes the time-independent function. The set of equations (3.142) becomes:

$$\frac{Eh}{1-\nu^2}\mathcal{A}^n(u_n, v_n, w_n; \delta u_n, \delta v_n, \delta w_n)$$

$$-\mu_s h\omega^2 \int_{\mathcal{L}} [u_n \delta u_n^* + v_n \delta v_n^* + w_n \delta w_n^*](s)\,\mathrm{d}s + \int_{\mathcal{L}} [\kappa_1^n(\omega; \varpi_n)\delta w_n^*](s)\,\mathrm{d}s$$

$$= -\int_{\mathcal{L}} [\,\mathrm{Tr}\, p_n^i \delta w_n^*](s)\,\mathrm{d}s$$

$$-\mu_0\omega^2 \int_{\mathcal{L}} [w_n \delta\psi_n^*](s)\,\mathrm{d}s + \int_{\mathcal{L}} [\kappa_2^n(\omega; \varpi_n)\delta\psi_n^*](s)\,\mathrm{d}s = -\int_{\mathcal{L}} [\,\mathrm{Tr}\,\partial_n p_n^i \delta\psi_n^*](s)\,\mathrm{d}s$$

$$\forall \delta u_n, \delta v_n, \delta w_n, \delta\psi_n \quad \text{and} \quad -\infty < n < +\infty \tag{3.144}$$

where the boundary operators κ_1^n and κ_2^n are defined by:

$$\kappa_1^n(\omega; \varpi_n) = \mathrm{Tr}\,\kappa^n(\omega; \mu_n) \quad , \quad \kappa_2^n(\omega; \varpi_n) = \mathrm{Tr}\,\partial_n \kappa^n(\omega; \varpi_n)$$

$$\text{with} \quad \kappa^n(\omega; \varpi_n)(M) = -2\pi \int_{\mathcal{L}} \varpi_n(s')\big[G_\omega^n(M; \rho'(s'), z'(s'))$$

$$+ \epsilon\, \partial_{n'} G_\omega^n(M; \rho'(s'), z'(s'))\big]\,\mathrm{d}s'$$

$$\text{and} \quad G_\omega^n(M; \rho'(s'), z'(s')) = -\int_0^{2\pi} \frac{\mathrm{e}^{\imath\omega r(M;\rho'(s'),z'(s'),\theta')/c_0}}{4\pi r(M; \rho'(s'), z'(s'), \theta')} \mathrm{e}^{\imath n\theta'}\,\mathrm{d}\theta' \tag{3.144'}$$

where a point M' on the shell surface is defined by its cylindrical coordinates $(\rho'(s'), z'(s'), \theta')$, the first two, $\rho'(s')$ and $z'(s')$, depending on its curvilinear coordinate s'.

The resonance modes $(U_n^m, V_n^m, W_n^m, \Pi_n^m)$ and the resonance angular-frequencies Ω_n^m are defined as the solutions to the homogenous equations:

$$\frac{Eh}{1-\nu^2}\mathcal{A}^n(U_n^m, V_n^m, W_n^m; \delta u_n, \delta v_n, \delta w_n) - \mu_s h {\Omega_n^m}^2 \Bigg[\int_{\mathcal{L}} [U_n^m \delta u_n^*$$

$$+ V_n^m \delta v_n^* + W_n^m \delta w_n^*](s)\,\mathrm{d}s - \frac{1}{{\Omega_n^m}^2 \mu_s h} \int_{\mathcal{L}} [\kappa_1^n(\Omega_n^m; \Pi_n^m)\delta w_n^*](s)\,\mathrm{d}s\Bigg] = 0$$

$$-\mu_0 {\Omega_n^m}^2 \int_{\mathcal{L}} [W_n^m \delta\psi_n^*](s)\,\mathrm{d}s + \int_{\mathcal{L}} [\kappa_2^n(\Omega_n^m; \Pi_n^m)\delta\psi_n^*](s)\,\mathrm{d}s = 0$$

$$\forall \delta u_n, \delta v_n, \delta w_n, \delta\psi_n \quad \text{and} \quad -\infty < n < +\infty \tag{3.145}$$

It can be shown that:

- ◇ the imaginary part of a resonance frequency is negative: this means that, because of the loss of energy at infinity in the fluid, the resonance modes are damped functions of the time variable;
- ◇ if Ω_n^m is a resonance frequency corresponding to the resonance mode $(U_n^m, V_n^m, W_n^m, \Pi_n^m)$, then $\Omega_n^{-m} = -\Omega_n^{m*}$ is also a resonance frequency which corresponds to the resonance mode $(U_n^{-m}, V_n^{-m}, W_n^{-m}, \Pi_n^{-m}) = (U_n^{m*}, V_n^{m*}, W_n^{m*}, \Pi_n^{m})^*$.

2. *Expression of the response of the system to an incident acoustic transient wave*

Let us denote by $\widetilde{\mathfrak{U}}_n$ the n-th Fourier component of the shell displacement vector with components $(\tilde{u}_n, \tilde{v}_n, \tilde{w}_n)$ and by $\mathcal{U}_n^m$ the resonance mode displacement vector with components (U_n^m, V_n^m, W_n^m). The solution $(\tilde{u}_n, \tilde{v}_n, \tilde{w}_n, \widetilde{\varpi}_n)$ of equations (3.137) is sought as a resonance modes series of the form:

$$\begin{pmatrix} \widetilde{\mathfrak{U}}_n \\ \widetilde{\varpi}_n \end{pmatrix} = Y(t) \sum_{m=1}^{+\infty} \left[\alpha_n^m \begin{pmatrix} \mathcal{U}_n^m \\ \varpi_n^m \end{pmatrix} e^{-\imath \omega_n^m t} + \alpha_n^{-m} \begin{pmatrix} \mathcal{U}_n^{-m} \\ \varpi_n^{-m} \end{pmatrix} e^{-\imath \omega_n^{-m} t} \right] \tag{3.146}$$

where $Y(t)$ stands for the Heaviside step function.

A direct solution is obtained as follows. This expression is introduced into equations (3.137) with $(u_n^{q*}, v_n^{q*}, w_n^{q*}, \varpi_n^{q*})$ as test functions. Thus, an infinite system of linear algebraic equations is obtained to determine the coefficients of the series expansion.

3.11.3. *Numerical Method and Comparison between Numerical Prediction and Experimental Results*

1. *Numerical method for the calculation of the resonance modes*

As already mentioned, the main difficulty is to compute the resonance frequencies and resonance modes of the fluid-loaded structure. The unknown functions can be approximated by various methods, the most popular being the Finite Element Method. The authors of [MAT 99, MAU 01, MAU 99] preferred to use an approximation based on Legendre polynomials. The advantage of such an approximation is to provide approximate resonance modes which are regular functions (indefinitely differentiable) and, thus, have the required regularity.

The layer density ϖ_n^m is approximated by a truncated series of Legendre polynomials of the variable s. The shell displacement components are approximated by a truncated series of polynomial functions built with a finite number of Legendre polynomials. Each polynomial function is chosen so that it satisfies the regularity conditions at the apexes. Thus, two kinds of approximations are adopted:

1. for the tangential displacement components u_n^m and v_n^m, the approximation functions involve a linear combination of two Legendre polynomials $P_q(s)$ and $P_{q+2}(s)$, with $q = 0, 1, \ldots$;

2. for the normal displacement component w_n^m, the approximation functions involve a linear combination of three Legendre polynomials $P_q(s)$, $P_{q+2}(s)$ and $P_{q+4}(s)$, with $q = 0, 1, \ldots$.

In both cases, the coefficients of the linear combination depend on the angular harmonic index n and are calculated analytically. These truncated series are introduced into the resonance modes equations, in which the approximation functions are used as test functions. A linear system of equations is thus obtained for the coefficients of the expansions of the resonance modes. The resonance angular frequencies Ω_n^m are approximated by the values of ω, for which the determinant of the system is numerically zero.

2. *Comparison between numerical and experimental resonance frequencies*

A first test of the validity of such a method is, of course, to check that the resonance frequencies of the structure are correctly predicted.

The first ten resonance frequencies of a *Line 2'* shell have been computed for the following data: $R = L = 27$ mm, $h = 0.81$ mm, $\mu_s = 7900$ kg m^{-2}, $E = 1.9973\ 10^{11}$ Pa, $\nu = 0.314$, $\mu_0 = 1000$ kg m^{-3}, $c_0 = 1470$ m s^{-1}.

Table 3.10 presents a comparison between the experimental results published in [DEC 93] and the values predicted by the method proposed here. It appears that the agreement is excellent: the relative error on the real part of the resonance frequencies is 1.8% on the first one and 0.12% on the tenth one. It must be recalled that the physical data, and, in particular the Young's modulus and the Poisson's ratio, have been measured very accurately by the authors of the experimental study. This explains why the numerical prediction can agree so well with the experiments.

Measured fluid-loaded	Computed fluid-loaded	Relative error	Computed *in vacuo*	Mode number
68.0	$69.3(1 - \imath 4.78\ 10^{-8})$	1.80%	72.6	3
88.0	$88.6(1 - \imath 2.48\ 10^{-2})$	0.70%	89.9	4
107.1	$107.9(1 - \imath 5.51\ 10^{-5})$	0.74%	108.0	5
124.0	$121.4(1 - \imath 4.26\ 10^{-2})$	2.00%	127.1	6
143.5	$143.4(1 - \imath 8.39\ 10^{-3})$	0.07%	146.1	7
163.8	$164.4(1 - \imath 4.32\ 10^{-2})$	0.36%	167.9	8
183.6	$183.3(1 - \imath 1.53\ 10^{-2})$	0.14%	184.7	9
198.6	$198.9(1 - \imath 3.02\ 10^{-3})$	0.15%	199.0	10
239.0	$239.9(1 - \imath 3.42\ 10^{-2})$	0.37%	240.2	11
257.0	$256.7(1 - \imath 2.45\ 10^{-2})$	0.12%	259.5	12

Table 3.10. *Comparison between computed and measured resonance frequencies (in kHz), for a steel-made Line 2' shell*

3. *Comparison between the predicted response of the shell to a transient incident wave and the experimental one*

The second test is to compare the experimental transient response of the system when excited by an incident transient wave to the numerical prediction.

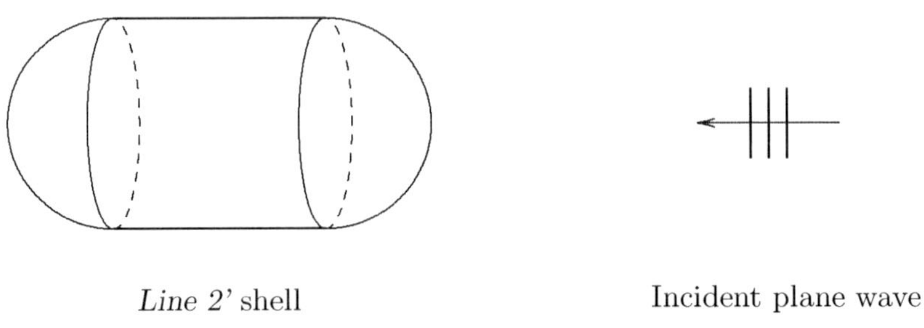

Figure 3.33. *Scheme of the experimental assembly*

The experimental assembly is as follows (see Figure 3.33). The *Line 2'* shell is made of an aluminum alloy. Its radius is $R = 0.020$ m, its total length is $2(L + R) = 0.080$ m and its thickness is $h = 0.001$ m. The material mechanical characteristics are: $\mu_s = 2611$ kg m^{-3}, $E = 0.81\ 10^{11}$ Pa, $\nu = 0.30$. It is immersed in a large water tank with anechoic boundaries. An acoustic transducer, located on the symmetry axis of the shell, 1 m away from the shell end cap, is used for the generation of a short signal with a spectrum centered around 250 kHz (4 oscillations, lasting 0.2 ms). The same transducer is used for recording the reflected wave in the backward direction.

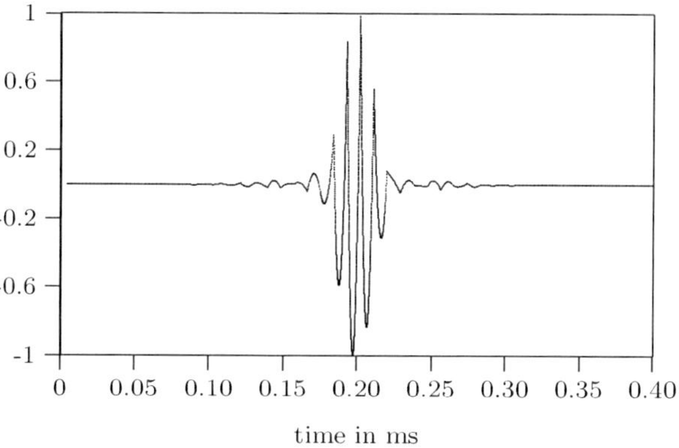

Figure 3.34. *Incident Sound Field (normalized)*

The incident wave is modeled as a plane wave. The signal which is emitted is modeled as an Ausher wavelet, which describes fairly well what is observed (see Figure 3.34): the experimental signal has a short duration and a quasi-finite bandwidth.

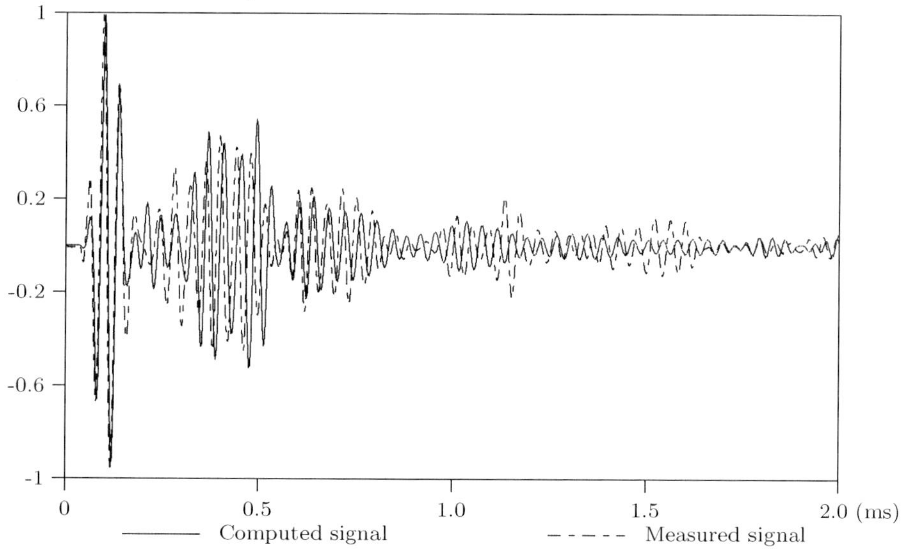

Figure 3.35. *Comparison between measurement and prediction of the time-dependent sound field diffracted in the axial incidence direction by a Line 2' shell*

In Figure 3.35, the experimental diffracted signal is compared to the predicted one, for a total time interval of 2.0 ms, that is ten times the duration of the incident signal. It must be noticed that the reflected wave is highly sensitive to the relative positions of the transducer and of the target: a small error in the orientation of the shell can change the diffracted wave by a rather large amount. Thanks to precise positioning of both the transducer and the target, it has been possible to obtain a good agreement between the two curves.

Although the reflected signal was computed as a whole, the specular reflection and the successive wave packets, which correspond roughly to creeping waves having travelled several times around the target, clearly appear. This shows the strong efficiency of the expansion of the diffracted acoustic transient pressure in terms of the resonance modes.

3.12. Exercises

1.Establish expressions (3.9) and (3.11).

2.Establish expressions (3.13).

3.Establish expressions (3.22) and (3.22′). Prove that P_2 given by expression (3.22) has a modulus equal to 1.

4.Establish expressions (3.40).

5.Establish expressions (3.43), (3.44) and (3.45).

6.Prove the orthogonality relationship (3.74).

7.Establish expressions (3.75).

8.Establish expressions (3.77) and (3.78).

9.Establish expression (3.97) and the corresponding expression of the acoustic pressure cross-spectrum density.

10. Establish the outer and inner expansions (first terms) for the shell problem of section 3.9.

11. Establish the expressions of $\mathcal{H}^n$, $\tilde{\kappa}_1^n$ and $\tilde{\kappa}_2^n$ which appear in equations (3.143).

Bibliography

[ABR 65] M. Abramowitz and I.A. Stegun, *Handbook of Mathematical Functions*. National Bureau of Standards, Applied Mathematics Series 55 (originally issued June 1964).

[BAR 89] C. Bardos, M. Concordel and G. Lebeau, *Journal d'Acoustique*, **2**, 31-38. Extension de la théorie de la diffusion pour un corps élastique immergé dans un fluide. Comportement asymptotique des résonances, 1989.

[BLA 86] W.K. Blake, *Mechanics of Flow-Induced Sound and Vibration, volumes 1 and 2*. Academic Press, 1986.

[BRU 06] M. Bruneau, *Manuel d'acoustique fondamentale*. Hermès Science Publication, Paris, 1998.

and

M. Bruneau, *Fundamentals of Acoustics*. ISTE Ltd, London, 2006.

[BUL 96] M.K. Bull, *Journal of Sound and Vibration*, **190**(3), 299-315. Wall-pressure fluctuation beneath turbulent boundary layers: some reflections on forty years of research, 1996.

[CHA 87] D.M. Chase, *Journal of Sound and Vibration*, **112**(1), 125-147. The character of the turbulent wall pressure spectrum at subconvective wavenumbers and a suggested comprehensive model, 1987.

[COR 63] G.M. Corcos, *Journal of the Acoustical Society of America*, **35**(2), 192-199. Resolution of pressure in turbulence, 1963.

[COU 57] R. COURANT and D. HILBERT, *Methods of Mathematical Physics*. Interscience, New York, 1957.

[CRI 89] D.G. CRIGHTON, *Journal of Sound and Vibration*, **133**:1–27. The 1988 Rayleigh medal lecture: fluid loading – the interaction between sound and vibration, 1989.

[CRI 92] D.G. CRIGHTON, A.P. DOWLING, J.E. FFOWCS WILLIAMS, M. HECKL and F.G. LEPPINGTON, *Modern Methods in Analytical Acoustics*. Lecture Notes. London: Springer-Verlag, 1992.

[DEC 93] D. DÉCULTOT, F. LECROQ, G. MAZE and J. RIPOCHE, *Journal of the Acoustical Society of America*. **94,** 2916-2923. Acoustic scattering from a cylindrical shell bounded by hemispherical end caps. Resonance interpretation with surface waves propagating in cylindrical and spherical shells, 1993.

[DUR 99-1] CH. DURANT, *Etude expérimentale de l'excitation et de la réponse vibroacoustique d'une conduite sollicitée par un écoulement interne*. Thesis, Ecole Centrale de Lyon. LMFA, UMR 5509, 1999.

[DUR 99-2] CH. DURANT and G. ROBERT, *Flow, Turbulence and Combustion*, **61**(1-4), 55-69. Vibro-Acoustic Response of a Pipe Excited by a Turbulent Internal Flow, 1999.

[ERD 54-1] A. ERDÉLYI, W. MAGNUS, F. OBERHETTINGER and F.G. TRICOMI, *Higher Transcendental Functions*, vol.1, 2 and 3. McGraw-Hill Book Company Inc, 1954.

[ERD 54-2] A. ERDÉLYI, W. MAGNUS, F. OBERHETTINGER and F.G. TRICOMI, *Tables of Integral Transforms*, vol. 1 and 2. McGraw-Hill Book Company Inc, 1954.

[FAH 85] F. FAHY, *Sound and Structural Vibration*, Academic Press, London, 1985.

[FIL 97] P.J.T. FILIPPI and D. MAZZONI, Response of a vibrating structure to a turbulent wall pressure: fluid-loaded structure modes series and boundary element method.

In *Acoustic Interactions with Submerged Elastic Structures*, Part III, 75-117. Edited by A. Guran, A. De Hoop, D. Guicking and F. Mainardi. *Series on Stability, Vibrations and Control of Systems, Series B: Vol. 5.* World Scientific Publishing Company, 1997.

and

In *Uncertainty Modeling in Finite Element, Fatigue and Stability of Structures*, 117-158. Edited by A. Hadar, A. Guran and B. M. Ayyub. *Series on Stability, Vibrations and Control of Systems, Series B: Vol. 9.* World Scientific Publisher Company, 1997.

[FIL 99] P.J.T. FILIPPI, D. HABAULT, J.-P. LEFEBVRE and A. BERGASSOLI, *Acoustics: Basic Physics, Theory and Methods.* Academic Press, London, 1999.

[FIL 01] P.J.T. FILIPPI, D. HABAULT, P.-O. MATTEI and C. MAURY, *Journal of Sound and Vibration*, **239**(4), 639-663. The role of the resonance modes in the response of a fluid-loaded structure, 2001.

[FLU 90] W. FLÜGGE, *Stresses in Shells.* Springer-Verlag, Berlin - Heidelberg - New York, 1990 (second edition).

[GUI 08] E. GUIBERT, *Etude vibro-acoustique et perceptive du bruit rayonné à l'intérieur d'une voiture de train.* Thèse de Mécanique – spécialité : Acoustique, Université de Provence, 2008.

[GER 73] P. GERMAIN, *Cours de Mécanique des Milieux Continus*, Tome 1 – Théorie Générale. Masson et Cie. Editeurs, Paris, 1973.

[GUY 02] J.-L. GUYADER, *Vibrations des milieux continus*, Hermès Science Publication, Paris, 2002.

and

J.-L. GUYADER, *Vibration in Continuous Media*, ISTE, London, 2006.

[HAB 99] D. HABAULT, *Fluid-Structure Interactions in Acoustics.* CISM Courses and Lecture Notes N° 396, Springer, Wien - New York, 1999.

[HAB 04] D. HABAULT and P.J.T. FILIPPI, *Journal of Sound and Vibration*, **270**(4), 207-231. A numerical method for the computation of the resonance modes of a fluid loaded plate: application to the transient response of the system, 2004.

[JOH 86] F. JOHN, *Partial Differential Equations*, 4th edn, Applied Mathematical Sciences Series, vol.1. Springer-Verlag, New York, 1986.

[JUN 93] M.C. JUNGER and D. FEIT, *Sound, Structures, and Their Interaction.* Acoustical Society of America. (1972 and 1986 by the Massachusetts Institute of Technology), 1993.

[KUN 61] J. KUNTZMANN, *Mathématiques de la physique et de la technique.* Hermann, Paris, 1961.

[LAN 67] L. LANDAU and E. LIFCHITZ, *Théorie de l'élasticité.* Editions MIR, Moscou, 1967.

[LEI 69] A.W. LEISSA *Vibrations of plates.* NASA Scientific and Technical Publications, Washington, D.C. 20402, 1969.
New edition:
Acoustical Society of America Publications, Sewickley, USA, 1993.

[LEI 73] A.W. LEISSA, *Vibrations of shells.* NASA Scientific and Technical Publications, Washington, D.C. 20546, 1973.
New edition:
Acoustical Society of America Publications, Sewickley, USA, 1993.

[LOV 44] A.E.H. LOVE, *A Treatise on the Mathematical Theory of Elasticity.* Dover Publications, New York, 1944 (fourth edition).

[MAR 77] N.C. MARTIN and P. LEEHEY, *Journal of Sound and Vibration,* **52**(1):95–120. Low wavenumber wall pressure measurements using a rectangular membrane as a spatial filter, 1977.

[SEI 97] S. MARTIN-SEIGLE, *Rayonnement acoustique de coques cylindriques immergées et couplées à un fluide interne léger.* Thèse de Mécanique – spécialité : Acoustique et Dynamique des Vibrations, Université de la Méditerranée, 1997.

[SEI 99] S. MARTIN-SEIGLE, M.-C. PÉLISSIER and P. FILIPPI, *Flow, Turbulence and Combustion,* **61**(1-4),71-84. Acoustic radiation from a finite length cylindrical shell excited by an internal acoustic source: solution based on a Boundary Element Method and a matched asymptotic expansions, 1999.

[MAT 99] P.-O. MATTEI, and P. FILIPPI, *Flow, Turbulence and Combustion,* **61**(1-4), 85-99. Response of a thin cylindrical shell excited by a turbulent internal flow, 1999.

[MAU 01] C. MAURY and P.J.T. FILIPPI, *Journal of Sound and Vibration,* **241**(459-483). Transient acoustic diffraction and radiation by an axisymmetrical elastic shell: a new statement of the basic equations and a numerical method based on polynomial approximations, 2001.

[MAU 99] C. MAURY, *Rayonnement acoustique de structures vibrantes : études en régimes temporels et fréquentiels.* Thèse de Mécanique – spécialité : Acoustique et Dynamique des Vibrations, Université de la Méditerranée, 1999.

[MOR 48] Ph. MORSE, *Vibration and Sound.* McGraw-Hill Book Company, New York, Toronto, London, 1948.

[MOR 53] Ph. MORSE and H. FESHBACH, *Methods of Theoretical Physics.* McGraw-Hill Book Company, New York, Toronto, London, 1953.

[MOR 68] Ph. MORSE and K.U. INGARD, *Theoretical Acoustics*. McGraw-Hill Book Company, New York, St. Louis, San Francisco, Toronto, London, Sydney, 1968.

[NAY 73] A. NAYFEH, *Perturbation Methods*. John Wiley and Sons, New York, London, Sydney, Toronto, 1973.

[NOR 84] A. NORRIS, G.A. KRIEGSMANN and E.L. REISS, *Journal of the Acoustical Society of America*, **75**(3):685–694. Acoustic scattering by baffled membranes, 1984.

[NOR 90] A. NORRIS, *Journal of the Acoustical Society of America*, **88**(1):505–514. Resonant acoustic scattering from solid targets, 1990.

[PIE 81] A.D. PIERCE, *Acoustics: An Introduction to its Physical Principles and Applications*. McGraw-Hill Book Company, New York, 1981.

[RUM 92] M.L. RUMERMAN, *Journal of the Acoustical Society of America*, **91**(2):907–910. Frequency-flow speed dependence of structural response to turbulent boundary layer excitation of a panel, 1992.

[SCH 61] L. SCHWARTZ, *Méthodes mathématiques pour les sciences physiques*. Hermann, Paris, 1961.

[TIM 70] S.P. TIMOSHENKO and J.N. GOODIER, *Theory of Elasticity*. McGraw-Hill Book Company, New York - San Francisco - London, 1970 (third edition).

[VIV 72] J. VIVOLI, *Vibrations de plaques et potentiels de couches*, Thèse de l'Université de Provence (Marseille, France), N° A.O. 7.868, 1972.

[VIV 74] J. VIVOLI and P. FILIPPI, *Journal of the Acoustical Society of America*, **55**(3): 562–567. Eigenfrequencies of thin plates and layer potentials, 1974.

Notations

Mechanical and geometrical data

μ_s .... solid density
$\bar{\mu}$ .... mass per unit length for a beam, or per unit area for a plate
E .... Young's modulus
ν .... Poisson's ratio
μ_0 .... fluid density
c_0 .... sound velocity in a fluid
h .... thickness (possibly variable) of shells and plates

Displacements, stresses, strains

$\mathcal{D}_{ij}$ .... strain tensor components
$\mathcal{E}_c$ .... kinetic energy
$\mathcal{E}_p$ .... potential energy
$\mathcal{S}_{ij}$ .... stress tensor components
U_i .... i^{th} component of the displacement of a 3-D solid
(u, v) .... tangent components of a shell displacement
w .... transverse component of a shell or plate displacement

Mathematical operators and symbols

$f_{,i}$ .... derivative of f with respect to the ith coordinate
$\dot{f}$ .... time derivative of f

$\vec{\nabla}$ .... gradient operator
$\partial_n = \vec{n} \cdot \vec{\nabla}$, $\partial_s = \vec{s} \cdot \vec{\nabla}$ .... normal and tangent derivation operators
$\partial_{\bar{s}}$ .... derivation operator with respect to the curvilinear abscissae
$d\bar{s}$ .... curvilinear abscissae element
Δ .... Laplace operator
Δ^2 .... iterated Laplace operator
$\text{Tr}\, f$ .... value of f along a boundary
δ_m^n .... Kronecker symbol : $= 1$ if $m = n$, $= 0$ if $m \neq n$
δ_{mn}^{pq} .... multi-index Kronecker symbol : $= 1$ if $m = p$ and $n = q$, $= 0$ else
$\begin{pmatrix} u \\ v \\ w \end{pmatrix} \cdot \begin{pmatrix} u' \\ v' \\ w' \end{pmatrix}$ .... scalar product of the vectors with components (u, v, w) and (u', v', w')
$f = \mathcal{O}(\varepsilon)$ .... means that f decreases as fast as ε
$f = o(\varepsilon)$ .... means that f decreases faster than ε
$\text{Tr}\, f(P)$ .... stands for $\lim_{M \in \Omega \to P \in \partial\Omega} f(M)$, with $\partial\Omega$ boundary of Ω

Mathematical functions

$J_n(z)$ .... Bessel function of the first kind and order n
$Y_n(z)$ .... Bessel function of the second kind and order n
$H_n^{(1)}(z) = J_n(z) + \imath Y_n(z)$ .... Hankel function of the first kind and order n
$H_n(z)$ .... simplified notation generally used instead of $H_n^{(1)}(z)$
$H_n^{(2)}(z) = J_n(z) - \imath Y_n(z)$ .... Hankel function of the second kind and order n
$I_n(z)$ .... modified Bessel function of the first kind and order n
$K_n(z)$ .... modified Bessel function of the third kind and order n

Index